Mathematics

MATHEMATICS

The Loss of Certainty

MORRIS KLINE
Professor Emeritus of Mathematics
Courant Institute of Mathematical Sciences
New York University

OXFORD UNIVERSITY PRESS
Oxford   New York

OXFORD UNIVERSITY PRESS
Oxford   London   Glasgow
New York   Toronto   Melbourne
Nairobi   Dar es Salaam   Cape Town
Kuala Lumpur   Singapore   Hong Kong   Tokyo
Delhi   Bombay   Calcutta   Madras   Karachi
and associate companies in
Beirut   Berlin   Ibadan   Mexico City

Library of Congress Cataloging in Publication Data
Kline, Morris, 1908–
Mathematics, the loss of certainty.
Bibliography: p.
1. Mathematics—History.   I. Title.
QA21.K525   510'.9   80-11050   ISBN 0-19-502754-X
ISBN 0-19-503085-0 (pbk.)

Printing (last digit): 15 14 13 12 11

Printed in the United States of America

To my wife
Helen Mann Kline

Preface

This book treats the fundamental changes that man has been forced to make in his understanding of the nature and role of mathematics. We know today that mathematics does not possess the qualities that in the past earned for it universal respect and admiration. Mathematics was regarded as the acme of exact reasoning, a body of truths in itself, and the truth about the design of nature. How man came to the realization that these values are false and just what our present understanding is constitute the major themes. A brief statement of these themes is presented in the Introduction. Some of the material could be gleaned from a detailed technical history of mathematics. But those people who are interested primarily in the dramatic changes that have taken place will find that a direct, non-technical approach makes them more readily accessible and more intelligible.

Many mathematicians would perhaps prefer to limit the disclosure of the present status of mathematics to members of the family. To air these troubles in public may appear to be in bad taste, as bad as airing one's marital difficulties. But intellectually oriented people must be fully aware of the powers of the tools at their disposal. Recognition of the limitations, as well as the capabilities, of reason is far more beneficial than blind trust, which can lead to false ideologies and even to destruction.

I wish to express my thanks to the staff of Oxford University Press for its thoughtful handling of this book. I am especially grateful to Mr. William C. Halpin and Mr. Sheldon Meyer for recognizing the importance of undertaking this popularization and to Ms. Leona Capeless and Mr. Curtis Church for valuable suggestions and criticisms. To my wife Helen I am indebted for many improvements in the writing and for her care in proofreading.

I wish to thank also the Mathematical Association of America for permission to use the quotations from articles in *The American Mathematical Monthly* reproduced in Chapter XI.

Brooklyn, N.Y.   M.K.
January 1980

Contents

Mathematics

The gods have not revealed all things from the beginning,
But men seek and so find out better in time.

. . .

Let us suppose these things are like the truth.

. . .

But surely no man knows or ever will know
The truth about the gods and all I speak of.
For even if he happens to tell the perfect truth,
He does not know it, but appearance is fashioned over everything.

XENOPHANES

Introduction: The Thesis

> To foresee the future of mathematics, the true method is
> to study its history and its present state.   HENRI POINCARÉ

There are tragedies caused by war, famine, and pestilence. But there
are also intellectual tragedies caused by limitations of the human mind.
This book relates the calamities that have befallen man's most effective
and unparalleled accomplishment, his most persistent and profound
effort to utilize human reason—mathematics.

Put in other terms, this book treats on a non-technical level the rise
and decline of the majesty of mathematics. In view of its present immense scope, the increasing, even flourishing, mathematical activity,
the thousands of research papers published each year, the rapidly
growing interest in computers, and the expanded search for quantitative relationships especially in the social and biological sciences, how
can we talk about the decline of mathematics? Wherein lies the tragedy? To answer these questions we must consider first what values won
for mathematics its immense prestige, respect, and glory.

From the very birth of mathematics as an independent body of
knowledge, fathered by the classical Greeks, and for a period of over
two thousand years, mathematicians pursued truth. Their accomplishments were magnificent. The vast body of theorems about number and
geometric figures offered in itself what appeared to be an almost endless vista of certainty.

Beyond the realm of mathematics proper, mathematical concepts
and derivations supplied the essence of remarkable scientific theories.
Though the knowledge obtained through the collaboration of mathematics and science employed physical principles, these seemed to be as
secure as the principles of mathematics proper because the predictions
in the mathematical theories of astronomy, mechanics, optics, and hydrodynamics were in remarkably accurate accord with observation and

experiment. Mathematics, then, provided a firm grip on the workings of nature, an understanding which dissolved mystery and replaced it by law and order. Man could pridefully survey the world about him and boast that he had grasped many of the secrets of the universe, which in essence were a series of mathematical laws. The conviction that mathematicians were securing truths is epitomized in Laplace's remark that Newton was a most fortunate man because there is just one universe and Newton had discovered its laws.

To achieve its marvelous and powerful results, mathematics relied upon a special method, namely, deductive proof from self-evident principles called axioms, the methodology we still learn, usually in high school geometry. Deductive reasoning, by its very nature, guarantees the truth of what is deduced if the axioms are truths. By utilizing this seemingly clear, infallible, and impeccable logic, mathematicians produced apparently indubitable and irrefutable conclusions. This feature of mathematics is still cited today. Whenever someone wants an example of certitude and exactness of reasoning, he appeals to mathematics.

The successes mathematics achieved with its methodology attracted the greatest intellectuals. Mathematics had demonstrated the capacities, resources, and strengths of human reason. Why should not this methodology be employed, they asked, to secure truths in fields dominated by authority, custom, and habit, fields such as philosophy, theology, ethics, aesthetics, and the social sciences? Man's reason, so evidently effective in mathematics and mathematical physics, could surely be the arbiter of thought and action in these other fields and obtain for them the beauty of truths and the truths of beauty. And so, during the period called the Enlightenment or the Age of Reason, mathematical methodology and even some mathematical concepts and theorems were applied to human affairs.

The most fertile source of insight is hindsight. Creations of the early 19th century, strange geometries and strange algebras, forced mathematicians, reluctantly and grudgingly, to realize that mathematics proper and the mathematical laws of science were not truths. They found, for example, that several differing geometries fit spatial experience equally well. All could not be truths. Apparently mathematical design was not inherent in nature, or if it was, man's mathematics was not necessarily the account of that design. The key to reality had been lost. This realization was the first of the calamities to befall mathematics.

The creation of these new geometries and algebras caused mathematicians to experience a shock of another nature. The conviction that they were obtaining truths had entranced them so much that they had rushed impetuously to secure these seeming truths at the cost of sound reasoning. The realization that mathematics was not a body of truths shook their confidence in what they had created, and they undertook to reexamine their creations. They were dismayed to find that the logic of mathematics was in sad shape.

In fact mathematics had developed illogically. Its illogical development contained not only false proofs, slips in reasoning, and inadvertent mistakes which with more care could have been avoided. Such blunders there were aplenty. The illogical development also involved inadequate understanding of concepts, a failure to recognize all the principles of logic required, and an inadequate rigor of proof; that is, intuition, physical arguments, and appeal to geometrical diagrams had taken the place of logical arguments.

However, mathematics was still an effective description of nature. And mathematics itself was certainly an attractive body of knowledge and in the minds of many, the Platonists especially, a part of reality to be prized in and for itself. Hence mathematicians decided to supply the missing logical structure and to rebuild the defective portions. During the latter half of the 19th century the movement often described as the rigorization of mathematics became the outstanding activity.

By 1900 the mathematicians believed they had achieved their goal. Though they had to be content with mathematics as an approximate description of nature and many even abandoned the belief in the mathematical design of nature, they did gloat over their reconstruction of the logical structure of mathematics. But before they had finished toasting their presumed success, contradictions were discovered in the reconstructed mathematics. Commonly these contradictions were referred to as paradoxes, a euphemism that avoids facing the fact that contradictions vitiate the logic of mathematics.

The resolution of the contradictions was undertaken almost immediately by the leading mathematicians and philosophers of the times. In effect four different approaches to mathematics were conceived, formulated, and advanced, each of which gathered many adherents. These foundational schools all attempted not only to resolve the known contradictions but to ensure that no new ones could ever arise, that is, to establish the consistency of mathematics. Other issues arose in the foundational efforts. The acceptability of some axioms and some principles of deductive logic also became bones of contention on which the several schools took differing positions.

As late as 1930 a mathematician might perhaps have been content with accepting one or another of the several foundations of mathematics and declared that his mathematical proofs were at least in accord with the tenets of that school. But disaster struck again in the form of a famous paper by Kurt Gödel in which he proved, among other significant and disturbing results, that the logical principles accepted by the several schools could not prove the consistency of mathematics. This, Gödel showed, cannot be done without involving logical principles so dubious as to question what is accomplished. Gödel's theorems produced a debacle. Subsequent developments brought further complications. For example, even the axiomatic-deductive method so highly regarded in the past as *the* approach to exact knowledge was seen to be flawed. The net effect of these newer developments was to add to the variety of possible approaches to mathematics and to divide mathematicians into an even greater number of differing factions.

The current predicament of mathematics is that there is not one but many mathematics and that for numerous reasons each fails to satisfy the members of the opposing schools. It is now apparent that the concept of a universally accepted, infallible body of reasoning—the majestic mathematics of 1800 and the pride of man—is a grand illusion. Uncertainty and doubt concerning the future of mathematics have replaced the certainties and complacency of the past. The disagreements about the foundations of the "most certain" science are both surprising and, to put it mildly, disconcerting. The present state of mathematics is a mockery of the hitherto deep-rooted and widely reputed truth and logical perfection of mathematics.

There are mathematicians who believe that the differing views on what can be accepted as sound mathematics will some day be reconciled. Prominent among these is a group of leading French mathematicians who write under the pseudonym of Nicholas Bourbaki:

> Since the earliest times, all critical revisions of the principles of mathematics as a whole, or of any branch of it, have almost invariably followed periods of uncertainty, where contradictions did appear and had to be resolved. . . . There are now twenty-five centuries during which the mathematicians have had the practice of correcting their errors and thereby seeing their science enriched, not impoverished; this gives them the right to view the future with serenity.

However, many more mathematicians are pessimistic. Hermann Weyl, one of the greatest mathematicians of this century, said in 1944:

> The question of the foundations and the ultimate meaning of mathematics remains open; we do not know in what direction it will find its final solution or even whether a final objective answer can be expected at all. "Mathematizing" may well be a creative activity of man, like language or music, of primary originality, whose historical decisions defy complete objective rationalization.

In the words of Goethe, "The history of a science is the science itself."

The disagreements concerning what correct mathematics is and the variety of differing foundations affect seriously not only mathematics proper but most vitally physical science. As we shall see, the most well-developed physical theories are entirely mathematical. (To be sure, the conclusions of such theories are interpreted in sensuous or truly physical objects, and we hear voices over our radios even though we have not the slightest physical understanding of what a radio wave is.) Hence scientists, who do not personally work on foundational problems, must nevertheless be concerned about what mathematics can be confidently employed if they are not to waste years on unsound mathematics.

The loss of truth, the constantly increasing complexity of mathematics and science, and the uncertainty about which approach to mathematics is secure have caused most mathematicians to abandon science. With a "plague on all your houses" they have retreated to specialties in areas of mathematics where the methods of proof seem to be safe. They also find problems concocted by humans more appealing and manageable than those posed by nature.

The crises and conflicts over what sound mathematics is have also discouraged the application of mathematical methodology to many areas of our culture such as philosophy, political science, ethics, and aesthetics. The hope of finding objective, infallible laws and standards has faded. The Age of Reason is gone.

Despite the unsatisfactory state of mathematics, the variety of approaches, the disagreements on acceptable axioms, and the danger that new contradictions, if discovered, would invalidate a great deal of mathematics, some mathematicians are still applying mathematics to physical phenomena and indeed extending the applied fields to economics, biology, and sociology. The continuing effectiveness of mathematics suggests two themes. The first is that effectiveness can be used as the criterion of correctness. Of course such a criterion is provisional. What is considered correct today may prove wrong in the next application.

The second theme deals with a mystery. In view of the disagreements about what sound mathematics is, why is it effective at all? Are we performing miracles with imperfect tools? If man has been deceived, can nature also be deceived into yielding to man's mathematical dictates? Clearly not. Yet, do not our successful voyages to the moon and our explorations of Mars and Jupiter, made possible by technology which itself depends heavily on mathematics, confirm mathematical theories of the cosmos? How can we, then, speak of the artificiality and varieties of

mathematics? Can the body live on when the mind and spirit are bewildered? Certainly this is true of human beings and it is true of mathematics. It behooves us therefore to learn why, despite its uncertain
foundations and despite the conflicting theories of mathematicians,
mathematics has proved to be so incredibly effective.

I

The Genesis of Mathematical Truths

Thrice happy souls! to whom 'twas given to rise
To truths like these, and scale the spangled skies!
Far distant stars to clearest view they brought,
And girdled ether with their chains of thought.
So heaven is reached—not as of old they tried
By mountains piled on mountains in their pride.
                                                  OVID

Any civilization worthy of the appelation has sought truths. Thoughtful people cannot but try to understand the variety of natural phenomena, to solve the mystery of how human beings came to dwell on this earth, to discern what purpose life should serve, and to discover human destiny. In all early civilizations but one, the answers to these questions were given by religious leaders, answers that were generally accepted. The ancient Greek civilization is the exception. What the Greeks discovered—the greatest discovery made by man—is the power of reason. It was the Greeks of the classical period, which was at its height during the years from 600 to 300 B.C., who recognized that man has an intellect, a mind which, aided occasionally by observation or experimentation, can discover truths.

What led the Greeks to this discovery is a question not readily answered. The initiators of the plan to apply reason to human affairs and concerns lived in Ionia, a Greek settlement in Asia Minor, and many historians have sought to account for the happenings there on the basis of political and social conditions. For example, the Ionians were rather freer to disregard the religious beliefs that dominated the European Greek culture. However, our knowledge of Greek history before about 600 B.C. is so fragmentary that no definitive explanation is available.

In the course of time the Greeks applied reason to political systems, ethics, justice, education, and numerous other concerns of man. Their chief contribution, and the one which decisively influenced all later cultures, was to undertake the most imposing challenge facing reason, learning the laws of nature. Before the Greeks made this contribution, they and the other civilizations of ancient times regarded nature as chaotic, capricious, and even terrifying. Acts of nature were either unexplained or attributed to the arbitrary will of gods who could be propitiated only by prayers, sacrifices, and other rituals. The Babylonians and Egyptians, who had notable civilizations as far back as 3000 B.C., did note some periodicities in the motions of the sun and moon and indeed based their calendars on these periodicities but saw no deeper significance in them. These few exceptional observations did not influence their attitude toward nature.

The Greeks dared to look nature in the face. Their intellectual leaders, if not the people at large, rejected traditional doctrines, supernatural forces, superstitions, dogma, and other trammels on thought. They were the first people to examine the multifarious, mysterious, and complex operations of nature and to attempt to understand them. They pitted their minds against the welter of seemingly haphazard occurrences in the universe and undertook to throw the light of reason upon them.

Possessed of insatiable curiosity and courage, they asked and answered the questions that occur to many, are tackled by few, and are resolved only by individuals of the highest intellectual caliber. Is there any plan underlying the workings of the entire universe? Are plants, animals, men, planets, light, and sound mere physical accidents or are they part of a grand design? Because they were dreamers enough to arrive at new points of view, the Greeks fashioned a conception of the universe which has dominated all subsequent Western thought.

The Greek intellectuals adopted a totally new attitude toward nature. This attitude was rational, critical, and secular. Mythology was discarded as was the belief that the gods manipulate man and the physical world according to their whims. The intellectuals eventually arrived at the doctrine that nature is orderly and functions invariably according to a grand design. All phenomena apparent to the senses, from the motions of the planets to the stirrings of the leaves on a tree, can be fitted into a precise, coherent, intelligible pattern. In short, nature is rationally designed and that design, though unaffected by human actions, can be apprehended by man's mind.

The Greeks were not only the first people with the audacity to conceive of law and order in the welter of phenomena but also the first with the genius to uncover some of the underlying patterns to which nature apparently conforms. Thus they dared to ask for, and found, design underlying the greatest spectacle man beholds, the motion of the brilliant sun, the changing phases of the many hued moon, the brightness of the planets, the broad panorama of lights from the canopy of stars, and the seemingly miraculous eclipses of the sun and moon.

It was the Ionian philosophers of the 6th century B.C. who also made the first attempts to secure a rational explanation of the nature and functioning of the universe. The famous philosophers of this period, Thales, Anaximander, Anaximenes, Heraclitus, and Anaxagoras, each fixed on a single substance to explain the constitution of the universe. Thales, for example, argued that everything is made up of water in either gaseous, liquid, or solid state, and he attempted explanations of many phenomena in terms of water—a not unreasonable choice because clouds, fog, dew, rain, and hail are forms of water and water is necessary to life, nourishes the crops, and supports much animal life. Even the human body, we now know, is 90 percent water.

The natural philosophy of the Ionians was a series of bold speculations, shrewd guesses, and brilliant intuitions rather than the outcome of extensive and careful scientific investigations. These men were perhaps a little over-eager to see the whole picture and so jumped to broad conclusions. But they did discard the older, largely mythical accounts and substituted materialistic and objective explanations of the design and operation of the universe. They offered a reasoned approach in place of fanciful and uncritical accounts and they defended their contentions by reason. These men dared to tackle the universe with their minds and refused to rely on gods, spirits, ghosts, devils, angels, and other mythical agents who might maintain or disrupt nature's happenings. The spirit of these rational explanations can be expressed in the words of Anaxagoras: "Reason rules the world."

The decisive step in dispelling the mystery, mysticism, and seeming chaos in the workings of nature and in replacing them by an understandable pattern was the application of mathematics. Here the Greeks displayed an insight almost as pregnant and as original as the discovery of the power of reason. The universe is mathematically designed, and through mathematics man can penetrate to that design. The first major group to offer a mathematical plan of nature was the Pythagoreans, a school led by Pythagoras (c.585–c.500B.C.) and rooted in southern Italy. While they did draw inspiration and doctrines from the prevailing Greek religion centering on purification of the soul and its redemption from the taint and prison of the body, Pythagorean natural philosophy was decidedly rational. The Pythagoreans were struck by the fact that phenomena most diverse from a qualitative point of view exhibit identical mathematical properties. Hence mathematical properties must be the essence of these phenomena. More specifically, the Pythagoreans found this essence in number and in numerical relationships. Number was the first principle in their explanation of nature. All objects were made up of elementary particles of matter or "units of existence" in combinations corresponding to the various geometrical figures. The total number of units represented, in fact, the material object. Number was the matter and form of the universe. Hence the Pythagorean doctrine, "All things are numbers." Since number is the "essence" of all objects, the explanation of natural phenomena could be achieved only through number.

This early Pythagorean doctrine is puzzling because to us numbers are abstract ideas, and things are physical objects or substance. But we have made an abstraction of number which the early Pythagoreans did not make. To them, numbers were points or particles. When they spoke of triangular numbers, square numbers, pentagonal numbers, and others, they were thinking of collections of points, pebbles, or point-like objects arranged in those shapes (Figs. 1.1–1.4).

Though historical fragments do not afford precise chronological data, there is no doubt that as the Pythagoreans developed and refined their own doctrines they began to understand numbers as abstract concepts, whereas objects were merely concrete realizations of numbers. With this later distinction we can make sense of the statement of Philolaus, a famous 5th-century Pythagorean: "Were it not for number and its nature, nothing that exists would be clear to anybody either in itself or in its relation to other things. . . . You can observe the power of number exercising itself . . . in all the acts and the thoughts of men, in all handicrafts and music."

The reduction of music, for example, to simple relationships among numbers became possible for the Pythagoreans when they discovered two facts: first that the sound caused by a plucked string depends upon the length of the string; and second that harmonious sounds are given off by equally taut strings whose lengths are to each other as the ratios of whole numbers. For example, a harmonious sound is produced by plucking two equally taut strings, one twice as long as the other. In our language the interval between the two notes is an octave. Another harmonious combination is formed by two strings whose lengths are in the ratio 3 to 2; in this case the shorter one gives off a note, called the fifth, above that given off by the first string. In fact, the relative lengths in every harmonious combination of plucked strings can be expressed as a ratio of whole numbers. The Pythagoreans also developed a famous musical scale. Though we shall not devote space to music of the Greek period, we would like to note that many Greek mathematicians, including Euclid and Ptolemy, wrote on the subject, especially on harmonious combinations of sounds and the construction of scales.

The Pythagoreans reduced the motions of the planets to number

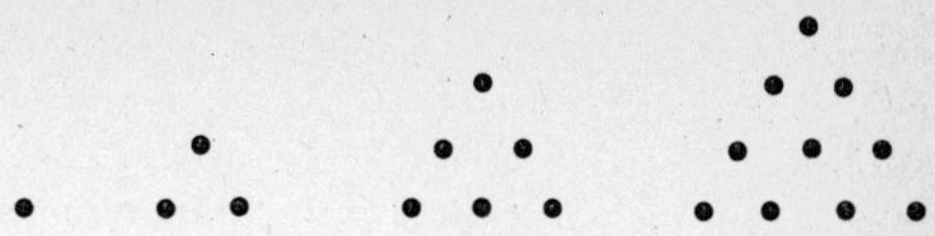

Figure 1.1
Triangular numbers

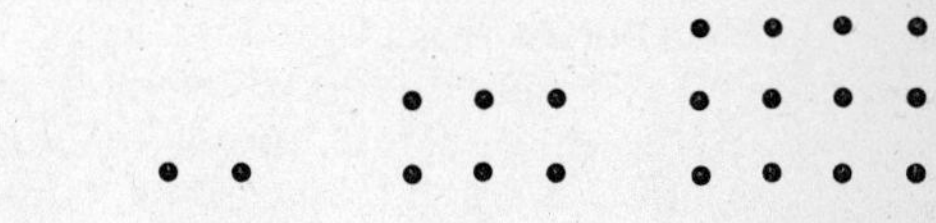

Figure 1.2
Square numbers

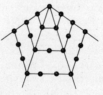

Figure 1.3
Pentagonal numbers

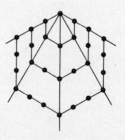

Figure 1.4
Hexagonal numbers

relations. They believed that bodies moving in space produce sounds. Perhaps this was suggested by the swishing of an object whirled on the end of a string. They believed, further, that a body which moves rapidly gives forth a higher note than one which moves slowly. Now according to their astronomy, the greater the distance of a planet from the earth the more rapidly it moved. Hence the sounds produced by the planets varied with their distances from the earth and these sounds all harmonized. But this "music of the spheres," like all harmony, reduced to no more than number relationships and hence so did the motions of the planets. We do not hear this music because we are accustomed to it from birth.

Other features of nature were "reduced" to number. The numbers 1, 2, 3, and 4, the *tetractys*, were especially valued. In fact, the Pythagorean oath is reported to be: "I swear in the name of the Tetractys which has been bestowed on our soul. The source and roots of the everflowing nature are contained in it." Nature was composed of fournesses such as the four geometric elements, point, line, surface, and solid; and the four material elements Plato later emphasized, earth, air, fire, and water.

The four numbers of the *tetractys* added up to ten and so ten was the ideal number and represented the universe. Because ten was ideal there must be ten bodies in the heavens. To fill out the required number the Pythagoreans introduced a central fire around which the earth, sun, moon, and the five planets then known revolved and a counter-earth on the opposite side of the central fire. We do not see this central fire and the counter-earth because the area of the earth on which we live faces away from them. The details are not worth pursuing; the main point is that the Pythagoreans tried to build an astronomical theory based on numerical relationships.

Because the Pythagoreans "reduced" astronomy and music to number, music and astronomy came to be linked with arithmetic and geometry, and all four subjects were regarded as mathematical. The four became part of the school curriculum and remained so even into medieval times, where they were labelled the quadrivium.

Aristotle in his *Metaphysics* sums up the Pythagorean identification of number and the real world:

> In numbers they seemed to see resemblances to things that exist and come into being—more than in fire and earth and water (such and such a modification of numbers being justice, another being soul and reason, another being opportunity—and similarly almost all other things being numerically expressible); since, again, that the modifications and the ratios of the musical scales were expressible in numbers;—since, then, all other things seemed in their whole nature

to be modelled on numbers, and numbers seemed to be the first things in the whole of nature, they supposed the elements of numbers to be the elements of all things, and the whole heaven to be a musical scale and a number.

The natural philosophy of the Pythagoreans is hardly substantial. Aesthetic considerations commingled with an obsession to find number relationships certainly led to assertions transcending observational evidence. Nor did the Pythagoreans develop any one branch of physical science very far. One could justifiably call their theories superficial. But, whether by luck or by intuitive genius, the Pythagoreans did hit upon two doctrines which proved later to be all-important: the first is that nature is built according to mathematical principles; the second that number relationships underlie, unify, and reveal the order in nature. Actually modern science adheres to the Pythagorean emphasis on number, though, as we shall see, the modern doctrines are a much more sophisticated form of Pythagoreanism.

The philosophers who chronologically succeeded the Pythagoreans were as much concerned with the nature of reality and the underlying mathematical design. Leuccipus (c. 440 B.C.) and Democritus (c. 460–c. 370 B.C.) are notable because they were most explicit in affirming the doctrine of atomism. Their common philosophy was that the world is composed of an infinite number of simple, eternal atoms. These differ in shape, size, hardness, order, and position. Every object is some combination of these atoms. Though geometrical magnitudes such as a line segment are infinitely divisible, the atoms are ultimate, indivisible particles. Properties such as shape, size, and the others just mentioned were properties of the atoms. All other properties such as taste, heat, and color were not in the atoms but in the effect of the atoms on the perceiver. This sensuous knowledge was unreliable because it varied with the perceiver. Like the Pythagoreans, the atomists asserted that the reality underlying the constantly changing diversity of the physical world was expressible in terms of mathematics. Moreover, the happenings in this world were strictly determined by mathematical laws.

After the Pythagoreans the most influential group to expound and propagate the doctrine of the mathematical design of nature was the Platonists, led, of course, by Plato. Though Plato (427–347 B.C.) took over some Pythagorean doctrines, he was a master who dominated Greek thought in the momentous 4th century B.C. He was the founder of the Academy in Athens, a center which attracted leading thinkers of his day and endured for nine hundred years.

Plato's belief in the rationality of the universe is perhaps best expressed in his dialogue the *Philebus:*

Protarchus: What question?

Socrates: Whether all this which they call the universe is left to the guidance of unreason and chance medley, or, on the contrary, as our fathers have declared, ordered and governed by a marvellous intelligence and wisdom.

Protarchus: Wide asunder are the two assertions, illustrious Socrates, for that which you were just now saying to me appears to be blasphemy, but the other assertion, that mind orders all things, is worthy of the aspect of the world, and of the sun, and of the moon, and of the stars and of the whole circle of the heavens; and never will I say or think otherwise.

The later Pythagoreans and the Platonists distinguished sharply between the world of things and the world of ideas. Objects and relationships in the material world were subject to imperfections, change, and decay and hence did not represent the ultimate truth, but there was an ideal world in which there were absolute and unchanging truths. These truths were the proper concern of the philosopher. About the physical world we can only have opinions. The visible and sensuous world is just a vague, dim, and imperfect realization of the ideal world. "Things are the shadows of ideas thrown on the screen of experience." Reality then was to be found in the ideas of sensuous, physical objects. Thus Plato would say that there is nothing real in a horse, a house, or a beautiful woman. The reality is in the universal type or idea of a horse, a house, or a woman. Infallible knowledge can be obtained only about pure ideal forms. These ideas are in fact constant and invariable, and knowledge concerning them is firm and indestructible.

Plato insisted that the reality and intelligibility of the physical world could be comprehended only through the mathematics of the ideal world. There was no question that this world was mathematically structured. Plutarch reports Plato's famous, "God eternally geometrizes." In the *Republic*, Plato said "the knowledge at which geometry aims is knowledge of the eternal, and not of aught perishing and transient." Mathematical laws were not only the essence of reality but eternal and unchanging. Number relations, too, were part of reality, and collections of things were mere imitations of numbers. Whereas with the earlier Pythagoreans numbers were immanent in things, with Plato they transcended things.

Plato went further than the Pythagoreans in that he wished not merely to understand nature through mathematics but to substitute mathematics for nature herself. He believed that a few penetrating glances at the physical world would suggest basic truths with which reason could then carry on unaided. From that point on there would be just mathematics. Mathematics would substitute for physical investigation.

Plutarch relates in his "Life of Marcellus" that Eudoxus and Archytas, famous contemporaries of Plato, used physical arguments to "prove" mathematical results. But Plato indignantly denounced such proofs as a corruption of geometry; they utilized sensuous facts in place of pure reasoning.

Plato's attitude toward astronomy illustrates his position on the knowledge to be sought. This science, he said, is not concerned with the movements of the visible heavenly bodies. The arrangement of the stars in the heavens and their apparent movements are indeed wonderful and beautiful to behold, but mere observations and explanation of the motions fall far short of true astronomy. Before we can attain to this true science we "must leave the heavens alone," for true astronomy deals with the laws of motion of true stars in a mathematical heaven of which the visible heaven is but an imperfect expression. He encouraged devotion to a theoretical astronomy whose problems please the mind and not the eye and whose objects are apprehended by the mind and not by vision. The varied figures the sky presents to the eye are to be used only as diagrams to assist the search for higher truths. We must treat astronomy, like geometry, as a series of problems merely suggested by visible things. The uses of astronomy in navigation, calendar-reckoning, and the measurement of time were of no interest to Plato.

Aristotle, though a student of Plato from whom he derived many ideas, had a quite different concept of the study of the real world and of the relation of mathematics to reality. He criticized Plato's other-worldliness and his reduction of science to mathematics. Aristotle was a physicist in the literal sense of the word. He believed in material things as the primary substance and source of reality. Physics, and science generally, must study the physical world and obtain truths from it. Genuine knowledge is obtained from sense experience by intuition and abstraction. These abstractions have no existence independent of human minds.

Aristotle did emphasize universals, general qualities that are abstracted from real things. To obtain these he said we "start with things which are knowable and observable to us and proceed toward those things which are clearer and more knowable by nature." He took the obvious sensuous qualities of objects, hypostatized them, and elevated them to independent, mental concepts.

Where was mathematics in Aristotle's scheme of things? The physical sciences were fundamental. Mathematics helped in the study of nature by describing formal properties such as shape and quantity. Also mathematics provided the reasons for facts observed in material phenomena. Thus geometry could provide the explanation of facts provided by optics and astronomy, and arithmetical ratios could give the basis for harmony. But mathematical concepts and principles are definitely abstractions from the real world. Because they are abstracted from the world, they are applicable to it. There is a faculty of the mind which enables us to arrive at these idealized properties of physical objects from sensations and these abstractions are necessarily true.

This brief survey of the philosophers who forged and molded the Greek intellectual world may serve to show that all of them stressed the study of nature for comprehension, understanding, and appreciation of the underlying reality. Moreover, from the time of the Pythagoreans practically all philosophers asserted that nature was designed mathematically. By the end of the classical period the doctrine of the mathematical design of nature was established and the search for mathematical laws had been instituted. Though this belief did not motivate all later mathematics, once accepted it was acted on by most of the great mathematicians, even those who had no contact with the belief. Of all the triumphs of the speculative thought of the Greeks, the most truly novel was their conception of the cosmos operating in accordance with mathematical laws discoverable by human thought.

The Greeks, then, were determined to seek truths and in particular truths about the mathematical design of nature. How does one go about seeking truths and guaranteeing that they are truths? Here, too, the Greeks provided the plan. Though this evolved gradually during the period from 600 to 300 B.C., and though there is some question as to when and by whom it was first conceived of, by 300 B.C. it was perfected.

Mathematics in a loose sense of the term, in the sense of utilizing numbers and geometrical figures, antedates the work of the classical Greeks by several thousand years. In this loose sense the term mathematics includes the contributions of many bygone civilizations among which the Egyptian and Babylonian are most prominent. In all of these, except the Greek civilization, mathematics was hardly a distinct discipline—it had no methodology nor was it pursued for other than immediate, practical ends. It was a tool, a series of disconnected, simple rules which enabled people to answer questions of daily life: calendar-reckoning, agriculture, and commerce. These rules were arrived at by trial and error, experience, and simple observation, and many were only approximately correct. About the best one can say for the mathematics of these civilizations is that it showed some vigor if not rigor of thought and more perseverance than brilliance. This mathematics is characterized by the word empirical. The empirical mathematics of the

Babylonians and the Egyptians also served as a prelude to the work of the Greeks.

Though the Greek culture was not entirely free of outside influences—Greek thinkers did travel and study in Egypt and Babylonia—and though mathematics in the modern sense of the word had to undergo a period of gestation even in the congenial intellectual atmosphere of Greece, what the Greeks created differs as much from what they took over as gold from tin.

Having decided to search for mathematical truths, the Greeks could not build upon the crude, empirical, limited, disconnected, and, in many instances, approximate results that their predecessors, notably the Egyptians and Babylonians, had compiled. Mathematics itself, the basic facts about number and geometrical figures, must be a body of truths, and mathematical reasoning, aimed at arriving at truths about physical phenomena, the motions of the heavens for example, must produce indubitable conclusions. How were these objectives to be attained?

The first principle was that mathematics was to deal with abstractions. For the philosophers who molded Greek mathematics, truth by its very meaning could pertain only to permanent, unchanging entities and relationships. Fortunately, the intelligence of man excited to reflection by the impressions of sensuous objects can rise to higher conceptions; these are the ideas, the eternal realities and the true object of thought. There was another reason for the preference for abstractions. If mathematics was to be powerful it must embrace in one abstract concept the essential feature of all the physical occurrences of that concept. Thus the mathematical straight line must embrace stretched strings, ruler's edges, boundaries of fields, and the paths of light rays. Accordingly, the mathematical line was to have no thickness, color, molecular structure, or tension. The Greeks were explicit in asserting that their mathematics dealt with abstractions. Speaking of geometricians, Plato said in *The Republic:*

> Do you not know also that although they make use of the visible forms and reason about them, they are thinking not of these, but of the ideals which they resemble; not of the figures which they draw, but of the absolute square and the absolute diameter . . . they are really seeking to behold the things themselves, which can be seen only with the eye of the mind?

Hence mathematics would deal first of all with abstract concepts such as point, line, and whole number. Other concepts such as triangle, square, and circle could then be defined in terms of the basic ones, which as Aristotle pointed out must be undefined or else there would be no starting point. The acuity of the Greeks is evident in the requirement that defined concepts must be shown to have counterparts in reality, either by demonstration or construction. Thus one could not define an angle trisector and prove theorems about it. It might not exist. And in fact, since the Greeks did not succeed in constructing an angle trisector under the limitations they imposed on constructions, they did not introduce this concept.

To reason about the concepts of mathematics the Greeks started with axioms, truths so self-evident that no one could doubt them. Surely such truths were available. Plato justified acceptance of the axioms by his theory of recollection or *anamnesis*. There was for him, as we noted earlier, an objective world of truths. Humans had experience as souls in another world before coming to earth and the soul had but to be stimulated to recall its prior experience in order to know that the axioms of geometry were truths. No experience on earth was necessary. Aristotle put it otherwise. The axioms are intelligible principles which appeal to the mind beyond possibility of doubt. The axioms, Aristotle said in *Posterior Analytics,* are known to be true by our infallible intuition. Moreover, we must have these truths on which to base our reasoning. If, instead, reasoning were to use some facts not known to be truths, further reasoning would be needed to establish these facts and this process would have to be repeated endlessly. There would then be an infinite regress. Among the axioms, he distinguished common notions and postulates. Common notions are true in all fields of thought and include statements such as "Equals added to equals give equals." Postulates apply to a specific subject such as geometry. Thus, "Two points determine a unique line." Aristotle did say that postulates need not be self-evident but when not must be supported by the consequences which follow from them. However, self-evidency was required by the mathematicians.

From the axioms, conclusions were to be derived by reasoning. There are many types of reasoning, for example, induction, reasoning by analogy, and deduction. Of the many types, only one guarantees the correctness of the conclusion. The conclusion that all apples are red because one thousand apples are found to be red is inductive and therefore not absolutely reliable. Likewise the argument that John should be able to graduate from college because his brother who inherited the same faculties did so, is reasoning by analogy and certainly not reliable. Deductive reasoning, on the other hand, though it can take many forms does guarantee the conclusion. Thus, if one grants that all men are mortal and Socrates is a man, one must accept that Socrates is mortal. The principle of logic involved here is one form of what Aristotle called syllogistic reasoning. Among other laws of deductive reasoning, Aristotle included the law of contradiction (a proposition cannot be both true and false) and the law of excluded middle (a proposition must be either true or false).

He and the world at large accepted unquestioningly that these deductive principles when applied to any premise yielded conclusions as reliable as the premise. Hence if the premises were truths, so would the conclusions be. It is worthy of note, especially in the light of what we shall be discussing later, that Aristotle abstracted the principles of deductive logic from the reasoning already practiced by mathematicians. Deductive logic is, in effect, the child of mathematics.

Though deductive reasoning was advocated by almost all the Greek philosophers as the only reliable method of obtaining truths, Plato's view was somewhat different. Though he would not object to deductive proof, he did regard it as superfluous, for the axioms and theorems of mathematics exist in some objective world independent of man, and in accordance with Plato's doctrine of anamnesis, man has but to recall them to recognize their indubitable truth. The theorems, to use Plato's own analogy in his *Theaetetus,* are like birds in an aviary. They exist and one has only to reach in to grasp them. Learning is but a process of recollection. In Plato's dialogue *Meno,* Socrates by skillful questioning elicits from a young slave the assertion that the square erected on the diagonal of an isosceles right triangle has twice the area of a square erected on a side. Socrates then triumphantly concludes that the slave, since he was not educated in geometry, recalled it under the proper suggestions.

It is important to appreciate *how radical the insistence on deductive proof was.* Suppose a scientist should measure the sum of the angles of a hundred different triangles in different locations and of different size and shape and find that sum to be 180° to within the limits of experimental accuracy. Surely he would conclude that the sum of the angles of any triangle is 180°. But his proof would be inductive, not deductive, and would therefore not be mathematically acceptable. Likewise, one can test as many even numbers as he pleases and find that each is a sum of two prime numbers. But this test is not a deductive proof and so the result is not a theorem of mathematics. Deductive proof is, then, a very stringent requirement. Nevertheless, the Greek mathematicians, who were in the main philosophers, insisted on the exclusive use of deductive reasoning because this yields truths, eternal verities.

There is another reason that philosophers favor deductive reasoning. Philosophers are concerned with broad knowledge about man and the physical world. To establish universal truths such as that man is basically good, that the world is designed, or that man's life has purpose, deductive reasoning from acceptable first principles is far more feasible than induction or analogy.

Still another reason for the classical Greeks' preference for deduction may be found in the organization of their society. Philosophical, mathematical, and artistic activities were carried on by the wealthier class. These people did no manual work. Slaves, metics (non-citizens), and free citizen-artisans were employed in business and in the household, and they even practiced the most important professions. Educated freemen did not use their hands and rarely engaged in commercial pursuits. Plato declared that the trade of a shopkeeper was a degradation to a freeman and wished that his engagement in such a trade be punished as a crime. Aristotle said that in the perfect state no citizen (as opposed to slaves) would practice any mechanical art. Among the Boeotians, one of the Greek tribes, those who defiled themselves with commerce were excluded from all state offices for ten years. To thinkers in such a society, experimentation and observation would be alien. Hence no results scientific or mathematical would be derived from such sources.

Though there are many reasons for the Greeks' insistence on deductive proof there is some question as to which philosopher or group of philosophers first laid down this requirement. Unfortunately our knowledge of the teachings and writings of the pre-Socratic philosophers is fragmentary and though various answers have been given there is no universally accepted one. By Aristotle's time the requirement was certainly in effect, for he is explicit about standards of rigor such as the need for undefined terms and the laws of reasoning.

How successful were the Greeks in executing their plan of obtaining mathematical laws of the universe? The cream of the mathematics created by such men as Euclid, Apollonius, Archimedes, and Claudius Ptolemy has fortunately come down to us. Chronologically these men belonged to the second great period of Greek culture, the Hellenistic or Alexandrian (300 B.C.–A.D. 600). During the 4th century B.C. King Philip of Macedonia undertook to conquer the Persians, who controlled the Near East and had been traditional enemies of the European Greeks. Philip was assassinated and was succeeded by his son Alexander. Alexander did defeat the Persians and moved the cultural center of the enlarged Greek empire to a new city which he modestly named after himself. Alexander died in 323 B.C. but his plan to develop the new center was continued by his successors in Egypt who adopted the royal title of Ptolemy.

It is quite certain that Euclid lived in Alexandria about 300 B.C. and trained students there, though his own education was probably acquired in Plato's Academy. This information, incidentally, is about all we have on Euclid's personal life. Euclid's work has the form of a systematic, deductive, and vast account of the separate discoveries of many classical Greeks. His chief work, the *Elements*, offers the laws of space and figures in space.

Euclid's *Elements* was by no means all of his contribution to the geometry of space. Euclid took up the theme of conic sections in a book no longer extant, and Apollonius (262–190 B.C.), a native of Pergamum in Asia Minor who learned mathematics in Alexandria, carried on this study of the parabola, ellipse, and hyperbola and wrote the classic work on the subject, the *Conic Sections*.

To this purely geometrical knowledge Archimedes (287–212 B.C.), who was educated in Alexandria but lived in Sicily, added several works, *On the Sphere and Cylinder, On Conoids and Spheroids,* and *The Quadrature of the Parabola,* all of which deal with the calculation of complex areas and volumes by a method introduced by Eudoxus (390–337 B.C.) and later known as the method of exhaustion. Nowadays these problems are solved by the methods of the calculus.

The Greeks made one more major addition to the study of space and figures in space—trigonometry. The originator of this work was Hipparchus, who lived in Rhodes and in Alexandria and died about 125 B.C. It was extended by Menelaus (*c.* A.D. 98) and given a complete and authoritative version by the Egyptian Claudius Ptolemy (d. A.D. 168), who worked in Alexandria. His major work was *Mathematical Composition,* known more popularly by the Arabic title, *Almagest.* Trigonometry concerns the quantitative relationships among the sides and angles of a triangle. The Greeks were concerned mainly with triangles on the surface of a sphere, the sides of which are formed by arcs of great circles (circles with centers at the center of the sphere) because the major application was to the motion of planets and stars, which in Greek astronomy moved along great circles. However, the same theory, when translated, readily applies to triangles in a plane, the form in which trigonometry is approached in our schools today. The introduction of trigonometry required of its users rather advanced arithmetic and some algebra. Just how the Greeks operated in these areas will be a later concern (Chapter V).

With these several creations mathematics emerged from obscure, empirical, disconnected fragments to brilliant, huge, systematic, and deep intellectual creations. However, the classics of Euclid, Apollonius, and Archimedes—Ptolemy's *Almagest* is an exception—that deal with the properties of space and of figures in space seem to be limited in scope and give little indication of the broader significance of their material. These works seem to have little relation to revealing truths about the workings of nature. In fact, these classics give only the formal, polished deductive mathematics. In this respect Greek mathematical texts are no different from modern mathematical textbooks and treatises. Such books seek only to organize and present the mathematical results that have been attained and so omit the motivations for the mathematics, the clues and suggestions for the theorems, and the uses to which the mathematical knowledge is put. Hence many writers on classical Greek mathematics assert that the mathematicians of the period were concerned only with mathematics for its own sake and they arrive at and defend this assertion by pointing to Euclid's *Elements* and Apollonius's *Conic Sections,* the two greatest compilations of work in that period. However, these writers have narrowed their focus. To look only at the *Elements* and the *Conic Sections* is like looking at Newton's paper on the binomial theorem and concluding that Newton was a pure mathematician.

The real goal was the study of nature. Insofar as the study of the physical world was concerned, even the truths of geometry were highly significant. It was clear to the Greeks that geometric principles were embodied in the entire structure of the universe, of which space was the primary component. Hence the study of space and figures in space was an essential contribution to the investigation of nature. Geometry was in fact part of the larger study of cosmology. For example, the study of the geometry of the sphere was undertaken when astronomy became mathematical, which happened in Plato's time. In fact, the Greek word for sphere meant astronomy for the Pythagoreans. And Euclids's *Phaenomena,* which was on the geometry of the sphere, was specifically intended for use in astronomy. With such evidence and with the fuller knowledge of how developments in mathematics took place in more recent times, we may be certain that the scientific investigations must have suggested mathematical problems and that the mathematics was part and parcel of the investigation of nature. But we need not speculate. We have only to examine what the Greeks accomplished in the study of nature and who were the men involved.

The greatest success in the field of physical science proper was achieved in astronomy. Plato, though fully aware of the impressive number of astronomical observations made by the Babylonians and Egyptians, emphasized that they had no underlying or unifying theory and no explanation of the seemingly irregular motions of the planets. Eudoxus, who was a student at the Academy and whose purely geometrical work is incorporated in Books V and XII of Euclid's *Elements,* took up the problem of "saving the appearances." His answer is the first reasonably complete astronomical theory known to history.

We shall not describe Eudoxus's theory except to state that it was thoroughly mathematical and involved the motions of interacting

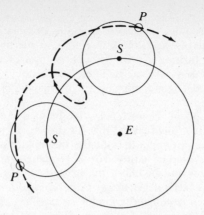

Figure 1.5

spheres. These spheres were, except for the "sphere" of fixed stars, not material bodies but mathematical constructions. Nor did he try to account for forces which would make the spheres rotate as he said they did. His theory is thoroughly modern in spirit, for today mathematical description and not physical explanation is the goal in science. This theory was superseded by the theory credited to the three greatest theoretical astronomers after Eudoxus, namely, Apollonius, Hipparchus and Ptolemy and incorporated in Ptolemy's *Almagest.*

Apollonius left no extant work in astronomy. However, his contributions are cited by Greek writers including Ptolemy in his *Almagest* (Book XII). He was so famous as an astronomer that he was nicknamed Ɛ (epsilon) because he had done much work on the motion of the moon and Ɛ was the symbol for the moon. Only one minor work of Hipparchus is known but he, too, is cited and credited in the *Almagest.*

The basic scheme of what is now referred to as Ptolemaic astronomy had entered Greek astronomy between the times of Eudoxus and Apollonius. In this scheme a planet P moves at a constant speed on a circle (Fig. 1.5) with center S while S itself moves with constant speed on a circle with center at the earth E. The circle on which S moves is called the deferent while the circle on which P moves is called an epicycle. The point S in the cases of some planets was the sun but in other cases it was just a mathematical point. The direction of the motion of P could agree with or be opposite to the direction of motion of S. The latter was the case for the sun and moon. Ptolemy also used a variation on this scheme to describe the motion of some of the planets. By properly selecting the radii of the epicycle and deferent, the speed of a body on its epicycle, and the speed of the center of the epicycle on the deferent, Hipparchus and Ptolemy were able to get descriptions of the motions which were quite in accord with the observations of their times. From the time of Hipparchus an eclipse of the moon could be predicted to within an hour or two, though eclipses of the sun were predicted somewhat less accurately. These predictions were possible because Ptolemy used trigonometry, which he said he created for astronomy.

From the standpoint of the search for truths, it is noteworthy that Ptolemy, like Eudoxus, fully realized that his theory was just a convenient mathematical description which fit the observations and was not necessarily the true design of nature. For some planets he had a choice of alternative schemes and he chose the mathematically simpler one. Ptolemy says in Book XIII of his *Almagest* that in astronomy one ought to seek as simple a mathematical model as possible. But Ptolemy's mathematical model was received as the truth by the Christian world.

Ptolemaic theory offered the first reasonably complete evidence of the uniformity and invariability of nature and is the final Greek answer to Plato's problem of rationalizing the apparent motions of the heavenly bodies. No other product of the entire Greek era rivals the *Almagest* for its profound influence on conceptions of the universe and none, except Euclid's *Elements,* achieved such unquestioned authority.

This brief account of Greek astronomy does not of course cover many other contributions to the subject nor does it reveal the depth and extent of the work even of the men treated. Greek astronomy was masterful and comprehensive and it employed a vast amount of mathematics. Moreover, almost every Greek mathematician devoted himself to the subject, including the masters Euclid and Archimedes.

The attainment of physical truths did not end with the mathematics of space and astronomy. The Greeks founded the science of mechanics. Mechanics deals with the motion of objects that may be considered as particles, the motion of extended bodies, and the forces that cause these motions. In his *Physics* Aristotle put together a theory of motion which is the high point of Greek mechanics. Like all of his physics, his mechanics is based on rational, seemingly self-evident principles, entirely in accord with observation. Though this theory held sway for almost two thousand years, we shall not review it because it was superseded by Newtonian mechanics. Notable additions to Aristotle's theory of motion were Archimedes' works on centers of gravity of bodies and his theory of the lever. What is relevant in all of this work is that mathematics played a leading role and thereby added to the conviction that mathematics was fundamental in penetrating the design of nature.

Next to astronomy and mechanics optics has been the subject most constantly pursued. This mathematical science, too, was founded by the Greeks. Almost all of the Greek philosophers, beginning with the Pythagoreans, speculated on the nature of light, vision, and color. Our concern, however, is with mathematical accomplishments in these areas. The first was the assertion on a priori grounds by Empedocles of Agrigentum (*c.* 490 B.C.)—Agrigentum was in Sicily—that light travels with finite velocity. The first systematic treatments of light that we have are Euclid's *Optics* and *Catoptrica*.* The *Optics* is concerned with the problem of vision and with the use of vision to determine sizes of objects. The *Catoptrica* (theory of mirrors) shows how light rays behave when reflected from plane, concave, and convex mirrors and the effect of this behavior on what we see. Like the *Optics* it starts with definitions which are really postulates. Theorem 1 (an axiom in modern texts) is fundamental in geometrical optics and is known as the law of reflection. It says that the angle *A* that a ray incident from point *P* makes with a mirror (Fig. 1.6) equals the angle *B* which the reflected ray makes with the mirror. Euclid also proves the law for a ray striking a convex or a concave mirror (Fig. 1.7). At the point of contact he substitutes the tangent *R* for the mirror. Both books are thoroughly mathematical not only in content but in organization. Definitions, axioms and theorems dominate as in Euclid's *Elements*.

From the law of reflection, the mathematician and engineer Heron (1st century A.D.) drew an important consequence. If *P* and *Q* in Figure 1.6 are any two points on one side of the line *ST*, then of all the paths one could follow in going from point *P* to the line and then to point *Q*, the shortest path is by way of the point *R* such that the two line segments *PR and QR* make equal angles with the line. And this is exactly the path a light ray takes. Hence, the light ray takes the shortest path in going from *P* to the mirror to *Q*. Apparently nature is well acquainted with geometry and employs it to full advantage. This proposition appears in Heron's *Catoptrica* which also treats concave and convex mirrors and combinations of mirrors.

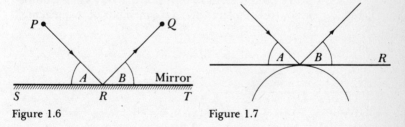

Figure 1.6   Figure 1.7

---

* The version we have today is probably a compilation of several works including Euclid's.

Any number of works were written on the reflection of light by mirrors of various shapes. Among these are the now lost works, Archimedes' *Catoptrica* and Apollonius's *On the Burning Mirror* (*c.*190 B.C.), and the extant work of Diocles, *On Burning-Mirrors* (*c.*190 B.C.). Burning mirrors were concave mirrors in the form of portions of a sphere, paraboloids of revolution (formed by revolving a parabola about its axis), and ellipsoids of revolution. Apollonius knew and Diocles' book contains the proof that a paraboloidal mirror will reflect light emanating from the focus into a beam parallel to the axis of the mirror (Fig. 1.8). Conversely, rays coming in parallel to the axis will after reflection be concentrated at the focus. The sun's rays thus concentrated produce great heat at the focus and hence the term burning mirror. This is the property of the paraboloidal mirror which Archimedes is reported to have used to concentrate the sun's rays on the Roman ships besieging his home city Syracuse and to set them afire. Apollonius also knew the reflection properties of the other conic sections, such as that all rays emanating from one focus of an ellipsoidal mirror will be reflected to the other focus. He gives the relevant geometrical properties of the ellipse and hyperbola in Book III of his *Conic Sections*.

The Greeks founded many other sciences, notably geography and hydrostatics. Eratosthenes of Cyrene (*c.*284–*c.*192 B.C.), one of the most learned men of antiquity and director of the library at Alexandria, made numerous calculations of distances between significant places on the portion of our earth known to the Greeks. He also made a now famous and quite accurate calculation of the circumference of the earth and wrote his *Geography*, in which beyond describing his mathematical methods he also gave his explanation of causes for the changes which had taken place on the earth's surface.

The most extensive work on geography was Ptolemy's *Geography*, in eight books. Ptolemy not only extended Eratosthenes' work but located eight thousand places on the earth in terms of the very same latitude and longitude we now use. Ptolemy also gave methods of mapmaking,

Figure 1.8

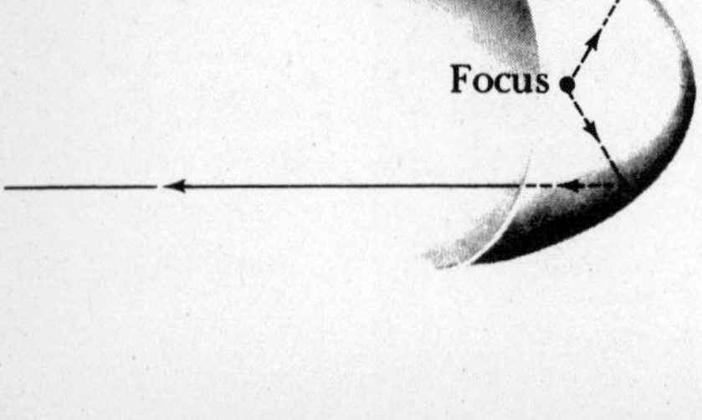

Focus

some of which are still used, particularly the method of stereographic projection. In all of this work in geography the geometry of figures on a sphere, applied from the 4th century B.C. onward, was basic.

As for hydrostatics, the subject which deals with the pressure on bodies which are placed in water, Archimedes' book *On Floating Bodies* is the foundational work. Like all of the works we have been describing it is thoroughly mathematical in approach and derivation of results. In particular it contains what is now known as Archimedes' principle, that a body immersed in water is buoyed up by a force equal to the weight of the water displaced. Thus we owe to Archimedes the explanation of how man can remain afloat in a world of forces that tend to submerge him.

Though the deductive approach to mathematics and the mathematical representation of the laws of nature dominated the Alexandrian Greek period, we should note that the Alexandrians, unlike the classical Greeks, also resorted to experimentation and observation. The Alexandrians took over and utilized the remarkably accurate astronomical observations which the Babylonians had made over a period of two thousand years. Hipparchus made a catalogue of the stars observable in his time. Inventions (notably by Archimedes and the mathematician and engineer Heron) included sun-dials, astrolabes, and uses for steam and water power.

Particularly famous was the Alexandrian Museum, which was started by Ptolemy Soter, the immediate successor of Alexander in Egypt. The Museum was a home for scholars and included a famous library of about 400,000 volumes. Since it could not house all the manuscripts an additional 300,000 were housed in the Temple of Serapis. The scholars also gave instruction to students.

With their mathematical work and many scientific investigations, the Greeks gave substantial evidence that the universe is mathematically designed. Mathematics is immanent in nature; it is the truth about nature's structure, or, as Plato would have it, the reality about the physical world. There is law and order in the universe and mathematics is the key to this order. Moreover, human reason can penetrate the plan and reveal the mathematical structure.

The impetus for the conception of a logical, mathematical approach to nature must be credited primarily to Euclid's *Elements*. Though this work was intended to be a study of physical space, its organization, ingenuity, and clarity inspired the axiomatic-deductive approach not only to other areas of mathematics such as the theory of numbers but to all of the sciences. Through this work the notion of a logical organization of all physical knowledge based on mathematics entered the intellectual world.

Thus the Greeks founded the alliance between mathematics and the study of nature's design which has since become the very basis of modern science. Until the latter part of the 19th century, the search for mathematical design was the search for truth. The belief that mathematical laws were the truth about nature attracted the deepest and noblest thinkers to mathematics.

II

The Flowering of Mathematical Truths

> The chief aim of all investigations of the external world
> should be to discover the rational order and harmony
> which has been imposed on it by God and which He re-
> vealed to us in the language of mathematics.
>
> JOHANNES KEPLER

The majestic Greek civilization was destroyed by several forces. The
first was the gradual conquest by the Romans of Greece, Egypt, and the
Near East. The Roman objective in extending its political power was
not to spread its materialistic culture. The subjugated areas became col-
onies from which great wealth was extracted by expropriation and by
taxation.

The rise of Christianity was another blow to pagan Greek culture.
Though Christian leaders adopted many Greek and Oriental myths
and customs with the intent of making Christianity more acceptable to
converts, they opposed pagan learning and even ridiculed mathematics, astronomy, and physical science. Despite cruel persecution by the
Romans, Christianity spread and became so powerful that the Roman
emperor Constantine the Great in his Edict of Milan of A.D. 313 recognized Christianity as the official religion of the Empire. Later, Theodosius (ruled A.D. 379–396) proscribed the pagan religions and in 392 ordered that their temples be destroyed.

Thousands of Greek books were burned by the Romans and the
Christians. In 47 B.C., the Romans set fire to the Egyptian ships in the
harbor of Alexandria; the fire spread and burned the library—the most
extensive of ancient libraries. In the year that Theodosius banned the
pagan religions, the Christians destroyed the temple of Serapis in Alexandria, which housed the only remaining sizable collection of Greek
works. Many other works written on parchment were expunged by the
Christians so that they could use the parchment for their own writings.

The late history of the Roman Empire is also relevant. The Emperor

Theodosius divided the extensive empire between his two sons, Honorius, who was to rule Italy and western Europe, and Arcadius, who was to rule Greece, Egypt, and the Near East. The western part was conquered by the Goths in the 5th century A.D. and its subsequent history belongs to the history of medieval Europe. The eastern part preserved its independence. Since the Eastern Roman Empire, known also as the Byzantine Empire, included Greece proper and Egypt, Greek culture and Greek works were to some extent preserved.

The final blow to the Greek civilization was the conquest of Egypt by the upsurging Moslems in A.D. 640. The remaining books were destroyed on the ground that, as Omar, the Arab conqueror, put it, "Either the books contain what is in the Koran, in which case we don't have to read them, or they contain the opposite of what is in the Koran, in which case we must not read them." And so for six months the baths of Alexandria were heated by burning rolls of parchment.

After the capture of Egypt by the Mohammedans the majority of scholars migrated to Constantinople, which had become the capital of the Eastern Roman Empire. Though no activity along the lines of Greek thought could flourish in the unfriendly Christian atmosphere of Byzantium, this inflow of scholars and their works to comparative safety increased the treasury of knowledge that was to reach Europe 800 years later.

India and Arabia contributed to the continuity of mathematical activity and introduced some ideas that were to play a larger role later.* During the years from A.D. 200 to about 1200 the Hindus, influenced somewhat by the Greek works, made some original contributions to arithmetic and algebra. The Arabs, whose empire at its height extended over all the lands bordering the Mediterranean and into the Near East and embraced many races united by Mohammedanism, absorbed the Greek and Hindu contributions and also made some advances of their own. These, in the spirit of the Alexandrian Greeks, commingled deductive reasoning and experimentation. The Arabs contributed to algebra, geography, astronomy, and optics. They also built colleges and schools for the transmission of knowledge. It is to the credit of the Arabs that though they were firm adherents of their own religion, they did not allow religious doctrines to restrict their mathematical and scientific investigations.

Despite the fact that both the Hindus and the Arabs were able to profit from the magnificent foundations erected by the Greeks and though they furthered Greek mathematics and science, they were not possessed as were the Greeks to understand the structure of the uni-

---

* We shall say more about the work of the Hindus and Arabs in Chapter V.

verse. The Arabs translated and commented extensively and even critically on Greek works but nothing of great moment or magnitude was added to the truths already known. By A.D. 1500 their empire was destroyed by the Christians in the West and by internal strife in the East.

While the Arabs were building and expanding their civilization, another civilization was being founded in Western Europe. A high level of culture in this region was attained in the medieval period, which extended from about A.D. 500 to 1500. This culture was dominated by the Catholic Church, and its teachings, however deep and meritorious, did not favor the study of the physical world. The Christian God ruled the universe and man's role was to serve and please Him and by so doing win salvation, whereupon the soul would live in an after-life of joy and splendor. The conditions of life on this earth were immaterial and hardship and suffering were not only to be tolerated but were in fact to be undergone as a test of man's faith in God. Understandably, interest in mathematics and science which had been motivated in Greek times by the study of the physical world was at a nadir. The intellectuals of medieval Europe were devoted seekers of truths but these they sought in revelation and in the study of the Scriptures. Hence medieval thinkers did not adduce additional evidence for the mathematical design of nature. However, late medieval philosophy did support the belief in the regularity and uniformity of nature's behavior, though this was thought to be subject to the will of God.

Late medieval Europe was shaken and altered by a number of revolutionary influences. Among the many which converted the medieval civilization into the modern, the most important for our present concern was the acquisition and study of Greek works. These became known through the Arabic translations and through Greek works which had been kept intact in the Byzantine Empire. In fact, when the Turks conquered this empire in 1453 many Greek scholars fled westward with their books. It was from Greek works that the leaders of the intellectual revitalization of Europe learned nature is mathematically designed and this design is harmonious, aesthetically pleasing, and the inner truth about nature. Nature not only is rational and orderly but acts in accordance with inexorable and immutable laws. European scientists began their study of nature as the children of ancient Greece.

That the revival of Greek ideals induced some to take up the study of nature is indubitable. But the speed and intensity of the revival of mathematics and science were due to many other factors. The forces which overthrew one culture and fostered the development of a new one are numerous and complicated. The rise of science has been studied by many scholars and much history has been devoted to pinpointing the causes. We shall not attempt here to do more than name them.

The rise of a class of free artisans, and a consequent interest in materials, skills, and technology, generated scientific problems. Geographical explorations, motivated by the search for raw materials and gold, introduced knowledge of strange lands and customs which challenged medieval European culture. The Protestant revolution rejected some Catholic doctrines, thereby fostering controversy and even scepticism concerning both religions. The Puritan emphasis on work and utility of knowledge to mankind, the introduction of gunpowder, which raised new military problems such as the motion of projectiles, and the problems raised by the navigations over thousands of miles of sea out of sight of land all motivated the study of nature. The invention of printing permitted the spread of knowledge which the Church had been able to control. Though authorities differ on the degree to which one or more of these forces may have influenced the investigation of nature, it suffices for our purposes to note their multitude and the universally accepted fact that the pursuit of science did become the dominant feature of modern European civilization.

The Europeans generally did not respond immediately to the new forces and influences. During the period often labelled humanistic the study and absorption of Greek works were far more characteristic than active pursuit of the Greek objectives. But by about A.D. 1500 minds infused with Greek goals—the application of reason to the study of nature and the search for the underlying mathematical design—began to act. However, they faced a serious problem. The Greek goals were in conflict with the prevailing culture. Whereas the Greeks believed in the mathematical design of nature, nature conforming invariably and unalterably to some ideal plan, late medieval thinkers ascribed all plan and action to the Christian God. He was the designer and creator, and all the actions of nature followed the plan laid down by this agency. The universe was the handiwork of God and subject to His will. The mathematicians and scientists of the Renaissance and several succeeding centuries were orthodox Christians and so accepted this doctrine. But Catholic teachings by no means included the Greek doctrine of the *mathematical* design of nature. How then was the attempt to understand God's universe to be reconciled with the search for the mathematical laws of nature? The answer was to add a new doctrine, that the Christian God had designed the universe mathematically. Thus the Catholic doctrine postulating the supreme importance of seeking to understand God's will and His creations took the form of a search for God's mathematical design of nature. Indeed the work of 16th-, 17th-, and most 18th-century mathematicians was, as we shall soon see more clearly, a religious quest. The search for the mathematical laws of nature was an act of devotion which would reveal the glory and grandeur of His handiwork. Mathematical knowledge, the truth about God's design of the universe, was as sacrosanct as any line of Scripture. Man could not hope to perceive the divine plan as clearly as God Himself understood it, but man could with humility and modesty seek at least to approach the mind of God and so understand God's world.

One can go further and assert that these mathematicians were sure of the existence of mathematical laws underlying natural phenomena and persisted in the search for them because they were convinced a priori that God had incorporated them into the construction of the universe. Each discovery of a law of nature was hailed as evidence of God's brilliance rather than the investigator's. The beliefs and attitudes of the mathematicians and scientists exemplify the larger cultural phenomenon that swept Renaissance Europe. The recently rediscovered Greek works confronted a deeply devout Christian world, and the intellectual leaders born in one and attracted to the other fused the doctrines of both.

Perhaps the most impressive evidence that the Greek doctrine of the mathematical design of nature coupled with the Renaissance belief in God's authorship of that design had taken hold in Europe is furnished by the work of Nicolaus Copernicus and Johannes Kepler. Up to the 16th century, the only sound and useful astronomical theory was the geocentric system of Hipparchus and Ptolemy. This was the theory accepted by professional astronomers and applied to calendar-reckoning and navigation. Work on a new astronomical theory was begun by Copernicus (1473–1543). At the University of Bologna, which he entered in 1497, he studied astronomy. In 1512 he assumed his duties as canon of the Cathedral of Frauenberg in East Prussia. This work left Copernicus with plenty of time to make astronomical observations and to think about the relevant theory. After years of reflection and observation, Copernicus evolved a new theory of planetary motions which he incorporated in a classic work, *On the Revolutions of the Heavenly Spheres.* He had written his first version in 1507 but feared to publish it because it would antagonize the Church. The book appeared in 1543, the year in which he died.

When Copernicus began to think about astronomy, the Ptolemaic theory had become somewhat more complicated. More epicycles had been added to those introduced by Ptolemy in order to make the theory fit the increased amount of observational data gathered largely by the Arabs. In Copernicus's time the theory required a total of seventy-seven circles to describe the motion of the sun, moon, and the five planets known then. To many astronomers the theory, as Copernicus says in his Preface, was scandalously complex.

Copernicus had studied the Greek works and had become convinced that the universe was mathematically and harmoniously designed. Harmony demanded a more pleasing theory than the complicated extension of Ptolemaic theory. He had read that some Greek authors, notably Aristarchus (3rd century B.C.), had suggested the possibility that the sun was stationary and that the earth revolved about the sun and rotated on its axis at the same time. He decided to explore this possibility.

The upshot of his reasoning was that he used the Ptolemaic scheme of deferent and epicycle (Chapter I) to describe the motions of the heavenly bodies, with, however, the all-important difference that the *sun* was at the center of each deferent. The earth itself became a planet moving on an epicycle while rotating on its axis. Thereby he achieved considerable simplification. He was able to reduce the total number of circles (deferents and epicycles) to thirty-four instead of the seventy-seven required in the geocentric theory.

The more remarkable simplification was achieved by Johannes Kepler (1571–1630), one of the most intriguing figures in the history of science. In a life beset by many personal misfortunes and hardships occasioned by religious and political events, Kepler had the good fortune in 1600 to become an assistant to the famous astronomer Tycho Brahe. Brahe was then engaged in making extensive new observations, the first major undertaking since Greek times. These observations and others which Kepler made himself were invaluable to him. When Brahe died in 1601 Kepler succeeded him as Imperial Mathematician to Emperor Rudolph II of Austria.

Kepler's scientific reasoning is fascinating. Like Copernicus he was a mystic, and like Copernicus he believed that the world was designed by God in accordance with some simple and beautiful mathematical plan. He said in his *Mystery of the Cosmos* (1596), the mathematical harmonies in the mind of the Creator furnish the cause "why the number, the size, and the motion of the orbs are as they are and not otherwise." This belief dominated all his thinking. But Kepler also had qualities which we now associate with scientists. He could be coldly rational. Though his fertile imagination triggered the conception of new theoretical systems, he knew that theories must fit observations and, in his later years, saw even more clearly that empirical data may indeed suggest fundamental principles of science. He therefore sacrificed even his most beloved mathematical hypotheses when he saw that they did not fit observational data, and it was precisely this incredible persistence in refusing to tolerate discrepancies any other scientist of his day would have disregarded that led him to espouse radical scientific ideas. He also had the humility, patience, and energy that enable great men to perform extraordinary labor.

Kepler's search for the mathematical laws of nature, which his beliefs assured him existed, led him to spend years in following up false trails. In the Preface to his *Mystery of the Cosmos,* we find him saying: "I undertake to prove that God, in creating the universe and regulating the order of the cosmos, had in view the five regular bodies of geometry as known since the days of Pythagoras and Plato, and that he has fixed according to those dimensions the number of heavens, their proportions, and the relations of their movements." However, the deductions from his attempt to build a theory based on the five regular polyhedra were not in accord with observations, and he abandoned this approach only after he had made extraordinary efforts to apply it in modified form.

But he was eminently successful in later efforts to find harmonious mathematical relations. His most famous and important results are known today as Kepler's three laws of planetary motion. The first two were published in a book of 1609 bearing a long title and often referred to by the first part, *The New Astronomy,* or by the last part, *Commentaries on the Motion of the Planet Mars.* The first of these laws is especially remarkable, for Kepler broke with the tradition of two thousand years that circles or spheres must be used to describe heavenly motions. Instead of resorting to deferent and several epicycles, which both Ptolemy and Copernicus had used to describe the motion of any one planet, Kepler found that a single ellipse would do. His first law states that each planet moves on an ellipse and that the sun is at one (common) focus of each of these elliptical paths (Fig. 2.1). The other focus of each ellipse is merely a mathematical point at which nothing exists. This law is of immense value in comprehending readily the paths of the planets. Of course Kepler, like Copernicus, added that the earth also rotates on its axis as it travels along its elliptical path.

But astronomy had to go much further if it was to be useful. It must tell us how to predict the positions of the planets. If one finds by observation that a planet is at a particular position, *P*, say, in Figure 2.1, he might like to know when it will be at some other position such as a solstice or an equinox, for example. What is needed is the velocity with which the planets move along their respective paths.

Here, too, Kepler made a radical step. Copernicus and the Greeks

Figure 2.1. Each planet moves in an ellipse about the sun.

Figure 2.2. Kepler's law of equal areas.

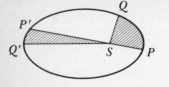

had always used constant velocities. A planet moved along its epicycle so as to cover equal arcs in equal times, and the center of each epicycle also moved at a constant velocity on another epicycle or on a deferent. But Kepler's observations told him that a planet moving on its ellipse does not move at a constant speed. A hard and long search for the correct law of velocity ended successfully. What he discovered was that if a planet moves from P to Q (Fig. 2.2) in, say, one month, then it will also move from P' to Q' in one month, provided that the *area PSQ* equals the *area P'SQ'*. Since P is nearer the sun than P' is, the arc PQ must be larger than the arc $P'Q'$ if the areas PSQ and $P'SQ'$ are equal. Hence the planets do not move at a constant velocity. In fact, they move faster when closer to the sun.

Kepler was overjoyed to discover this second law. Although it is not so simple to apply as a law of constant velocity, it nonetheless confirmed his fundamental belief that God had used mathematical principles to design the universe. God had chosen to be just a little more subtle, but, nevertheless, a mathematical law clearly determined how fast the planets moved.

Another major problem remained. What law described the distances of the planets from the sun? The problem was now complicated by the fact that a planet's distance from the sun was not constant. Hence Kepler searched for a new principle which would take this fact into account. He believed that nature was not only mathematically but harmoniously designed and he took this word "harmony" very literally. Thus he believed that there was a music of the spheres, which produced a harmonious tonal effect, not in actual sound but discernible by some translation of the facts about planetary motions into musical notes. He followed this lead and after an amazing combination of mathematical and musical arguments arrived at the law that, if T is the period of·revolution of any planet and D is its mean distance from the sun, then $T^2 = kD^3$, where k is a constant which is the same for all the planets. This is Kepler's third law of planetary motion and the one he triumphantly announced in his book *The Harmony of the World* (1619).

After stating his third law Kepler broke forth into a paean to God:

"Sun, moon, and planets glorify Him in your ineffable language! Celestial harmonies, all ye who comprehend His marvelous works, praise Him. And thou, my soul, praise thy Creator! It is by Him and in Him that all exists. That which we know best is comprised in Him, as well as in our vain science."

The strength of Copernicus's and Kepler's conviction that God must have designed the world harmoniously and simply can be judged by the objections with which they had to contend. That the other planets were in motion even according to Ptolemaic theory was explained by the Greek doctrine that these were made of special, light matter and therefore easily moved, but how could the heavy earth be put into motion? Neither Copernicus nor Kepler could answer this question. An argument against the earth's rotation maintained that objects on the surface would fly off into space just as an object on a rotating platform will fly off. Neither man could rebut this argument. To the further objection that a rotating earth would itself fly apart, Copernicus replied weakly that the earth's motion was natural and could not destroy the planet. Then he countered by asking why the sky did not fall apart under the very rapid daily motion which the geocentric theory called for. Yet another objection was that, if the earth rotated from west to east, an object thrown up into the air would fall back to the west of its original position, because the earth moved while the object was in the air. If, moreover, the earth revolved about the sun, then, since the velocity of an object is proportional to its weight, or so at least Greek and Renaissance physics maintained, lighter objects on the earth should be left behind. Even the air should be left behind. To the last argument Copernicus replied that air is earthy and so moves in sympathy with the earth. The substance of these objections is that a rotating and revolving earth did not fit in with the theory of motion due to Aristotle and commonly accepted in Copernicus's and Kepler's time.

Another class of scientific arguments against a heliocentric theory came from astronomy proper. The most serious one stemmed from the fact that the heliocentric theory regarded the stars as fixed. However, in six months the earth changes its position in space by about 186 million miles. Hence if one notes the direction of a particular star at one time and again six months later, he should observe a difference in direction. But this difference was not observed in Copernicus's or Kepler's time. Copernicus argued that the stars are so far away that the difference in direction was too small to be observed. His explanation did not satisfy his critics, who countered that, if the stars were that distant, then they should not be clearly observable. In this instance, Copernicus's answer was correct. The change in direction over a six-month period for the nearest star is an angle of $0.31''$; and this was first detected in 1838 by the mathematician Friedrich Wilhelm Bessel, who by that time had a good telescope at his disposal.

The traditionalists asked further, why do we not feel any motion if the earth is moving around the sun at about 18 miles per second and is rotating at the equator at about 0.3 miles per second? Our senses, in fact, tell us that the sun is moving in the sky. To people of Kepler's time, the argument that we do not feel ourselves moving at the very high speeds called for by the new astronomy was incontrovertible. All of these scientific objections to a moving earth were weighty and could not be dismissed as the stubborness of die-hards who refused to see the truth.

Copernicus and Kepler were deeply religious and yet both denied one of the prevailing doctrines of Christianity. This doctrine affirmed that man was at the center of the universe; he was the chief concern of God. The heliocentric theory, in contrast, by putting the sun at the center of the universe, undermined this comforting dogma of the Church. It caused man to appear to be one of a possible host of wanderers drifting through a cold sky. It seemed less likely that he was born to live gloriously and to attain paradise upon his death. Less likely, too, was it that he was the object of God's ministrations. Copernicus attacked the doctrine that the earth is at the center of the universe by pointing out that the size of the universe is so immense that to speak of a center is meaningless. But this counter-argument carried little weight with his contemporaries.

Against all these objections to a heliocentric theory, Copernicus and Kepler had but one overwhelming retort. Each had achieved mathematical simplification and a more harmonious and aesthetically superior theory. If a better mathematical account could be given, then, in view of the belief that God had designed the world and would clearly have used the superior theory, the heliocentric theory must be correct.

There are many passages in Copernicus's *On the Revolutions of the Heavenly Spheres* and in Kepler's numerous writings which bear unmistakable testimony to their conviction that they had found the right theory. Kepler, for example, remarked of his elliptical theory of motion, "I have attested it as true in my deepest soul and I contemplate its beauty with incredible and ravishing delight." The very title of Kepler's work of 1619, *The Harmony of the World,* and endless paeans to God, expressing satisfaction with the grandeur of God's mathematical design, exhibit his conviction.

At first only mathematicians supported the new theory. This is not surprising. Only a mathematician convinced that the universe was mathematically and simply designed would have had the mental fortitude to disregard the prevailing philosophical, religious, and scientific counter-arguments and to appreciate the mathematics of such a revolutionary astronomy. Only one possessed of unshakable convictions about the importance of mathematics in the design of the universe would have dared to affirm the new theory against the mass of powerful opposition it met.

Support for the new theory came from an unexpected development. Early in the 17th century the telescope was invented and Galileo, upon hearing of this invention, built one himself. He then proceeded to make observations of the heavens which startled his contemporaries. He detected four moons of Jupiter (we can now observe twelve) and this discovery showed that a moving planet can have satellites. Galileo saw irregular surfaces and mountains on the moon, spots on the sun, and a bulge around the equator of Saturn (which we now call the rings of Saturn). Here was further evidence that the planets were like the earth and certainly not perfect bodies composed of some special ethereal substance, as Greek and medieval thinkers had believed. The Milky Way, which had hitherto appeared to be just a broad band of light, could be seen with the telescope to be composed of thousands of stars. Thus there were other suns and presumably other planetary systems suspended in the heavens. Copernicus had predicted that if human sight could be enhanced, then man would be able to observe phases of Venus and Mercury, just as the naked eye can discern phases of the moon. With his telescope Galileo did view the phases of Venus. His observations convinced him that the Copernican system was correct and in the classic *Dialogue on the Great World Systems* (1632) he defended it strongly. The new theory won acceptance also because it was simpler for calculations made by astronomers, geographers, and navigators. By the middle of the 17th century the scientific world was willing to proceed on a heliocentric basis and the claim of mathematical laws to truth was immeasurably strengthened.

To maintain the doctrines of the revolution of the earth around the sun and the daily rotation of the earth in the intellectual atmosphere of the early 17th century was by no means a casual matter. Galileo's trial before the Inquisition is well known. The devout Catholic Pascal found his works on the Index of Prohibited Books because he had the temerity to defy the Jesuits by declaring in his *Provincial Letters:* "It was also in vain that you did obtain the decree of Rome against Galileo which condemned his opinion touching the motion of the earth, for that will not prove that the earth is standing still. . . ."

Copernicus and Kepler accepted unquestioningly the fusion of the Greek doctrine that nature is designed mathematically and the Catholic doctrine that God created and designed the universe. René Descartes (1596–1650) set about erecting the new philosophy of science systematically, clearly, and forcefully. Descartes was primarily a philosopher, second a cosmologist, third a physicist, fourth a biologist, and only fifth a mathematician, though he is regarded as one of the gems in the diadem of mathematics. His philosophy is important because it dominated 17th-century thought and influenced such giants as Newton and Leibniz. His primary goal, to find the method of establishing truths in all fields, he pursued in his basic work, *Discourse on the Method of Rightly Conducting the Reason and Seeking for Truth in the Sciences* (1637).

Descartes began the construction of his philosophy by accepting only those facts that were to him beyond doubt. How then did he differentiate between acceptable and unacceptable evidence? In his *Rules for the Direction of the Mind* (written in 1628 but published posthumously), he stated: "Concerning the objects we propose to study, we should investigate not what others have thought nor what we ourselves conjecture, but what we can intuit clearly and evidently or deduce with certainty, for there is no other way to acquire knowledge." The mind's immediate apprehension of basic, clear, and distinct truths—an intuitive power— and the deduction of consequences are the essence of his philosophy of knowledge. There are thus, according to Descartes, only two mental acts that enable us to arrive at knowledge without any fear of error: intuition and deduction. However, in the *Rules* he gave greater credence to intuition: "Intuition is the undoubting conception of a pure and attentive mind, which arises from the light of reason alone, and is more certain than deduction."

In the *Discourse,* he defended the existence of mind and the sure and indubitable knowledge it possesses. By relying upon fundamental intuitions Descartes in the *Discourse* hastened to prove that God exists. And then with an argument that surely involves circular reasoning he reassured himself that our intuitions and methods of deduction must be sound because God would not deceive us. God he stated is "a substance that is infinite, eternal, immutable, independent, all-knowing, all-powerful and by which I myself and everything else . . . have been created."

As for truths in mathematics proper, he said in his *Meditations* (1641), "I counted as the most certain the truths which I conceived clearly as regards figures, numbers, and other matters which pertain to arithmetic and geometry, and, in general to pure and abstract mathematics." "Only the mathematicians contrived to reach certainty and evidence since they started from what is easiest and simplest." The concepts and truths of mathematics do not come from the senses. They are innate in our minds from birth and are placed there by God. Sense perception of material triangles could never give the mind the concept of an ideal triangle. It is equally clear to the mind that the sum of the angles of a triangle must be 180°.

Descartes turned next to the physical world. We can be sure, he said, that the intuitions which the mind recognizes clearly and the deductions from them apply to the physical world. It was clear to him that God designed the world mathematically. In his *Discourse* he affirmed the existence of "certain laws which God has so established in nature and of notions which he has impressed in our souls, that once we have reflected sufficiently upon them, we can no longer doubt their being accurately observed in all that exists or happens in the world."

Descartes affirmed further that the laws of nature are invariable since they are but part of a predetermined mathematical pattern. Even before publishing his *Discourse,* Descartes wrote to Father Marin Mersenne, a theologian and close associate of mathematicians, on April 15, 1630:

> Do not be afraid to proclaim everywhere that God established these laws in nature just as a sovereign establishes laws in his kingdom. . . . And just as a king has more majesty when he is less familiarly known by his subjects, we judge God's greatness as incomprehensible and do not think that we are without a king. One will tell you that if God established these truths, He would be able to change them just as a king changes his laws; to which one should reply that it is possible if His will can change. But I understand these truths as eternal and unvarying in the same way that I judge God.

Here Descartes denied the prevailing belief that God continually intervenes in the functioning of the universe.

To study the physical world Descartes wished to employ only mathematics, for, he said in the *Discourse,* "Of all those who have hitherto sought for truth in the Sciences, it has been the mathematicians alone who have been able to succeed in making any demonstrations, that is to say, producing reasons which are evident and certain." In the study of the physical world, Descartes was sure mathematics would suffice. He said in the *Principles of Philosophy* (1644),

> I frankly confess that in respect to corporeal things I know of no other matter than that . . . which the geometers entitle quantity and take as being the object of their demonstrations. In treating it I consider only the divisions, the shapes and the movements, and in short admit nothing as true save what can be deduced from those common notions (the truth of which cannot be doubted) with the same evidence as is secure in mathematical demonstration. And since, in this manner we can explain all the phenomena of nature, . . . I do not think that we should admit any additional physical principles, or that we have the right to look for any other.

Descartes was explicit in his *Principles* that the essence of science was mathematics. He says that he "neither admits nor hopes for any principles in Physics other than those which are in Geometry or in Abstract Mathematics, because thus all phenomena of nature are explained and sure demonstrations of them can be given." The objective world is space solidified or geometry incarnate. Its properties should therefore be deducible from the first principles of geometry (the term that he and others of his time used as practically synonymous with mathematics because the bulk of mathematics was then geometry).

Descartes elaborated on why the world must be accessible to mathematics. He insisted that the most fundamental and reliable properties of matter are shape, extension, and motion in space and time, all of which are mathematically describable. Since shape reduces to extension, Descartes asserted, "Give me extension and motion and I shall construct the universe." He did add that all physical phenomena are the result of the mechanical action of molecules moved by forces. But forces too obeyed invariable mathematical laws.

Since Descartes regarded the external world as consisting only of matter in motion, how could he account for tastes, smells, colors, and the qualities of sounds? Here Descartes adopted an older Greek doctrine, Democritus's doctrine of primary and secondary qualities. The primary qualities, matter and motion, exist in the physical world; the secondary qualities, taste, smell, color, warmth, and the pleasantness or harshness of sounds, are only effects which the primary qualities induce in the sense organs of human beings by the impact of external atoms on these organs. The real world is the totality of mathematically expressible motions of objects in space and time and the entire universe is a great, harmonious, and mathematically designed machine. Science and in fact any discipline that sought to establish order and measurement were subject to mathematics. He said in Rule IV of his *Rules for the Direction of the Mind:*

> All the sciences, which have for their end investigations concerning order and measure, are related to mathematics, it being of small importance whether this measure be sought in numbers, forms, stars, sounds, or any other object; that accordingly, there ought to exist a general science which should explain all that can be known about order and measure, considered independently of any application to a particular subject, and that, indeed, this science has its own proper name consecrated, by long usage, to wit, mathematics. . . . And a proof that it far surpasses in facility and importance the sciences which depend upon it is that it embraces at once all the objects to which these are devoted and a great many others besides. . . .

Descartes's contributions to mathematics proper did not offer new truths but rather a powerful methodology which we now call analytic geometry (Chapter V). From the technical standpoint, analytic geometry revolutionized mathematical methodology.

In science, too, Descartes's contributions, though not of the magnitude and broad significance of the work of Copernicus, Kepler, or Newton, were certainly not negligible. His theory of vortices (Chapter III) was the dominant cosmological theory of the 17th century. He was the founder of the philosophy of mechanism, that is, the philosophy that all natural phenomena including the human body, but excepting the soul, reduce to the motions of particles obeying the laws of mechanics. In mechanics proper, he formulated the law of inertia, now known as Newton's first law of motion: If no force acts on a body and the body is at rest, it will remain at rest, and if it is in motion it will continue to move in a straight line at a constant speed.

Optics—the design of lenses in particular—was another major interest. Indeed part of his *Geometry* and all of the *Dioptrics,* which he appended to his *Discourse,* are devoted to optics. He shares with Willebrord Snell the discovery of the correct law of refraction of light, that is, how light behaves when it passes through an abrupt change in medium, as from air to glass or water. The Greeks began the mathematization of optics but Descartes's work established the subject as a mathematical science. He made important contributions also to geography, meteorology, botany, anatomy (in the dissection of animals), zoology, psychology, and even medicine.

Though Descartes's philosophical and scientific doctrines subverted Aristotelianism and medieval scholasticism, in one fundamental respect he was a scholastic: he drew from his own mind propositions about the nature of being and reality. He believed in a priori truths and that the intellect by its own power may arrive at a perfect knowledge of all things. Thus he stated laws of motion on the basis of a priori reasoning. (Actually in his biological work and in some other fields he did experiment and drew significant conclusions from his experiments.) However, by reducing natural phenomena to purely physical happenings, he did much to rid science of mysticism and occult forces.

Though not as influential in his philosophy, one of the great mathematicians of the 17th century, Blaise Pascal (1623–1662), readily added his support to the belief that mathematics and the mathematical laws of science are truths. Unlike Descartes, who spoke of intuitions clearly acceptable to the mind, Pascal spoke of acceptability to the heart. Truths must either appeal clearly and distinctly to the heart or be logical consequences of such truths. In his *Pensées* he tells us:

> Our knowledge of the first principles, such as space, time, motion, number, is as certain as any knowledge we obtain by reasoning. As a matter of fact, this knowledge provided by our hearts and instinct is necessarily the basis on which our reasoning has to build its conclusions. It is just as pointless and absurd for reason to demand proof of first principles from the heart before agreeing to accept them as it

would be for the heart to demand an intuition of all the propositions
demonstrated by reason before agreeing to accept them.

For Pascal, science is the study of God's world. The pursuit of science
for mere enjoyment is wrong. To make enjoyment the chief end of
science is to corrupt research, for then one acquires "a greed or lust for
learning, a profligate appetite for knowledge." "Such a study of science
springs from a priori concern for self as the center of things rather
than a concern for seeking out, amid all surrounding natural phenomena, the presence of God and His glory."

Of the seminal thinkers who forged modern mathematics and
science, Galileo Galilei (1564–1642) ranks with Descartes. Of course,
he, too, was certain that nature is mathematically designed, and designed by God. His statement in *The Assayer* of 1610 is famous:

> Philosophy [nature] is written in that great book which ever lies before
> our eyes—I mean the universe—but we cannot understand it if we do
> not first learn the language and grasp the symbols in which it is writ-
> ten. The book is written in the mathematical language, and the sym-
> bols are triangles, circles and other geometrical figures, without whose
> help it is impossible to comprehend a single word of it; without which
> one wanders in vain through a dark labyrinth.

Nature is simple and orderly and its behavior is regular and necessary.
It acts in accordance with perfect and immutable mathematical laws.
Divine reason is the source of the rational in nature. God put into the
world that rigorous mathematical necessity which men reach only la-
boriously, even though their reason is related to God's. Mathematical
knowledge is therefore not only absolute truth but as sacrosanct as any
line of the Scriptures. Moreover, the study of nature is as devout as the
study of the Bible. "Nor does God less admirably reveal himself to us in
Nature's actions than in the Scriptures' sacred dictions."

Galileo asserted in his *Dialogue on the Great World Systems* (1632) that
in mathematics man reaches the pinnacle of all possible knowledge—a
knowledge not inferior to that possessed by the divine intellect. Of
course the divine intellect knows and conceives an infinitely greater
number of mathematical truths than man does but with regard to objective certainty the few verities known by the human mind are known
as perfectly by man as by God.

Though Galileo was a professor of mathematics and a court mathematician, his major contribution was his many innovations in scientific
method. Of these the most notable was his injunction to abandon physical explanation, which Aristotle had regarded as the true goal of
science, and to seek instead mathematical description. The difference
in these two goals is readily illustrated. A body which is dropped falls to

earth, and in fact falls with increasing velocity. Aristotle and the medieval scientists who followed his methodology sought to account for the cause of the fall, which presumably was mechanical. Instead, Galileo merely described the fall with the mathematical law which, stated in modern notation, is $d = 16t^2$, where d is the number of feet the body falls in t seconds. This formula says nothing about why the body falls and would seem to offer far less than what one would want to know about the phenomenon. But Galileo was sure that the knowledge of nature we should seek was descriptive. He wrote in his *Two New Sciences*, "The cause of the acceleration of the motion of falling bodies is not a necessary part of the investigation." More generally, he pointed out that he was going to investigate and demonstrate some of the properties of motion without regard to what the causes might be. Positive scientific inquiries were to be separated from questions of ultimate causation, and speculation about physical causes was to be abandoned. Galileo may well have put it to scientists: theirs not to reason why, theirs but to quantify.

First reactions to this plank of Galileo's program are likely even today to be negative. Descriptions of phenomena in terms of formulas hardly seem to be more than a first step. It would seem that the true function of science had really been grasped by the Aristotelians, namely, to explain why phenomena happened. Even Descartes protested Galileo's decision to seek descriptive formulas. He said, "Everything that Galileo says about bodies falling in empty space is built without foundation: he ought first to have determined the nature of weight." Further, said Descartes, Galileo should reflect about ultimate reasons. But we now know, in the light of subsequent developments, that Galileo's decision to aim for description was the deepest and the most fruitful innovation that anyone has made about scientific methodology. Its significance, which will be more fully apparent later, is that it placed science far more squarely under the aegis of mathematics.

Galileo's next principle was that any branch of science should be patterned on the model of mathematics. Two essential steps are implied here. Mathematics starts with axioms, that is, clear, self-evident truths. From these it proceeds by deductive reasoning to establish new truths. So any branch of science should start with axioms or principles and then proceed deductively. Moreover, one should pull out from the axioms as many consequences as possible. Of course, this plan was advanced by Aristotle, who also aimed at deductive structure in science with the mathematical model in mind.

However, Galileo departed radically from the Greeks, the medievalists, and Descartes in the method of obtaining first principles. The pre-Galileans and Descartes had believed that the mind supplies the basic principles. The mind had but to think about any class of phenomena and it would immediately recognize fundamental truths. This power of the mind was clearly evidenced in mathematics. Axioms such as equals added to equals give equals and two points determine a line suggested themselves immediately in thinking about number or geometrical figures and were indubitable truths. So too did the Greeks find some physical principles equally appealing. That all objects in the universe should have a natural place was no more than fitting. The state of rest seemed clearly more natural than the state of motion. It seemed indubitable, too, that to put and keep bodies in motion force must be applied. To believe that the mind supplies fundamental principles does not deny that observation might help us to reach these principles. But observation merely evokes the correct principles just as the sight of a familiar face may call to mind facts about that person.

These savants, as Galileo put it, first decided how the world should function in accordance with their preconceived principles. Galileo decided that in physics as opposed to mathematics first principles must come from experience and experimentation. The way to obtain correct and basic principles is to pay attention to what nature says rather than what the mind prefers. He openly criticized scientists and philosophers who accepted laws which conformed to their preconceived ideas as to how nature must behave. Nature did not first make men's brains, he said, and then arrange the world so that it would be acceptable to human intellects. To the medievalists who kept repeating Aristotle and debating what he meant, Galileo addressed the criticism that knowledge comes from observation and not from books. It was useless to debate about Aristotle. Those who did he called paper scientists who fancied that science was to be studied like the *Aeneid* or the *Odyssey* or by collation of texts. "When we have the decree of nature, authority goes for nothing."

Of course, some Renaissance thinkers and Galileo's contemporary Francis Bacon had also arrived at the conclusion that experimentation was necessary. In this particular plank of his new method Galileo was not ahead of all others. Yet the modernist Descartes did not grant the wisdom of Galileo's reliance upon experimentation. The facts of the senses, Descartes said, can only lead to delusion. Reason penetrates such delusions. From the innate general principles supplied by the mind we can deduce particular phenomena of nature and understand them. Actually, as we noted earlier, in much of his scientific work Descartes did experiment and require that theory fit facts, but in his philosophy he was still tied to truths of the mind.

A few mathematical physicists agreed with Galileo that reason alone could not ensure correct physical principles. Christian Huygens actually criticized Descartes. The English physicists also attacked pure rationalism. Robert Hooke (1635–1703) said that the members of the Royal Society of London "having before their eyes so many fatal instances of the errors and falsehoods in which the greater part of mankind had wandered because they relied upon the strength of human reason alone, have now begun to correct all hypotheses by *sense.*"

Of course, Galileo realized that one may glean an incorrect principle from experimentation and that as a consequence the deductions from it could be incorrect. Hence he proposed and presumably did use experiments to check the conclusions of his reasonings as well as to acquire basic principles. However, the extent to which Galileo experimented is certainly open to question. Some of his presumed experiments are sometimes called *Gedanken* (German for "thoughts") experiments, that is, he imagined what must happen were one to experiment. Nevertheless, his doctrine that *physical* principles must rest on experience and experiments is revolutionary and crucial. Galileo himself had no doubt that some true principles among those used by God to fashion the universe could still be reached by the mind, but by opening the door to the role of experience he allowed the devil of doubt to slip in. If the basic principles of science must come from experience, why not the axioms of mathematics? This question did not trouble Galileo or his successors until 1800. Mathematics still enjoyed a privileged position.

In his scientific work Galileo concentrated on matter and motion. He recognized clearly and independently of Descartes the principle of inertia, now called Newton's first law of motion. He was successful also in obtaining the laws of motion of bodies that rise straight up and fall, bodies sliding along inclined planes, and projectiles. The motion of this last class he showed was parabolic. In sum, he obtained the laws of motion of terrestrial objects. Though, as in every major innovation, predecessors can be found, no one was so clear as Galileo about the concepts and principles that were to guide scientific inquiry, and no one demonstrated their application in so simple and effective a manner.

Radically innovative for his time, Galileo's philosophy and methodology of science were prefatory to the accomplishments of Isaac Newton, who was born in the year that Galileo died.

III

The Mathematization of Science

> In any particular theory there is only as much real science
> as there is mathematics.   IMMANUEL KANT

If the conviction that the mathematical laws of science were truths incorporated by God in His design of the universe needed any reinforcement, it was superbly provided by Sir Isaac Newton (1642–1727). Though Newton was a professor of mathematics at Cambridge University and is ranked as one of the great mathematicians, he has a still higher rank as a physicist. His work inaugurated a new era and a new methodology in science which increased and deepened the role of mathematics.

The work of Copernicus, Kepler, Descartes, Galileo, and Pascal did indeed demonstrate that some phenomena of nature operated in accordance with mathematical laws, and all of these men were convinced not only that God had so designed the universe but that the mathematical thinking of humans was in accord with God's design. However, the philosophy or methodology of science predominant in the 17th century was formulated and advanced by Descartes. Descartes did indeed say that all of physics reduces to geometry, the word he and others often used as a synonym for mathematics. But Descartes's methodology, adopted by most pre-Newtonians, especially Huygens, proposed an additional function of science, namely, to provide a *physical* explanation of the action of natural phenomena.

The Greeks, notably Aristotle, had also sought to explain in physical terms the behavior of natural phenomena, and their chief theory was that all matter was composed of four elements, earth, air, fire, and water, and these possessed one or more qualities, heaviness, lightness, dryness, and moisture. These qualities explained why matter behaved as it seemingly does: fire rises because it is light whereas earthy matter falls because it possesses the quality of heaviness. To these qualities the medieval scholars added many others, such as sympathy, which accounted for the attraction of one body to another as iron to a magnet, and antipathy, to account for the repulsion of one body by another.

Descartes, on the other hand, discarded all these qualities and insisted that all physical phenomena could be explained in terms of matter and motion. The essential attribute of matter was extension and this was measurable and so reducible to mathematics. Further, there could be no extension without matter. Hence a vacuum was impossible. Space was filled with matter. Moreover, matter acted on matter only by direct contact. However, matter was composed of small, invisible particles that differed in size, shape, and other properties. Because these particles were too small to be visible it was necessary to make hypotheses about how they behaved in order to account for the larger phenomena that man can observe. In this view all of space was filled with particles which swept the larger accumulations, the planets, around the sun. This is the essence of Descartes's vortex theory.

Descartes was the founder of the mechanistic theory, and he was followed in this by the French philosopher and priest Pierre Gassendi (1592–1655), the English philosopher Thomas Hobbes (1588–1679), and the Dutch mathematician and physicist Christian Huygens (1629–1695). Thus Huygens in his *Treatise on Light* (1690) accounted for various phenomena of light by the assumption that space was filled with ether particles which transmitted the motion of light from one to the next. In fact the subtitle of his book reads, "In which are explained the causes of that which occurs in Reflexion and in Refraction." In his Introduction, Huygens said that in the true philosophy, "One conceives the causes of all natural effects in terms of mechanical motion. This, in my opinion, we must necessarily do, or else renounce all hopes of ever comprehending anything in Physics." In one respect Gassendi differed: he believed that atoms moved in a void.

Physical hypotheses about how the minute particles acted did account, at least in the large, for the gross behavior of nature. But they were creations of the mind. Moreover, the physical hypotheses of Descartes and his followers were qualitative. They could explain but because they were qualitative they could not predict precisely what observation and experimentation would reveal. Leibniz called this collection of physical hypotheses a beautiful romance.

A counter-philosophy of science was inaugurated by Galileo. Science must seek mathematical description rather than physical explanation. Moreover, the basic principles must be derived from experiments and induction from experiments. In accordance with this philosophy, Newton, influenced also by his teacher Isaac Barrow, changed the course of science by adopting *mathematical* premises in place of physical hypotheses and could consequently predict with the certainty that Bacon had called for. Moreover, the premises were inferred from experiments and observations.

Galileo anticipated Newton in that he treated the fall of a body and the flight of a projectile. Newton tackled a far broader problem, one which was uppermost in the minds of the scientists about 1650. Could one establish a connection between Galileo's laws of terrestrial motions and Kepler's laws of celestial motions? The thought that *all* the phenomena of motion should follow from one set of principles might seem grandiose and inordinate, but it occurred very naturally to the religious mathematicians of the 17th century. God had designed the universe, and it was to be expected that all phenomena of nature would follow one master plan. One mind designing a universe would almost surely have employed one set of basic principles to govern related phenomena. To the mathematicians and scientists of the 17th century engaged in the quest for God's design of nature, it seemed very reasonable that they should seek the unity underlying the diverse earthly and heavenly motions.

In the course of executing his program of producing universal laws of motion, Newton made many contributions to algebra, geometry, and above all the calculus (Chapter VI). But these were merely aids to his scientific goals. In fact he considered mathematics proper to be dry and barren, and only a tool for the expression of natural laws. He concentrated on finding the scientific principles that would lead to a unifying theory of earthly and celestial motions. Fortunately, as Denis Diderot put it, nature admitted Newton into her confidence.

Newton was, of course, familiar with the principles established by Galileo. But these were not sufficiently encompassing. It was clear from the first law of motion that the planets must be acted on by a force which pulls them toward the sun, for, if no force were acting, each planet would move in a straight line. The idea of a force which constantly pulls each planet toward the sun had occurred to many men, Copernicus, Kepler, the famous experimental physicist Robert Hooke, the physicist and renowned architect Christopher Wren, Edmund Halley, and others, even before Newton set to work. It had also been conjectured that this force exerted on a distant planet must be weaker than that exerted on a nearer one and that this force must decrease as the square of the distance between sun and planet increased. But, before Newton's work, none of these thoughts about a gravitational force advanced beyond speculation.

Newton adopted the conjecture already made by his contemporaries, namely, that the force of attraction, F, between *any* two bodies of masses m and M, respectively, separated by a distance r is given by the formula

$$F = G\,\frac{mM}{r^2}.$$

In this formula, G is a constant, that is, it is the same number no matter what m, M, and r may be. The value of this constant depends upon the units used for mass, force, and distance. He also generalized Galileo's laws of terrestrial motion; these generalizations are now known as Newton's three laws of motion. The first is the law Descartes and Galileo had already affirmed: that if no force acts on a body, then if the body is at rest it will remain at rest and if in motion it will move along a straight line at a constant speed. The second states that, if a force acts on a body of mass m, then it will give that body an acceleration; moreover, the force equals the mass times the acceleration. In symbols, $F = ma$. The third law affirms that if a body A exerts a force F on body B then B exerts a force on A that is equal in magnitude but opposite in direction to the force F. From these three laws and the law of gravitation Newton deduced readily the behavior of earthly bodies.

As for heavenly motions, Newton's truly great triumph was his demonstration that the three Keplerian laws, which Kepler had obtained only after years of observation and trial and error, were also mathematical consequences of the law of gravitation and the three laws of motion. Hence the laws of planetary motion, which before Newton's work seemed to have no relation to earthly motions, were shown to follow from the same basic principles as did the laws of earthly motions. In this sense, Newton "explained" the laws of planetary motion. Further, since the Keplerian laws agree with observations, their derivation from the law of gravitation constituted superb evidence for the correctness of that law.

The few deductions from the laws of motion and the law of gravitation just described are but a sample of what Newton was able to accomplish. He applied the law of gravitation to explain a phenomenon which theretofore had not been understood, namely, the oceans' tides. These were due to the gravitational forces exerted by the moon, and to a lesser extent the sun, on large bodies of water. From the data collected on the height of lunar tides, that is, tides due to the moon, Newton calculated the moon's mass. Newton and Christian Huygens calculated the bulge of the earth around the equator. Newton and others also showed the paths of comets to be in conformity with the law of gravitation. Hence the comets, too, were recognized as lawful members of our solar system and ceased to be viewed as accidental occurrences or visitations from God intended to wreak destruction. Newton then showed that the attraction of the moon and the sun on the earth's equatorial bulge causes the axis of the earth's rotation to describe a cone over a period of 26,000 years instead of always pointing to the same star in the sky. This periodic change in the direction of the earth's axis causes a slight change each year in the time of the spring and fall equinoxes, a fact which had been observed by Hipparchus 1800 years earlier. Thus Newton explained the precession of the equinoxes.

Finally, Newton, using approximate methods, solved a number of problems involving the motion of the moon. For example, the plane in which the moon moves is inclined somewhat to the plane in which the earth moves. He was able to show that this phenomenon follows from the interaction of the sun, earth, and moon under the law of gravitation. Newton and his immediate successors deduced so many and such weighty consequences about the motions of the planets, the comets, the moon, and the seas that their accomplishments were viewed for the next two hundred years as the "explication of the System of the World."

In all of this work, Newton adopted Galileo's proposal to seek mathematical description rather than physical explanation. Newton not only unified a vast number of experimental and theoretical results of Kepler, Galileo, and Huygens but he placed *mathematical* description and deduction at the forefront of all scientific accounts and prediction. In the Preface to his aptly titled major work, *Mathematical Principles of Natural Philosophy* (1687), he said,

> Since the ancients (as we are told by Pappus) esteemed the science of mechanics of greatest importance in the investigation of natural things, and the moderns, rejecting substantial forms and occult qualities, have endeavored to subject the phenomena of nature to the laws of mathematics, I have in this treatise cultivated mathematics as far as it relates to philosophy [science] . . . and therefore I offer this work as the mathematical principles of philosophy, for the whole burden in philosophy seems to consist in this—from the phenomena of motions to investigate the forces of nature, and then from these forces to demonstrate the other phenomena; and to this end the general propositions in the first and second Books are directed. . . . Then from these forces, by other propositions which are also mathematical, I deduce the motions of the planets, the comets, the moon, and the sea.

Clearly, mathematics was to play the major role.

Newton had good reason to emphasize quantitative mathematical laws as opposed to physical explanation because the central physical concept in his celestial mechanics was the force of gravitation and the action of this force could not be explained at all in physical terms. The concept of a gravitational force that attracted any two masses to each other even when separated by hundreds of millions of miles of empty space seemed as incredible as many of the qualities that the Aristotelians and medieval scholars had invented to account for scientific phenomena. The concept was especially repugnant to Newton's contemporaries who insisted on mechanical explanations and had come to see force as the effect of contact between bodies wherein one body "pushes" another. The abandonment of physical mechanism in favor of mathematical description shocked even the greatest scientists. Huygens regarded the idea of gravitation as "absurd" because its action through empty space precluded any mechanism. He expressed surprise that Newton should have taken the trouble to make such a number of laborious calculations with no foundation but the mathematical principle of gravitation. Many others, including Leibniz, objected to the purely mathematical account of gravitation. Leibniz began his critique in 1690 after reading Newton's *Principia* and he kept it up until he died. Voltaire returning from Newton's funeral in 1727 remarked that he had left a vacuum in London and found a plenum (space filled with matter) in France, where Descartes's philosophy still held sway. The attempts to explain "action at a distance" persisted until 1900.

But Newton's amazing contributions were made possible by his reliance on mathematical description even where physical understanding was completely lacking. In lieu of physical explanation, Newton did have a quantitative formulation of how gravity acted, which was significant and usable. And this is why he says at the beginning of his *Principia*, "For I here design only to give a mathematical notion of these forces, without considering their physical causes and seats." And toward the end of the book he repeats this thought:

> But our purpose is only to trace out the quantity and properties of this force from the phenomena, and to apply what we discover in some simple cases as principles, by which, in a mathematical way, we may estimate the effects thereof in more involved cases. . . . We said in a *mathematical way* [the italics are Newton's], to avoid all questions about the nature or quality of this force, which we would not be understood to determine by any hypothesis. . . .

In his letter of February 25, 1692 to the Reverend Dr. Richard Bentley, Newton did write:

> That gravity should be innate, inherent, essential to matter, so that one body may act upon another at a distance through a *vacuum*, without the mediation of anything else, by and through which the action and force may be conveyed from one to another, is to me so great an absurdity that I believe no man who has in philosophical matters a

competent faculty of thinking can ever fall into it. Gravity must be caused by an agent acting constantly according to certain laws, but whether this agent be material or immaterial I have left to the consideration of my readers.

Despite Newton's mathematical successes the absence of a physical mechanism continued to bother scientists but their efforts to supply one were not successful. Bishop George Berkeley in his dialogue *Alciphron* (1732) made this point:

> Euphranor: . . . Let me entreat you, Alciphron, be not amused by terms: lay aside the *word* force, and exclude every other thing from your thoughts, and then see what precise *idea* you have of force.
>
> Alciphron: Force is that in bodies which produceth motion and other sensible effects.
>
> Euphranor: It is then something distinct from those effects?
>
> Alciphron: It is.
>
> Euphranor: Be pleased now to exclude the consideration of its subject and effects, and contemplate force itself in its own precise idea.
>
> Alciphron: I profess I find it no such easy matter.
>
> And that, replied Euphranor, which it seems that neither you nor I can frame an idea of, by your own remark of men's minds and faculties being made much alike, we may suppose others have no more an idea of than we.

Newton did hope that the nature of the force of gravity would be investigated and learned. Contrary to Newton's hope and popular belief, no one has ever explained how gravity acts; the physical reality of this force has never been demonstrated. It is a scientific fiction suggested by the human ability to exert force. However, the mathematical deductions from the quantitative law proved so effective that this approach has been accepted as an integral part of physical science. What science has done, then, is to sacrifice physical intelligibility for mathematical description and mathematical prediction.

The work of the 17th century is often summarized by the statement that the mathematical physicists built up a mechanical world picture, a world operating like a machine. Of course the physics of Aristotle and of medieval scientists was also mechanical if one means by that word an account that describes motion by forces such as heaviness, lightness, sympathy, and others noted earlier, acting on particles and extended bodies. However, the men of the 17th century, the Cartesians especially, discarded the multiplicity of qualities that their predecessors had hypothesized to account for motion and restricted force to the material and obvious: weight or the force required to throw an object. One could call this pre-Newtonian physics a material physics. Mathematics could describe but was not fundamental.

The essential difference between Newtonian mechanics and the older mechanics, then, was not the introduction of mathematics to describe the behavior of bodies. Mathematics was not just an aid to physics in the sense of a convenient, briefer, clearer, and more general language; rather it provided the fundamental concepts. Gravitational force is merely a name for a mathematical symbol. Likewise in Newton's second law of motion ($F = ma$, force equals mass times acceleration) force is anything that gives acceleration to a mass. The nature of the force itself may be physically unknowable. Thus Newton spoke of and used centripetal and centrifugal forces though the mechanism of these forces is not known.

Even the notion of mass is a fiction in Newtonian mechanics. To be sure, mass is matter and matter is real as Samuel Johnson "demonstrated" by kicking a stone. But for Newton the primary property of mass is inertia whose meaning is expressed in Newton's first law of motion, namely, that if no force acts on a mass and if it is at rest it will remain at rest, and if it is in motion it will move in a straight line at a constant velocity. Why a straight line? Why not a circle? Indeed Galileo did believe that inertial motion is circular. And why should it move at a constant velocity? Why if no force acts does not the mass always remain at rest or move with a constant acceleration? The property of inertia is a fictitious concept and not an experimental fact. No mass is free of all forces. The only element of physical reality in Newton's laws of motion is acceleration. We can observe and measure the acceleration of objects.

However reluctantly he abandoned physical explanation, Newton recast the entire body of 17th-century physics by employing mathematical concepts, their quantitative formulation, and mathematical deductions from the resulting formulas.* Newton's crowning work presented mankind with a new world order, a universe controlled throughout by a common set of only mathematically expressible physical principles. Here was a majestic scheme which embraced the fall of a stone, the tides of the oceans, the motions of the planets and their moons, the defiant sweep of the comets, and the brilliant, stately motion of the sphere of stars. The Newtonian scheme was decisive in convincing the world that nature is mathematically designed and that the true laws of nature are mathematical. Newton's *Principles* is an epitaph to physical explanation. Laplace once remarked that Newton was a most fortunate

---

* In his *Opticks* Newton did give physical explanations. However, they were not adequate to explain all of the behavior of light.

man because there was just one universe and he had succeeded in finding its laws.

Throughout the 18th century, the mathematicians, who were also the
major scientists, pursued Newton's plan. Lagrange's *Analytical Mechanics*
of 1788 may be regarded as the prime example of Newton's mathematical approach. In this book, mechanics was treated entirely mathematically. There was little reference to physical processes. In fact, Lagrange boasted that he had no need of these or even of geometrical
diagrams. As newer branches of physics—hydrodynamics, elasticity,
electricity, and magnetism—were tackled, it was Newton's approach to
mechanics and astronomy that was adopted. The quantitative, mathematical method became the essence of science, and truth resided most
securely in mathematics.

The insurgent 17th century found a qualitative world whose study
was aided by mathematical description. It bequeathed to posterity a
mathematical, quantitative world which substituted mathematical formulas for the concreteness of the physical world. It was the beginning
of the mathematization of nature which flourishes to this very day.
When Sir James Jeans remarked in *The Mysterious Universe* (1930), "The
Great Architect of the Universe now begins to appear as a pure mathematician," he was at least two centuries behind the times.

Though, as we have noted, Newton himself was not at ease about
relying solely upon mathematical formulas unsupported by physical explanations, he nevertheless not only advocated his mathematical principles of natural philosophy (physical science) but certainly believed it to
be a true account of the phenomena he described. Whence came the
conviction? The answer is that Newton like all the mathematicians and
scientists of his day believed that God designed the world in accordance
with mathematical principles. Most eloquent is Newton's statement in
his *Opticks* (1704) of the classic argument for the existence of God as
the framer of the universe:

> The main business of natural philosophy is to argue from phenomena
> without feigning hypotheses, and to deduce causes from effects, till we
> come to the very first, which certainly is not mechanical. . . . What is
> there in places almost empty of matter, and whence is it that the sun
> and planets gravitate towards one another, without dense matter be-
> tween them? Whence is it that nature doth nothing in vain; and
> whence arises all the order and beauty we see in the world? To what
> end are comets, and whence is it that planets move all one and the
> same way in orbs concentric, and what hinders the fixed stars from
> falling upon one another? How came the bodies of animals to be con-
> trived with so much art, and for what ends were their several parts?
> Was the eye contrived without skill in optics, or the ear without knowl-

edge of sounds? How do the motions of the body follow from the will, and whence is the instinct in animals? . . . And these things being rightly dispatched, does it not appear from phenomena that there is a being incorporeal, living, intelligent, omnipresent, who, in infinite space, as it were in his sensory, sees the things themselves intimately, and thoroughly perceives them; and comprehends them wholly by their immediate presence to himself?

In the third edition of his *Mathematical Principles of Natural Philosophy*, Newton answers his own questions:

> This most beautiful system of sun, planets, and comets could only proceed from the counsel and dominion of an intelligent and powerful Being. . . . This Being governs all things, not as the Soul of the world, but as Lord over all. . . .

Newton was convinced, too, that God was a skilled mathematician and physicist. He said in a letter to the Reverend Richard Bentley of December 10, 1692,

> To make this [solar] system, therefore, with all its motions, required a cause which understood, and compared together the quantities of matter in the several bodies of the sun and planets, and the gravitating powers resulting from thence; the several distances of the primary planets from the sun, and of the secondary ones [i.e., moons] from Saturn, Jupiter, and the earth; and the velocities with which these planets could revolve about those quantities of matter in the central bodies; and to compare and adjust all these things together in so great a variety of bodies, argues that cause to be not blind or fortuitous, but very skilled in mechanics and geometry.

Science should uncover God's glorious designs. Newton began the same letter to Bentley with this thought: "When I wrote my treatise about our system [*The Mathematical Principles of Natural Philosophy*], I had an eye on such principles as might work with considering men for the belief in a Deity; and nothing can rejoice me more than to find it useful for that purpose." There are many other such letters in Newton's correspondence.

Newton's religious interests were the true motivation of his mathematical and scientific work. He believed first of all that the doctrines of Christianity were a revelation from God. God was the cause of all natural forces, of everything that exists and happens. Divine intent, guidance, and control were present in all phenomena. From his youth on, Newton made critical studies and interpretations of religious writings and late in life devoted himself entirely to theology. His books *Observations on the Prophecies of Daniel and the Apocalypse of St. John* (published in 1733) and *The Chronology of Ancient Kingdoms Amended* (not published)

are extant, as are hundreds of manuscript papers in which he attempted to fix the chronology of Biblical events. Science was a form of
worship though the scientific work proper was to be kept free of mystical or supernatural forces. Newton himself rejoiced that his work
revealed the hand of an omnipotent God. He considered strengthening
the foundations of religion far more important than scientific and
mathematical achievements, for the latter were restricted to displaying
God's design of the natural world only. He often justified the hard and
at times dreary scientific work because he felt it supported religion by
providing evidence of God's order in the universe. It was as pious a
pursuit as the study of the Scriptures. The wisdom of God could be
manifested by disclosing the structure of the universe. God was also the
cause of whatever happened. Thus miracles were only God's occasional
intervention in the usual functioning of the universe. God might also
occasionally have to intervene to correct some malfunctioning, much as
a watchmaker might repair clocks.

If the belief that God has designed the universe and that the role of
mathematics and science was to uncover that design needed any reinforcement, it was provided by Gottfried Wilhelm Leibniz (1646–1716).
Like Descartes, Leibniz was primarily a philosopher and even more
versatile; his contributions to mathematics, science, history, logic, law,
diplomacy, and theology were first-rate. Like Newton, Leibniz
regarded science as a religious mission which scientists were duty
bound to undertake. In an undated letter of 1699 or 1700 he wrote, "It
seems to me that the principal goal of the whole of mankind must be
the knowledge and development of the wonders of God, and that this
is the reason that God gave him the empire of the globe."

In his *Essais de Théodicée* (1710), he affirmed the by that time familiar
thought that God is the intelligence who created this carefully designed
world. His explanation of the concord between the real and the mathematical worlds, and his ultimate defense of the applicability of mathematics to the real world, was the unity of the world and God. *Cum Deus
calculat, fit mundus* (as God calculates, so the world is made). There is a
pre-established harmony between mathematics and nature. The universe is the most perfect conceivable, the best of all possible worlds, and
rational thinking discloses its laws.

True knowledge is innate in our minds though not, contrary to Plato,
because of some prior existence. The senses could never teach us necessary truths such as that God exists or that all right angles are equal.
Thus the axioms of mathematics are innate truths as are the fundamental principles of the deductive sciences such as mechanics and
optics, "in which to be sure the senses are very necessary, in order to
have certain ideas of sensible things, and experiments are necessary to

establish certain facts. . . . But the force of the demonstrations depends upon intelligible notions and truths, which alone are capable of making us discern what is necessary. . . ."

Leibniz's mathematical and scientific work was extensive and valuable and we shall have occasion to say more about it later. However, somewhat like Descartes's, his contributions were to technique. His work in the calculus and in the beginnings of differential equations and his recognition of the importance of some emerging concepts such as what we now call kinetic energy were of the highest order. But Leibniz did not contribute any new fundamental laws of nature. Rather it was his philosophy of science, in which the role of mathematics was fundamental, that was most important in spurring on man's search for truths.

Though the men of the 18th century extended vastly both mathematics and the mathematical sciences, insofar as convincing the intellectuals that mathematics and the mathematical laws of science are truths is concerned, their work was largely an encore. The several Bernoullis—especially the brothers James (1665–1705) and John (1667–1748) and John's son Daniel (1700–1782)—Leonhard Euler (1707–1783), Jean Le Rond d'Alembert (1717–1783), Joseph-Louis Lagrange (1736–1813), Pierre-Simon Laplace (1749–1827), and many others continued the mathematical investigation of nature. In mathematics proper all of these men continued the development of the techniques of the calculus and created totally new branches, notably ordinary and partial differential equations, differential geometry, the calculus of variations, infinite series, and functions of a complex variable. These were in themselves accepted as truths but also furnished more powerful tools for the investigation of nature. As Euler put it in 1741: "The usefulness of mathematics, commonly allowed to its elementary parts, not only does not stop in higher mathematics but is in fact so much greater, the further that science is developed."

The goal of this mathematical work was the attainment of more laws of nature, a deeper penetration of nature's design. And the successes were numerous and profound. The greatest effort was in astronomy, to carry further the work of Newton in describing and predicting the motions of the heavenly bodies. Newton's chief theoretical derivation, that the path of a planet is an ellipse, is, as he well knew, correct only were the sun and just one planet in the heavens. But in Newton's days and during most of the 18th century, six planets were known; each attracts the others and all are attracted by the sun. Moreover, some—the earth, Jupiter and Saturn—have moons. Hence the elliptical path is perturbed. What are the precise paths? All the great mathematicians of the 18th century worked on this problem.

The core of this problem is the question of the mutual gravitational effects of three bodies on each other. If one could devise a procedure to determine the perturbing effect of a third body, this procedure could be used to determine the perturbing effect of a fourth body, and so on. However, exact solution of the general problem of the motion of even three bodies has not been obtained even today. Instead, better and better approximate procedures were devised.

Even with approximate methods the successes of the 18th century were remarkable. One of the most dramatic proofs of the accuracy of the mathematical work in astronomy is the prediction by Alexis-Claude Clairaut (1713–1765) of the return of what had been known as Halley's comet. This comet had been observed by several men, and Halley in 1682 attempted to determine its orbit. He predicted that it would return in 1758. On November 14, 1758, Clairaut announced at a session of the Paris Academy of Sciences that Halley's comet would return to its position nearest the sun in the middle of April 1759, with a possible error of thirty days. The comet appeared one month ahead of schedule. The error of one month may seem enormous but comets can be observed for only a few days at most and this one had not been seen for seventy-seven years.

Another of the great successes in astronomy was due to the work of Lagrange and Laplace. A number of irregularities had been observed in the motions of the earth's moon and the planets. These irregularities could mean that a planet might move farther and farther from the sun or move into the sun. Lagrange and Laplace proved that the irregularities that had been observed in the velocities of Jupiter and Saturn were periodic so that basically their motions were stable. The work of the century is embodied in one of the masterpieces of science, Laplace's *Celestial Mechanics,* which appeared in five volumes from 1799 to 1825.

In fact Laplace devoted his entire life to astronomy, and every branch of mathematics he investigated he intended for application to astronomy. The story is well known that in his writings he often omitted difficult mathematical steps and said, "It is easy to see. . . ." The real point of this story is that he was impatient with mathematical details and wanted to get on with the application. His many, basic contributions to mathematics were by-products of his great work in natural philosophy and were developed by others.

Equally dramatic is the oft-told account of the discovery of the planet Neptune. Though it took place in 1846, the discovery was based on 18th-century mathematical work. In 1781 William Herschel, using a powerful new telescope, discovered the planet Uranus. But the path of Uranus did not adhere to predictions. The hypothesis was advanced by Alexis Bouvard that still another unknown planet was perturbing the motion of Uranus. Many attempts were made to locate such a planet by observation or by calculations of its possible size and path. In 1845 John Couch Adams (1819–1892), a twenty-six-year-old student at Cambridge University, was able to make a close estimate of the mass, position, and path of the conjectured planet. When informed of this work, the eminent Sir George Airy, director of the Royal Astronomical Observatory at Greenwich, refused to pay attention to it. But another young astronomer, the Frenchman Urbain J. J. Leverrier (1811–1877), having independently done about the same thing as Adams, sent a set of directions for locating the planet to the German astronomer Johann Galle. Galle received the information on September 23, 1846, and that very evening discovered Neptune only 55 minutes of arc away from the direction predicted by Leverrier. How could one doubt the truth of an astronomical theory which made possible such remarkable predictions, indeed predictions accurate to within one ten-thousandth of a percent?

Beyond astronomy, the science which had already yielded somewhat to mathematical treatment even in Greek times was optics. The early 17th-century inventions of the microscope and the telescope stimulated enormous interest in optics and, as in Greek times, every mathematician of the 17th and 18th centuries worked in this area. We have already noted that in the 17th century Snell and Descartes had found what Ptolemy had sought in vain, the law of refraction of light: how light behaves when it moves through an abrupt change in media, as from air to water. The fact that light has a finite velocity was noted by Olaus Roemer (1644–1710), and Newton's discovery that white light is a composite of all colors from red to violet stimulated enormous additional interest in optics. Newton's *Opticks* (1704) advanced this subject immensely and contributed to the improvement of microscopes and telescopes. Here, too, mathematics was the major instrument. In the 18th century the subject was pursued intensively, and Euler's three-volume work on optics is another landmark.

The physical nature of light was by no means clear. Whereas Newton thought that light was a motion of particles and Huygens spoke of wave motion—though hardly in the sense of waves—Euler was the first to treat the vibrations of light mathematically and to derive the equations of motion. In his advocacy of the wave nature of light he was the only one to oppose Newton and was vindicated by the early 19th-century work of Augustin-Jean Fresnel and Thomas Young. But the nature of light did not become any clearer even then and mathematical laws remained the mainstay. The currently accepted theory of light, the electromagnetic theory, was still fifty years in the future.

During the 18th century several new fields of investigation were opened and at least partial successes achieved. The first of these was the mathematical description and analysis of musical sounds. The story here is somewhat lengthy. It began with the study of sounds given off by a vibrating string, a violin string. Daniel Bernoulli, d'Alembert, Euler, and Lagrange each made contributions but also had sharp disagreements on some of the mathematical analysis. Although the controversies were not settled until Joseph Fourier (1768–1830) did his work in the early 19th century, enormous progress was made in the 18th century nevertheless. Our present understanding that every musical sound consists of a fundamental or first harmonic and higher harmonics whose frequencies (pitches, in musical terminology) are all integral multiples of the frequency of the fundamental was established conclusively in the work of the 18th-century masters. This understanding is basic in the design today of all recording and transmitting instruments such as the telephone, phonograph, radio, and television.

Still another branch of mathematical physics was at least initiated in the 18th century, the study of the flow of fluids (liquids and gases) and the motion of bodies in fluids. Newton had already considered and treated the problem of what shape a body should have in order to move forward in a fluid with least resistance. The founding classic was Daniel Bernoulli's *Hydrodynamica* (1738) in which, incidentally, he noted that the theory might be used to describe the flow of blood in human arteries and veins. It was followed by a basic paper of Euler's (1755) in which he derived the equations for the motion of compressible fluids. In this paper Euler said:

> If it is not permitted to us to penetrate to a complete knowledge concerning the motion of fluids, it is not to mechanics, or to the insufficiency of the known principles of motion, that we must attribute the cause. It is [mathematical] analysis itself which abandons us here, since all the theory of the motion of fluids has just been reduced to the solution of analytic formulas.

Actually, there is much more to the theory of fluids than Euler envisaged, which was added some seventy years later. Euler neglected viscosity. (Water flows freely and is non-viscous, whereas oil, for example, flows sluggishly and is somewhat viscous.) Nevertheless, Euler can be said to have founded the science of hydrodynamics which applies to the motion of ships and airplanes.

If men of the 18th century needed any additional evidence that the world was mathematically and indeed most efficiently designed and that God was surely the architect of all the workings of nature, they found it in another mathematical discovery. Heron had proved (Chapter I) that light in going from a point P to a point Q via reflection from a mirror takes the shortest path. Since the light travels in such a situation at a constant velocity, the shortest path also means least time.

In the 17th century Pierre de Fermat (1601–1665), one of the giants of mathematics, had asserted on the basis of rather limited evidence his Principle of Least Time, which affirms that light in traveling from one point to another always takes the path requiring least time. Apparently God had arranged that light not only obey mathematical laws but travel most efficiently. Fermat became all the more convinced of the correctness of his principle when he succeeded in deriving from it the law of refraction of light previously discovered by Willebrord Snell and Descartes.

By the early 18th century mathematicians had several impressive examples of the fact that nature does attempt to maximize or minimize some important quantities. Christian Huygens, who had at first objected to Fermat's principle, showed that it does hold for the propagation of light in media with continuously changing character. Even Newton's first law of motion, which states that an object in motion and undisturbed by any force follows a straight line—the shortest path—is an example of nature's economy.

The men of the 18th century were convinced that, since a perfect universe would not tolerate waste, the action of nature should be the least required to achieve its purposes and a search for a general principle was undertaken. The first formulation of such a principle was offered by Pierre-Louis Moreau de Maupertuis (1698–1759), who was primarily a mathematician and had led an expedition to Lapland to measure the length of one degree along a meridian. His measurement showed that the earth is indeed flattened at the poles as Newton and Huygens had concluded from theoretical arguments. Maupertuis's findings disposed of the contrary arguments of Jean-Dominique Cassini and his son Jacques. Maupertuis was dubbed the "earth-flattener," or, as Voltaire put it, Maupertuis flattened the earth and the Cassinis.

In 1744, while working with the theory of light, he propounded his famous principle of least action in a paper entitled "Accord of different laws of nature which hitherto had appeared incompatible." He worked from Fermat's principle but, in view of disagreements at that time as to whether the velocity of light was greater in water, say, than in air (as Descartes and Newton believed) or smaller (as Fermat believed), Maupertuis abandoned least time and substituted the concept of action. Action, Maupertuis said, is the integral (in the calculus sense) of the product of mass, velocity, and distance traversed, and any happenings in nature are such as to make the action least. Maupertuis was somewhat vague in failing to specify the time interval over which the product was to be taken and in assigning a different meaning to action in each of the applications he made to optics and some problems of mechanics.

Though he had some physical examples to support his principle,

Maupertuis advocated it also for theological reasons. The laws of the behavior of matter had to possess the perfection worthy of God's creation and the least action principle seemed to satisfy the criterion because it showed that nature was economical. Maupertuis proclaimed his principle to be a universal law of nature and the first scientific proof of the existence and wisdom of God.

Leonhard Euler, the greatest of 18th-century mathematicians, who had corresponded with Maupertuis on this subject between 1740 and 1744, agreed with Maupertuis that God must have constructed the universe in accordance with some such basic principle and that the existence of the principle evidenced the hand of God. He expressed this conviction in these words: "Since the fabric of the universe is most perfect and the work of a most wise Creator, nothing at all takes place in the universe in which some rule of maximum or minimum does not appear."

Euler went further than Maupertuis in his belief that all natural phenomena behave so as to maximize or minimize some function and so the basic physical principles should contain a function that is maximized or minimized. God was surely a wiser mathematician than 16th- and 17th-century men had appreciated. Euler's religious convictions also assured him that God has entrusted man with the task of using his faculties to understand His laws. The book of nature was open before us but written in a language that we do not understand at once but can, with persistence, love, and suffering, learn. The language is mathematics. Since the world is the best possible, the laws must also be beautiful.

The Principle of Least Action was clarified and generalized by Lagrange. Action became essentially energy and from this generalized principle it became possible to deduce solutions for many more mechanical problems. (This principle is at the heart of the subject of the calculus of variations, a new branch of mathematics which Lagrange, using preliminary work by Euler, founded.) The principle was further generalized by William R. Hamilton (1805–1865), Britain's "second Newton." It is today the most comprehensive principle underlying mechanics and has served as the paradigm for similar principles, called variational principles, in other branches of physics. However, as we shall see, by Hamilton's time the inferences that Maupertuis and Euler drew as to God's incorporation of this principle in the design of the universe were abandoned. Some indication that a change in the interpretation of its significance would occur can be obtained from Voltaire's *History of Doctor Akakia,* in which he mocked this argument for the existence of God. However, the 18th-century men were still firmly convinced that such an all-embracing principle could mean only that the world had been designed, by God of course, to conform to it.

The reign of mathematics was affirmed in no uncertain terms by the greatest intellects of the 18th century. As the renowned mathematician Jean Le Rond d'Alembert, chief collaborator of Denis Diderot (1718–1784) in the writing of the famous French *Encyclopedia,* put it, "The true system of the world has been recognized, developed and perfected." Natural law clearly was mathematical law.

More famous is the statement of Laplace:

> We may regard the present state of the universe as the effect of its past and the cause of its future. An intellect which at any given moment knew all the forces that animate nature and the mutual positions of the beings that compose it, if this intellect were vast enough to submit the data to analysis, could condense into a single formula the movement of the greatest bodies of the universe and that of the lightest atom: for such an intellect nothing could be uncertain; and the future just like the past would be present before its eyes.

William James in his *Pragmatism* described the attitude of the mathematicians of this period thus:

> When the first mathematical, logical and natural uniformities, the first *laws,* were discovered, men were so carried away by the clearness, beauty and simplification that resulted that they believed themselves to have deciphered authentically the eternal thoughts of the Almighty. His mind also thundered and reverberated in syllogisms. He also thought in conic sections, squares and roots and ratios, and geometrized like Euclid. He made Kepler's laws for the planets to follow; he made velocity increase proportionally to the time in falling bodies, he made the law of sines for light to obey when refracted. . . . He thought the archetypes of all things and devised their variations; and when we rediscover any one of these wondrous institutions, we seize his mind in its very literal intention.

The conviction that nature was mathematically designed, designed by God, was expressed even by poets, Joseph Addison, for instance, in his *Hymn:*

> The spacious firmament on high,
> With all the blue ethereal sky,
> And spangled heavens, a shining frame,
> Their great Original proclaim.
> Th' unwearied Sun from day to day
> Does his Creator's power display;
> And publishes to every land
> The work of an Almighty hand.
>   . . .
> And all the planets in their turn,

Confirm the tidings as they roll,
And spread the truth from pole to pole.

By the end of the 18th century, mathematics was like a tree firmly grounded in reality, with roots already two thousand years old, with majestic branches, and towering over all other bodies of knowledge. Surely such a tree would live forever.

IV

The First Debacle:
The Withering of Truth

> Every age has its myths and calls them higher truths.
>
> ANONYMOUS

The 19th century opened auspiciously. Lagrange was still active; Laplace was at the height of his power; Fourier (1768–1830) was working hard on his manuscript of 1807, which was later incorporated in his classic *Theory of Heat* (1822); Karl Friedrich Gauss (1777–1855) had just published a landmark in the theory of numbers, his *Disquisitiones arithmeticae* (Arithmetical Dissertations, 1801), and was soon to burst forth with a mass of contributions that earned for him the title Prince of Mathematics; and Augustin-Louis Cauchy (1789–1857), the French counterpart of Gauss, had begun to show his extraordinary strength in a paper of 1814.

A few words about the work of these men may give some indication of the immense progress made in the first half of the 19th century toward uncovering more of nature's design. Though Gauss made magnificent contributions to mathematics proper—we shall be discussing one of these shortly—he devoted most of his life to physical studies. In fact he was not a professor of mathematics but for almost fifty years a professor of astronomy and director of the observatory at Göttingen. Astronomy occupied him most and his interest in it dates back to his student days of 1795–1798 at Göttingen. He achieved his first notable success in 1801. On January 1 of that year Giuseppi Piazzi (1746–1826) discovered the minor planet Ceres. Although it was possible to observe it for only a few weeks, Gauss, then only twenty-four years old, applied new mathematical theory to the observations and predicted the planet's path. It was observed at the end of the year very nearly in accord with Gauss's predictions. When Wilhelm Olbers discovered another minor planet, Pallas, in 1802, Gauss again successfully determined its path. All of this early work on astronomy was summarized in one of Gauss's major works, the *Theory of Motion of Heavenly Bodies* (1809).

Later, at the request of the Duke of Hannover, Gauss led a survey of Hannover, founded the science of geodesy, and derived therefrom seminal ideas in differential geometry. During the years from 1830 to 1840 Gauss also earned great distinction in his physical research on theoretical and experimental magnetism. He established a method of measuring the earth's magnetic field. James Clerk Maxwell, the founder of electromagnetic theory, says in his *Electricity and Magnetism* that Gauss's studies of magnetism reconstituted the whole science, the instruments used, the methods of observation, and the calculation of results. Gauss's papers on terrestrial magnetism are models of physical research. In honor of this work the unit of magnetism is called the gauss.

Though Gauss and Wilhelm Weber (1804–1891) did not originate the idea of telegraphy (there were many earlier attempts), in 1833 they designed a practical device which made a needle rotate right or left depending upon the direction of electric current sent over a wire. This was but one of Gauss's several inventions. He also worked in optics, which had been neglected since Euler's days, and his investigations of 1838–1841 gave a totally new basis for the handling of optical problems.

Leadership on a level with Gauss in the 19th-century mathematical world must be accorded to Cauchy. He had universal interests. In mathematics he wrote over seven hundred papers, second only to Euler in number. His works in a modern edition fill twenty-six volumes and embrace all branches of mathematics. He was the founder of the theory of functions of a complex variable (Chapers VII and VIII). But Cauchy devoted at least as much of his energy to physical problems. In 1815 he won a prize awarded by the French Academy of Sciences for a paper on water waves. He wrote fundamental works on the equilibrium of rods and elastic membranes such as metal sheets and on waves in elastic media and was one of the founders of that branch of mathematical physics. He took up the theory of light waves, which Augustin-Jean Fresnel had initiated, and extended the theory to the dispersion and polarization of light. Cauchy was a superb mathematical physicist.

Though Fourier was not quite in the same class as Gauss and Cauchy, his work is especially noteworthy because he brought one more natural phenomenon under the dominion of mathematics, the conduction of heat. Fourier regarded this subject as one of the most important cosmological studies because study of the conduction of heat within the earth might show that the earth had cooled down from a molten state. Thereby some estimate of the age of the earth might be made. In the course of this work he so advanced the theory of infinite trigonometric series, now called Fourier series, that it became applicable to many other areas of applied mathematics. It would be unjust to praise Fourier's work in measured terms.

These achievements of Gauss, Cauchy, Fourier, and hundreds of others were additional, seemingly incontestable evidence that more and more true laws of nature were being uncovered. And indeed throughout the century the giants of mathematics continued to pursue the course set by their predecessors and to create more powerful mathematical machinery and employ it successfully in the further investigation of nature. They forged ahead in the search for the mathematical laws of nature as though hypnotized by the belief that they, the mathematicians, were the anointed ones to discover God's design.

Had they paid attention to the words of some of their own brethren and kindred spirits, they might have been prepared for the disaster they soon would face. Early in the modern period Francis Bacon in his *Novum organum* (New Instrument [of reasoning], 1620) had noted:

> The idols of the tribe are inherent in human nature, and the very tribe or race of man. For man's sense is falsely asserted to be the standard of things. On the contrary, all the perceptions, both of the senses and the mind, bear reference to man, and not to the universe, and the human mind resembles those uneven mirrors, which impart their own properties to different objects, from which rays are emitted, and distort and disfigure them.

In this same work Bacon, calling for experience and experimentation as the basis for all knowledge, also said:

> It cannot be that axioms established by argumentation can suffice for the discovery of new works, since the subtlety of nature is greater many times over than the subtlety of argument.

Unintentionally, even the most faithful began to make distinctions that led gradually to the elimination of God's role in the design of the universe.

The work of Copernicus and Kepler on the heliocentric theory, hailed by both as evidence of God's mathematical wisdom, nevertheless contradicted statements in the Scriptures as to the importance of man. Galileo, Robert Boyle, and Newton affirmed that the objective of their scientific work was to evidence God's design and presence, but their scientific work proper did not involve God. In fact Galileo says in one of his letters, "For my part any discussion of the Sacred Scriptures might have lain dormant forever; no astronomer or scientist who remained within proper bounds has ever gotten into such things." Of course, Galileo, as we have seen, believed in God's mathematical design; all he meant by this statement was that no mystic or supernatural forces should be invoked to explain the workings of nature. In Galileo's time the belief that the omnipotent God could change the design still held sway. It was Descartes, however devout, who asserted the invariability of the laws of nature and implicitly limited God's power. Newton too believed in a fixed order but banked on God to keep the world functioning according to plan. He used the analogy of a watchmaker keeping a watch in repair. Newton had good reason to call upon God's intervention. Though he was well aware that the path of any one planet was not a perfect ellipse because other planets were perturbing its path, he could not demonstrate mathematically that the irregularities or the departures from the elliptical path were due to the attractions exerted by the other planets. Hence he feared that the stability of the solar system would be disrupted unless God continually intervened to keep it moving according to plan.

Leibniz objected. In a November 1715 letter to the philosopher and champion of Newton, Samuel Clarke, he said of Newton's view that God needs to repair and wind up the watch occasionally, "He [God] had not, it seems, sufficient foresight to make it a perpetual motion. . . . According to my opinion, the same force and vigour remain constant in the world, and only pass from one part of matter to another, in conformity with the laws of nature." Thus Leibniz charged Newton with deprecating God's power. Leibniz actually accused Newton of contributing to the decline of religion in England.

Leibniz was not off the mark. Newton's work unintentionally initiated a divorce or emancipation of natural philosophy from theology. Galileo, as we have noted, had insisted that physical science must be kept separate from theology. Newton adhered to this doctrine in his *Principles* and made much greater progress toward a purely mathematical account of natural phenomena. Thus God was excluded more and more from mathematical accounts of scientific theories. Indeed, the aberrations that Newton could not explain were essentially accounted for by later investigators.

Universal laws embracing heavenly and earthly motions began to dominate the intellectual scene, and the continued agreement between predictions and observations bespoke the perfection of the laws. Though belief that this perfection was evidence of God's design continued after Newton's time, God sank into the background and the mathematical laws of the universe became the focus. Leibniz saw some of the implications of Newton's *Principles*, a world functioning according to plan with or without God, and attacked the book as anti-Christian. The concern for the attainment of purely mathematical results gradually replaced the concern for God's design. Though many a mathematician after Euler continued to believe in God's presence, His design of the world, and mathematics as the science whose main function was to provide the tools to decipher God's design, the further the development of mathematics proceeded in the 18th century and the more numerous its successes, the more the religious inspiration for mathematical work receded and God's presence became dim.

Lagrange and Laplace though of Catholic parentage were agnostics. Laplace completely rejected any belief in God as the mathematical designer of the universe. There is a well-known story that when Laplace gave Napoleon a copy of his *Celestial Mechanics* Napoleon remarked, "Monsieur Laplace, they tell me you have written this large book on the system of the universe and have never even mentioned its Creator." Laplace is said to have replied, "I have no need of this hypothesis." Nature replaced God. As Gauss put it, "Thou, nature, are my goddess, to thy laws my services are bound." Gauss did in fact believe in an eternally existing, omniscient, and omnipotent God but thoughts about God had nothing to do with mathematics or the search for mathematical laws of nature.

Hamilton's remark about the Principle of Least Action (Chapter III) also reveals the change in intellectual outlook. He said in an article of 1833:

> Although the law of least action has thus attained a rank among the highest theorems of physics, yet its pretensions to a cosmological necessity, on the ground of economy in the universe, are now generally rejected. And the rejection appears just, for this, among other reasons, that the quantity pretended to be economized is in fact often lavishly expended. . . . We cannot, therefore, suppose the economy of this quantity to have been designed in the divine idea of the universe: though a simplicity of some high kind may be believed to be included in that idea.

In retrospect one can see that the doctrine of God's mathematical design of nature was being undermined by the very work of the mathematicians themselves. The intellectuals became more and more convinced that human reason was a most powerful faculty and the best evidence for it was the successes of the mathematicians. Could not then reason be applied to justify prevailing doctrines of religion and ethics if for no less salutary a reason than to reinforce them? Fortunately, or unfortunately, the application of reason to the foundation of religious beliefs undermined the orthodoxy of many. Religious beliefs drifted from orthodoxy to various intermediate stages such as rationalistic supernaturalism, deism, agnosticism, and outright atheism. These movements had their effect on mathematicians, who, in the 18th century, were men of broad culture. As Denis Diderot, the rationalistic, anti-clerical, leading intellectual of the age put it, "If you want me to believe in God, you must make me touch him." Not all 19th-century mathematicians denied God's role. Cauchy, a devout Catholic, said that man "rejects without hesitation any hypothesis which is in contradiction with revealed truth." Nevertheless the belief in God as the mathematical designer of the universe began to fade.

The weakening of this belief soon entailed the question of why the mathematical laws of nature were necessarily truths. Among the first to deny truths was Diderot in his *Thoughts on the Interpretation of Nature* (1753). He said the mathematician was like a gambler: both played games with abstract rules that they themselves had created. Their subjects of study were only matters of convention that had no foundation in reality. Equally critical was the intellectual Bernard Le Bovier de Fontenelle (1657–1757) in his *Plurality of Worlds*. He attacked the belief in the immutability of the laws of the heavenly motions with the observation that as long as roses could remember no gardener ever died.

Mathematicians prefer to believe that they create the nourishment on which philosophers feed. But in the 18th century the philosophers were in the van in denying truths about the physical world. We shall pass over the doctrines of Thomas Hobbes (1588–1679), John Locke (1632–1704) and of Bishop George Berkeley (1685–1753) not because they are easily refutable but because they were not as influential as those of the radical David Hume (1711–1776) who, in effect, not only endorsed Berkeley but went even further. In his *Treatise of Human Nature* (1739–40), Hume maintained that we know neither mind nor matter; both are fictions. We perceive sensations; simple ideas such as images, memories, and thoughts are but faint effects of sensations. Any complex idea is but a collection of simple ideas. The mind is *identical* with our collection of sensations and ideas. We should not assume the existence of any substances other than those which can be tested by immediate experience; but experience yields only sensations.

Hume was equally dubious about matter. Who guarantees that there is a permanently existing world of solid objects? All we *know* are our own sensations of such a world. Repeated sensations of a chair do not prove that a chair actually exists. Space and time are but a manner and order in which ideas occur to us. Similarly, causality is but a customary connection of ideas. Neither space nor time nor causality is an objective reality. We are deluded by the force and firmness of our sensations into believing in such realities. The existence of an external world with fixed properties is really an unwarranted inference; the origin of our sensations is inexplicable. Whether they arise from external objects or the mind itself or God, we cannot say.

Man himself is but an isolated collection of perceptions, that is, sensations and ideas. He exists only as such. The ego is a bundle of different perceptions. Any attempt to perceive one's self leads only to a perception. All other men and the supposed external world are just the perceptions of any one man, and there is no assurance that they exist.

Hence there can be no scientific laws concerning a permanent, objective physical world; such laws signify merely convenient summaries of sensations. Moreover, since the idea of causality is based not on scientific proof but merely on a habit of mind resulting from the frequent occurrence of the usual order of "events," we have no way of knowing that sequences perceived in the past will recur in the future. Thus Hume stripped away the inevitability of the laws of nature, their eternality and their inviolability.

By destroying the doctrine of an external world following fixed mathematical laws, Hume had destroyed the value of a logical deductive structure which represented reality. But mathematics also contains theorems about number and geometry which follow indubitably from supposed truths about numbers and geometrical figures. Hume did not reject the axioms, but chose to deflate them and the results obtained by deduction. As for the axioms, they come from sensations about the presumed physical world. The theorems are indeed necessary consequences of the axioms but they are no more than elaborated repetitions of the axioms. They are deductions but deductions of statements implicit in the axioms. They are tautologies. Thus there are no truths in the axioms or the theorems.

Hume, then, answered the fundamental question of how man obtains truths by denying their existence; man cannot arrive at truths. Hume's work not only deflated the efforts and results of science and mathematics, but challenged the value of reason itself. Such a denial of man's highest faculty was revolting to most 18th-century thinkers. Mathematics and other manifestations of human reason had accomplished too much to be cast aside as useless. Hume's philosophy was so contrary and repugnant to most intellectuals of the 18th century and so much at odds with the remarkable successes of mathematics and science that a refutation was called for.

The most revered and perhaps the deepest philosopher of all times, Immanuel Kant, took up the challenge. But the outcome of Kant's cogitations when carefully examined was not much more comforting. In his *Prolegomena to Any Future Metaphysics* (1783), Kant did seem to take the side of the mathematicians and scientists: "We can say with confidence that certain pure a priori synthetical cognitions, pure mathematics and pure physics, are actual and given; for both contain propositions which are thoroughly recognized as absolutely certain . . . and yet as independent of experience." In his *Critique of Pure Reason* (1781), Kant started out with even more reassuring words. He affirmed that all axioms and theorems of mathematics were truths. But why, Kant asked himself, was he willing to accept such truths? Surely experience itself did not validate them. The question could be answered if one could answer the larger question of how the very science of mathematics is possible. Kant's answer was that our minds possess the forms of space and time. Space and time are modes of perception—Kant called them intuitions—in terms of which the mind views experience. We perceive, organize, and understand experience in accordance with these mental forms. Experience fits into them like dough into a mold. The mind imposes these modes on the received sense impressions and causes these sensations to fall into built-in patterns. Since the intuition of space has its origin in the mind, the mind automatically accepts certain properties of this space. Principles such as that a straight line is the shortest path between two points, that three points determine a plane, and the parallel axiom of Euclid, which Kant called a priori synthetic truths, are part of our mental equipment. The science of geometry merely explores the logical consequences of these principles. The very fact that the mind views experience in terms of the "spatial structure" of the mind means that experience will conform to the basic principles and the theorems. The order and rationality that we think we perceive in the external world are imposed on it by our minds and our modes of thinking.

Since Kant manufactured space from the cells of the human brain he could see no reason not to make it Euclidean. His inability to conceive of another geometry convinced him that there could be no other. Thus the laws of Euclidean geometry were not inherent in the universe nor was the universe so designed by God: the laws were man's mechanism for organizing and rationalizing his sensations. As for God, Kant said that the nature of God fell outside of rational knowledge but we should believe in Him. Kant's boldness in philosophy was surpassed by his rashness in geometry, for despite never having been more than forty miles from his home city of Königsberg in East Prussia he presumed he could decide the geometry of the world.

What about the mathematical laws of science? Since all experience is grasped in terms of the mental framework of space and time, mathematics must be applicable to all experience. In his *Metaphysical Foundations of Natural Science* (1786) Kant accepted Newton's laws and their consequences as self-evident. He claimed to have demonstrated that Newton's laws of motion can be derived from pure reason and that these laws are the only assumptions under which nature is understandable. Newton, he said, gave us so clear an "insight into the structure of the universe that it will remain unchanged for all time."

More generally, Kant argued that the world of science is a world of sense impressions arranged and controlled by the mind in accordance with innate categories such as space, time, cause and effect, and substance. The mind contains furniture into which the guests must fit. The sense impressions do originate in a real world, but *unfortunately this world is unknowable.* Actuality can be known only in terms of the subjective categories supplied by the perceiving mind. Hence there never would be a way other than Euclidean geometry and Newtonian mechanics to organize experience. As experience broadens, as new sciences are formed, the mind does not formulate new principles by generalizing from these new experiences; rather, unused compartments of the mind are called into use to interpret these new experiences. The mind's power of vision is lit up by experience. This accounts for the relatively late recognition of some truths, such as, for example, the laws of mechanics, compared with others known for many centuries.

Kant's philosophy, barely intimated here, glorified reason; however, he assigned to it the role of exploring not nature but the recesses of the human mind. Experience received due recognition as a necessary element in knowledge, since sensations from the external world supply the raw material which the mind organizes. And mathematics retained its place as the discloser of the necessary laws of the mind.

That mathematics was a body of a priori truths the mathematicians were attuned to hear. But most did not pay enough attention to how Kant arrived at this conclusion. His doctrine that what mathematics asserts is not inherent in the physical world but comes from man's mind should have given pause to all mathematicians. Are all our minds prefabricated so as to make the same organization of our sensations, and is that organization of spatial sensations necessarily Euclidean? How do we know this? Unlike Kant, mathematicians and physicists still believed in an external world subject to laws independent of human minds. The world was rationally designed and man merely uncovered that design and used it to predict what would happen in that external world.

Kant's doctrines and influence had both a liberating and a restricting effect. By emphasizing the power of the mind to organize experiences in a world we shall never truly know, he paved the way for new structures contrary to those so firmly maintained in his time. But by insisting that the mind necessarily organizes spatial sensations in accordance with the laws of Euclidean geometry he hindered the acceptance of any contrary views. Had Kant been paying more attention to contemporary work in mathematics, he might at least have been less insistent that the mind must organize spatial sensations in the Euclidean format.

Indifference to and even dismissal of God as the law-maker of the universe, as well as the Kantian view that the laws were inherent in the structure of the human mind, brought forth a reaction from the Divine Architect. God decided that He would punish the Kantians and especially those egotistic, proud, and overconfident mathematicians. And He proceeded to encourage non-Euclidean geometry, a creation that devastated the achievements of man's presumably self-sufficient, all-powerful reason.

Though by 1800 God's presence was less and less felt and though a few radical philosophers such as Hume had denied all truths, the mathematicians of that time still believed in the truth of mathematics proper and of the mathematical laws of nature. Among the branches of mathematics, Euclidean geometry was most venerated, not only because it was the first to be deductively established but also because for over two thousand years its theorems had been found to be in perfect accord with the physical facts. It was precisely Euclidean geometry that "God" attacked.

One axiom of Euclidean geometry had bothered mathematicians somewhat, not because there was in their minds any doubt of its truth but because of its wording. This was the parallel axiom or, as it is often referred to, Euclid's fifth postulate. As Euclid worded it, it states:

> If a straight line [Fig. 4.1] falling on two straight lines makes the interior angles on the same side less than two right angles, then the two straight lines if extended will meet on that side of the straight line on which the angles are less than two right angles.

That is, if angles 1 and 2 add up to less than 180° then lines a and b if produced sufficiently far will meet.

Euclid had good reasons to word his axiom in this manner. He could have asserted instead that if the sum of the angles 1 and 2 is 180° then the lines a and b will never meet, that is, the lines a and b would be parallel. But Euclid apparently feared to assume that there could be *infinite* lines which never meet. Certainly experience did not vouch for the behavior of infinite straight lines, whereas axioms were supposed to be self-evident truths about the physical world. However, he did *prove* the existence of parallel lines on the basis of his parallel axiom and the other axioms.

The parallel axiom in the form stated by Euclid was thought to be somewhat too complicated. It lacked the simplicity of the other axioms. Apparently even Euclid did not like his version of the parallel axiom because he did not call upon it until he had proved all the theorems he could without it.

A related problem, which did not bother many people but ultimately came to the fore as a vital concern, is whether one can be sure of the

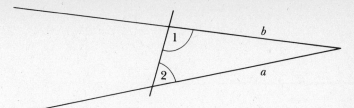

Figure 4.1

existence of infinite straight lines in physical space. Euclid was careful
to postulate only that one can extend a finite line segment as far as necessary, so that even the extended segment was still finite. Nevertheless
Euclid did imply the existence of infinite straight lines for, were they finite, they could not be extended as far as necessary in any given context.

Even in Greek times the mathematicians began efforts to resolve the
problem presented by Euclid's parallel axiom. Two types of attempts
were made. The first was to replace the parallel axiom by a seemingly
more self-evident statement. The second was to try to deduce it from
the other nine axioms of Euclid; were this possible Euclid's statement
would be a theorem and so no longer questionable. Over two thousand
years many dozens of major mathematicians, to say nothing of minor
ones, engaged in both types of efforts. The history is long and technical
and most of it will not be reproduced here because it is readily available
and not especially relevant.*

Of the many substitute axioms perhaps one should be noted because
it is the one we usually learn in high school today. This version of the
parallel axiom, which is due to John Playfair (1748–1819) and was
proposed by him in 1795, states: Through a given point P not on a line
l (Fig. 4.2) there is one and only one line in the plane of P and l which

Figure 4.2

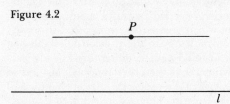

* This history can be found in Roberto Bonola: *Non-Euclidean Geometry,* first published in
Italian in 1906, Dover Publications reprint of the English translation, 1955.

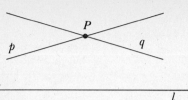

Figure 4.3

does not meet l. All of the substitute axioms that were proposed, though seemingly simpler than Euclid's version, proved on closer examination no more satisfactory than Euclid's. Many of them, including Playfair's, made assertions about what happens indefinitely far out in space. On the other hand the substitute axioms which did not involve "infinity" directly, for example, the axiom that there exist two similar but unequal triangles, were seen to be rather complex assumptions and by no means preferable to Euclid's own parallel axiom.

Of the second group of efforts to solve the problem of the parallel axiom, those which sought to deduce Euclid's assertion from the other nine axioms, the most significant was made by Gerolamo Saccheri (1667–1733), a Jesuit priest and professor at the University of Pavia. His thought was that if one adopted an axiom that differed essentially from Euclid's then one might arrive at a theorem that contradicted another of Euclid's theorems. Such a contradiction would mean that the axiom denying Euclid's parallel axiom, the only axiom in question, is false and so the Euclidean parallel axiom must be true, that is, a consequence of the other nine axioms.

In view of Playfair's axiom, which is equivalent to Euclid's, Saccheri assumed first* that through the point P (Fig. 4.3) there are no lines parallel to l. And from this axiom and the other nine that Euclid adopted Saccheri did deduce a contradiction. Saccheri tried next the second and only other possible alternative, namely, that through the point P there are at least two lines p and q that no matter how far extended do not meet l. Saccheri proceeded to prove many interesting theorems until he reached one which seemed so strange and so repugnant that he decided it was contradictory to the previously established results. Saccheri therefore felt justified in concluding that Euclid's parallel axiom was really a consequence of the other nine axioms and so entitled his book, *Euclides ab omni naevo vindicatus* (Euclid Vindicated from All Fault, 1733). However, later mathematicians realized that Saccheri did not really obtain a contradiction in the second case and the

---

* The description that follows is a minor modification of Saccheri's procedure.

problem of the parallel axiom was still open. The efforts to find an acceptable substitute for the Euclidean axiom on parallels or to prove that the Euclidean assertion must be a consequence of Euclid's nine other axioms were so numerous and so futile that in 1759 D'Alembert called the problem of the parallel axiom "the scandal of the elements of geometry."

Gradually the mathematicians began to approach the correct understanding of the status of Euclid's parallel axiom. In his doctoral dissertation of 1763 Georg S. Klügel (1739–1812), later a professor at the University of Helmstädt, who knew Saccheri's book and many other attempts to vindicate the parallel axiom, made the remarkable observation that the certainty with which men accepted the truth of the Euclidean parallel axiom was based on experience. This observation introduced for the first time the thought that experience rather than self-evidence substantiated the axioms.* Klügel expressed doubt that the Euclidean assertion could be proved. Moreover, he realized that Saccheri had not arrived at a contradiction but merely at results that were strange.

Klügel's paper suggested work on the parallel axiom to Johann Heinrich Lambert (1728–1777). In his book *Theory of Parallel Lines* (written in 1766 and published in 1786), Lambert, somewhat like Saccheri, considered the two alternative possibilities. He, too, found that the assumption of no lines through P parallel to l (Fig. 4.3) led to a contradiction. However, unlike Saccheri, Lambert did not conclude that he had obtained a contradiction from the assumption of at least two parallels through P. Moreover, he realized that any collection of hypotheses which did not lead to contradictions offered a possible geometry. Such a geometry would be a valid logical structure even though it might have little to do with physical figures.

The work of Lambert and other men such as Abraham G. Kästner (1719–1800), a professor at Göttingen who was Gauss's teacher, warrants emphasis. They were convinced that Euclid's parallel axiom could not be proven on the basis of the other nine Euclidean axioms; that is, it is independent of Euclid's other axioms. Further, Lambert was convinced that it is possible to adopt an alternative axiom contradicting Euclid's and build a logically consistent geometry, though he made no assertions about the applicability of such a geometry. Thus all three recognized the existence of a non-Euclidean geometry.

The most distinguished mathematician who worked on the problem posed by Euclid's parallel axiom was Gauss. Gauss was fully aware of the vain efforts to establish Euclid's parallel axiom because this was

*Newton had also made this assertion but he did not stress it and it was ignored.

common knowledge in Göttingen. In fact the history of these efforts was thoroughly familiar to Gauss's teacher Kästner. Many years later, in 1831, Gauss told his friend Schumacher that as far back as 1792 (Gauss was then 15) he had already grasped the idea that there could be a logical geometry in which Euclid's parallel postulate did not hold. But even up to 1799 Gauss still tried to deduce Euclid's parallel postulate from other more plausible assumptions and he still believed Euclidean geometry to be the geometry of physical space though he could conceive of other logical non-Euclidean geometries. However, on December 17, 1799, Gauss wrote to his friend and fellow mathematician Wolfgang Bolyai:

> As for me I have already made some progress in my work. However, the path I have chosen does not lead at all to the goal which we seek [deduction of the parallel axiom], and which you assure me you have reached. It seems rather to compel me to doubt the truth of geometry itself. It is true that I have come upon much which by most people would be held to constitute a proof [of the deduction of the Euclidean parallel axiom from the other axioms]; but in my eyes it proves as good as nothing. For example, if we could show that a rectilinear triangle is possible, whose area would be greater than any given area, then I would be ready to prove the whole of [Euclidean] geometry absolutely rigorously.
>   Most people would certainly let this stand as an axiom; but I, no! It would indeed be possible that the area might always remain below a certain limit, however far apart the three angular points of the triangle were taken.

From about 1813 on Gauss developed his non-Euclidean geometry, which he first called anti-Euclidean geometry, then astral geometry, and finally non-Euclidean geometry. He became convinced that it was logically consistent and rather sure that it might be applicable.

In a letter to his friend Franz Adolf Taurinus (1794–1874) of November 8, 1824, Gauss wrote:

> The assumption that the angle sum [of a triangle] is less than 180° leads to a curious geometry, quite different from ours [Euclidean] but thoroughly consistent, which I have developed to my entire satisfaction. The theorems of this geometry appear to be paradoxical, and, to the uninitiated, absurd, but calm, steady reflection reveals that they contain nothing at all impossible.

In his letter to the mathematician and astronomer Friedrich Wilhelm Bessel of January 27, 1829, Gauss reaffirmed that the parallel postulate could not be proved on the basis of the other axioms in Euclid.

We shall not discuss the specific non-Euclidean geometry that was created by Gauss. He did not finish a full deductive presentation, and the theorems he did prove are much like those we shall encounter shortly in the work of Lobatchevsky and Bolyai. He said in his letter to Bessel that he probably would never publish his findings on this subject because he feared ridicule or, as he put it, he feared the clamor of the Boeotians, a reference to a proverbially dull-witted Greek tribe. One must remember that though a few mathematicians had been gradually reaching the denouement of the work in non-Euclidean geometry the intellectual world at large was still dominated by the conviction that Euclidean geometry was the only possible geometry. What we do know about Gauss's work in non-Euclidean geometry is gleaned from his letters to friends, two short reviews in the *Göttingische gelehrte Anzeigen* of 1816 and 1822, and some notes of 1831 found among his papers after his death.

The two men who receive more credit than Gauss for the creation of non-Euclidean geometry are Lobatchevsky and Bolyai. Actually their work was an epilogue to prior innovative ideas but because they published systematic deductive works they are usually hailed as the creators of non-Euclidean geometry. Nikolai Ivanovich Lobatchevsky (1793–1856), a Russian, studied at the University of Kazan and from 1827 to 1846 was professor and rector there. From 1825 on he presented his views on the foundations of geometry in many papers and two books. Johann Bolyai (1802–1860), son of Wolfgang Bolyai, was a Hungarian army officer. He published a twenty-six-page paper, "The Science of Absolute Space," on non-Euclidean geometry, which he called absolute geometry, as an appendix to the first volume of his father's two-volume book *Tentamen juventutem studiosam in elementa Matheseos* (Essay on the Elements of Mathematics for Studious Youths). Though this book appeared in 1832–33 and therefore after publications by Lobatchevsky, Bolyai seems to have worked out his ideas on non-Euclidean geometry by 1825 and was convinced by that time that the new geometry was not self-contradictory. In a letter to his father dated November 23, 1823, Johann wrote, "I have made such wonderful discoveries that I am myself lost in astonishment."

Gauss, Lobatchevsky, and Bolyai had realized that the Euclidean parallel axiom could not be proved on the basis of the other nine axioms and that some additional axiom about parallel lines was needed to found Euclidean geometry. Since this last was an independent fact it was then at least logically possible to adopt a contradictory statement and develop the consequences of the new set of axioms.

The technical content of what these men created is rather simple. It is just as well to note Lobatchevsky's work since all three did about the same thing. Lobatchevsky boldly rejects Euclid's parallel axiom and makes the assumption which, in effect, Saccheri had made. Given a line

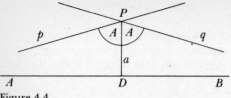

Figure 4.4

AB and a point *P* (Fig. 4.4) then all lines through *P* fall into two classes with respect to *AB*, namely, the class of lines which meet *AB* and the class of lines which do not. To the latter belong two lines *p* and *q* which form the boundary between the two classes. More precisely, if *P* is a point at a perpendicular distance *a* from the line *AB*, then there exists an acute angle *A* such that all lines which make with the perpendicular *PD* an angle less than *A* will intersect *AB*, while lines which make an angle greater than or equal to *A* do not intersect *AB*. The two lines *p* and *q* which make the angle *A* with *PD* are the parallels and *A* is called the angle of parallelism. Lines through *P* (other than the parallels) which do not meet *AB* are called non-intersecting lines, though in Euclid's sense they would be parallel lines. In this sense Lobatchevsky's geometry admits an infinite number of parallels through *P*.

He then proved several key theorems. If angle *A* equals $\pi/2$ then the Euclidean parallel axiom results. If angle *A* is acute, then it follows that it increases and approaches $\pi/2$ as the perpendicular *a* decreases to 0; further, angle *A* decreases and approaches 0 as *a* becomes infinite. The sum of the angles of a triangle is always less than 180° and approaches 180° as the area of the triangle decreases. Moreover, two similar triangles are necessarily congruent.

No major branch of mathematics or even a major specific result is the work of one man. At best some decisive step or assertion may be credited to an individual. This cumulative development of mathematics certainly applies to non-Euclidean geometry. If non-Euclidean geometry means the development of the consequences of a system of axioms containing an alternative to Euclid's parallel axiom, then most credit must be accorded to Saccheri and even he benefited by the work of many men who tried to find an acceptable substitute axiom for Euclid's. If the creation of non-Euclidean geometry means the recognition that there can be geometries alternative to Euclid's, then Klügel and Lambert created it. However, the most significant fact about non-Euclidean geometry is that *it can be used to describe the properties of physical space as accurately as Euclidean geometry does.* Euclidean geometry is not the necessary geometry of physical space; its physical truth cannot be guaranteed on any a priori grounds. This realization, which did not call for any technical mathematical development, because this had already been done, was first attained by Gauss.

According to one of his biographers Gauss attempted to test this conviction. He noted that the sum of the angles of a triangle is 180° in Euclidean geometry and less in non-Euclidean geometry. He had spent some years making a survey of the kingdom of Hannover and he had recorded the data. It is possible that he used these to measure the sum of the angles of a triangle. In a famous paper he wrote in 1827 Gauss noted that the angle sum of the triangle formed by three mountain peaks, Brocken, Hohehagen, and Inselberg, exceeded 180° by about 15″. This proved nothing because the experimental error was much larger than the excess, so that the correct sum could have been 180° or less. Gauss must have realized that this triangle was too small to be decisive because in his non-Euclidean geometry the departure from 180° is proportional to the area. Only a large triangle, such as one involved in astronomical investigations, could possibly reveal any significant departure from an angle sum of 180°. Nevertheless Gauss was convinced that this new geometry was as applicable as Euclidean geometry.

Lobatchevsky also considered the applicability of his geometry to physical space and did give an argument to show that it could be applicable to very large geometrical figures. Thus by the 1830s not only was a non-Euclidean geometry accepted by a few men but its applicability to physical space was considered at least possible.

The problem of which geometry fits physical space, raised primarily by the work of Gauss, stimulated another creation, a new geometry that gave the mathematical world further inducement to believe that the geometry of physical space could be non-Euclidean. The creator was Georg Bernhard Riemann (1826–1866), a student of Gauss and later professor of mathematics at Göttingen. Though the details of Lobatchevsky's and Bolyai's work were unknown to Riemann, they were known to Gauss, and Riemann certainly knew Gauss's doubts as to the necessary applicability of Euclidean geometry.

Gauss assigned to Riemann the subject of the foundations of geometry for the lecture he was required to deliver to qualify for the title of *Privatdozent.* Riemann gave this lecture in 1854 to the philosophical faculty at Göttingen with Gauss present. It was published in 1868 under the title "On the Hypotheses which Lie at the Foundation of Geometry." In it Riemann reconsidered the entire problem of the structure of space. He took up first the question of just what is certain about physical space. What conditions or facts are presupposed in the very concept of space before we determine *by experience* whatever properties may hold in physical space? From these conditions or facts, treated as axioms, he planned to deduce further properties. The axioms and their logical consequences would be a priori and necessarily true. Any other properties of space would then have to be learned empirically. One of Riemann's objectives was to show that Euclid's axioms were indeed empirical rather than self-evident truths. He adopted the analytical approach (calculus and its extensions) because in geometrical proofs we may be misled by our perceptions to assume facts not explicitly recognized.

Riemann's approach to the structure of space is very general and for present purposes we need not examine it in detail. While studying what can be asserted a priori, he made a distinction which later became more important, the distinction between boundlessness and infiniteness of space. (Thus the surface of a sphere is boundless but not infinite.) Unboundedness, he pointed out, has a greater empirical credibility than infinite extent.

Riemann's idea about the possibility of a space being unbounded rather than infinite suggested another elementary non-Euclidean geometry now called double elliptic geometry. At first Riemann himself and Eugenio Beltrami (1835–1900) thought of this new geometry as one which applies to certain surfaces, such as the surface of a sphere where great circles are the "straight lines." But later work by Cayley and others obliged mathematicians to accept the fact that double elliptic geometry, like the geometry of Gauss, Lobatchevsky, and Bolyai, can describe our three-dimensional physical space in which the straight line is the ruler's edge.

In double elliptic geometry the straight line is in fact unbounded but not infinite in length. Moreover, there are *no* parallel lines. Since this new geometry does retain some of the axioms of Euclidean geometry, some of the theorems are the same. Thus the theorem that two triangles are congruent when two sides and the included angle of one equal the corresponding parts of the other is a theorem of the new geometry as are other familiar congruence theorems. However, the major theorems of this geometry differ from those of Euclidean geometry and the geometry of Gauss, Lobatchevsky, and Bolyai. One theorem asserts that all straight lines have the same finite length and meet in two points. Another asserts that all perpendiculars to a line meet in one point. The sum of the angles of a triangle is always *greater* than 180° but decreases to 180° as the area of the triangle approaches zero. Two similar triangles are necessarily congruent. As to the applicability of this double elliptic geometry, all of the arguments for the applicability of the previously created non-Euclidean geometry, now called hyperbolic geometry, apply with equal force.*

---

* Later Felix Klein pointed out that there is another elementary non-Euclidean geometry in which any two lines meet in one point. This he called single elliptic.

At first thought the idea that any of these strange geometries could possibly compete with or even supersede Euclidean geometry seems absurd. But Gauss faced this possibility. Whether or not he used the measurements recorded in his paper of 1827 to test the applicability of non-Euclidean geometry, he was the first man to assert not only its applicability but to recognize that we could no longer be sure of the *truth* of Euclidean geometry. That Gauss was influenced directly by the writings of Hume cannot be ascertained. Kant's rebuttal of Hume he disdained. Nevertheless he lived at a time when the truth of mathematical laws was being challenged, and he must have absorbed the intellectual atmosphere as surely as all of us breathe the air about us. New intellectual outlooks take hold even if imperceptibly. Had Saccheri been born one hundred years later, perhaps he too would have arrived at Gauss's conclusions.

At first Gauss seems to have concluded that there is no truth in all of mathematics. In a letter to Bessel of November 21, 1811, he said, "One should never forget that the functions [of a complex variable], like all mathematical constructions, are only our own creations, and that when the definition with which one begins ceases to make sense, one should not ask, What is, but what is it convenient to assume in order that it remain significant." But no one yields up treasures readily. Gauss apparently reconsidered the matter of the truth of mathematics and saw the rock to which he could anchor it. In a letter to Heinrich W. M. Olbers (1758–1840) written in 1817 he said, "I am becoming more and more convinced that the [physical] necessity of our [Euclidean] geometry cannot be proved, at least not by human reason nor for human reason. Perhaps in another life we will be able to obtain insight into the nature of space which is now unattainable. Until then we must not place geometry in the same class with arithmetic, which is purely a priori, but with mechanics." Gauss, unlike Kant, did not accept the laws of mechanics as truths. Rather he and most men followed Galileo in believing that these laws are founded on experience. On April 9, 1830, Gauss wrote to Bessel:

> According to my most sincere conviction the theory of space has an entirely different place in knowledge from that occupied by pure mathematics [the mathematics built on number]. There is lacking throughout our knowledge of it the complete persuasion of necessity (also of absolute truth) which is common to the latter; we must add in humility, that if number is exclusively the product of our mind, space has a reality outside our mind and we cannot completely prescribe its laws.

Gauss was asserting that truth lies in arithmetic and consequently in algebra and analysis (calculus and its extensions), which are built on arithmetic, because the truths of arithmetic are clear to our minds.

That Euclidean geometry is the geometry of physical space, that it is the truth about space, was so ingrained in people's minds that for many years any contrary thoughts such as Gauss's were rejected. The mathematician Georg Cantor spoke of a law of conservation of ignorance. A false conclusion once arrived at and widely accepted is not easily dislodged and the less it is understood the more tenaciously it is held. For thirty or so years after the publication of Lobatchevsky's and Bolyai's works all but a few mathematicians ignored the non-Euclidean geometries. They were regarded as a curiosity. Some mathematicians did not deny their logical coherence. Others believed that they must contain contradictions and so were worthless. Almost all mathematicians maintained that the geometry of physical space, *the* geometry, must be Euclidean.

Unfortunately, mathematicians had abandoned God and so the Divine Geometer refused to reveal which of the several competing geometries He had used to design the universe. Mathematicians were thrown upon their own resources. However, the material in Gauss's notes became available after his death in 1855 when his reputation was unexcelled and the publication in 1868 of Riemann's 1854 paper convinced many mathematicians that a non-Euclidean geometry could be the geometry of physical space and that we could no longer be sure which geometry was true. The mere fact that there can be alternative geometries was in itself a shock. But the greater shock was that one could no longer be sure which geometry was true or whether any one of them was true. It became clear that mathematicians had adopted axioms for geometry that seemed correct on the basis of limited experience and had been deluded into thinking that these were self-evident truths. Mathematicians were in the position described by Mark Twain: "Man is the religious animal. He's the only one who's got the true religion—several of them."

Non-Euclidean geometry and its implication about the truth of geometry were accepted gradually by mathematicians, but not because the arguments for its applicability were strengthened in any way. Rather the reason was given in the early 1900s by Max Planck, the founder of quantum mechanics: "A new scientific truth does not triumph by convincing its opponents and making them see light, but rather because its opponents eventually die, and a new generation grows up that is familiar with it."

As to the truth of mathematics as a whole, some mathematicians took the position of Gauss. Truth resides in number, which is the basis of arithmetic, algebra, calculus, and higher branches of analysis. As Karl Gustav Jacob Jacobi (1804–1851) put it, "God ever arithmetizes." He does not, as Plato had maintained, eternally geometrize.

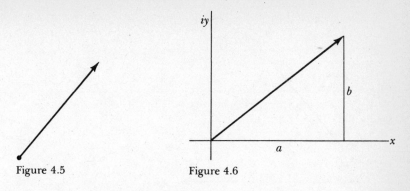

Figure 4.5   Figure 4.6

It would seem then that mathematicians had managed to rescue and maintain as truths the portion of mathematics built on arithmetic, which by 1850 was far more extensive and more vital for science then the several geometries. Unfortunately, other shattering events were to follow. To understand these we must backtrack a bit.

Since the 16th century, mathematicians had been using the concept of vectors. A vector, usually pictured as a directed line segment, has both direction and magnitude (Fig. 4.5). It is used to represent a force, a velocity, or any other quantity for which both direction and magnitude are important. Vectors in one plane can be combined geometrically under the usual operations of addition, subtraction, multiplication, and division to yield a resultant vector. The same century saw the introduction of complex numbers, that is, numbers of the form $a + bi$, where $i = \sqrt{-1}$ and a and b are real numbers. These were, even to mathematicians, rather mysterious quantities. It was therefore a boon when, about 1800, several mathematicians, Caspar Wessel (1745–1818), Jean-Robert Argand (1786–1822), and Gauss, realized that one can represent complex numbers as directed line segments in a plane (Fig. 4.6). These same men saw at once that complex numbers could be used not only to represent vectors in a plane but that the operations of addition, subtraction, multiplication, and division of vectors could be performed by using the complex numbers. That is, complex numbers serve as the algebra of vectors, just as the whole numbers and fractions serve to represent, for example, commercial transactions. Hence one need not carry out operations with vectors geometrically but can work with them algebraically. Thus the addition of the two vectors OA and OB (Fig. 4.7), which according to the parallelogram law for the addition of vectors must yield the sum or resultant vector OC, can be carried out algebraically by representing OA as the complex number $3 + 2i$, and OB

Figure 4.7

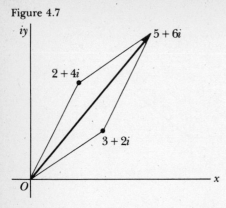

as the complex number $2 + 4i$. The sum, $5 + 6i$, then represents the resultant vector OC.

This use of complex numbers to represent vectors and operations with vectors that lie in a plane had become somewhat well known by 1830. However, if several forces act on a body, these forces and their vector representations need not and generally will not lie in one plane. If for the sake of convenience we call ordinary real numbers one-dimensional numbers and complex numbers two-dimensional, then what would be needed to represent and work algebraically with vectors in space is some sort of three-dimensional number. The desired operations with these three-dimensional numbers, as in the case of complex numbers, would have to include addition, subtraction, multiplication, and division and moreover obey the usual properties of real and complex numbers so that algebraic operations could be applied freely and effectively. Hence mathematicians began a search for what was called a three-dimensional complex number and its algebra.

Many mathematicians took up this problem. The creation in 1843 of a useful spatial analogue of complex numbers is due to William R. Hamilton. For fifteen years Hamilton was baffled. All of the numbers known to mathematicians at this time possessed the commutative property of multiplication, that is $ab = ba$, and it was natural for Hamilton to believe that the three-dimensional or three-component numbers he sought should possess this same property as well as the other properties that real and complex numbers possess. Hamilton succeeded but only by making two compromises. First, his new numbers contained four components and, second, he had to sacrifice the commutative law of multiplication. Both features were revolutionary for algebra. He called the new numbers quaternions.

Whereas a complex number is of the form $a + bi$ wherein $i = \sqrt{-1}$, a quaternion is a number of the form

$$a + bi + cj + dk$$

wherein the i, j, and k possess the same property as $\sqrt{-1}$, namely,

$$i^2 = j^2 = k^2 = -1.$$

The criterion for equality of two quaternions is that the coefficients a, b, c, and d be equal.

Two quaternions are added by the addition of the respective coefficients to form new coefficients. Thus the sum of two quaternions is itself a quaternion. To define multiplication Hamilton had to specify what the products of i and j, i and k, and j and k should be. To ensure that the product be a quaternion and to secure for his quaternions as many of the properties of real and complex numbers as possible, he arrived at

$$jk = i, \; kj = -i, \; ki = j, \; ik = -j, \; ij = k, \; ji = -k.$$

These agreements mean that multiplication is not commutative. Thus if p and q are quaternions, pq does not equal qp. Division of one quaternion by another can also be effected. However, the fact that multiplication is not commutative means that to divide the quaternion p by the quaternion q one can intend to find r such that $p = qr$ or such that $p = rq$. The quotient r need not be the same in the two cases. Though quaternions did not prove to be as broadly useful as Hamilton expected, he was able to apply them to a large number of physical and geometrical problems.

The introduction of quaternions was another shock to mathematicians. Here was a physically useful algebra which failed to possess a fundamental property of all real and complex numbers, namely, that $ab = ba$.

Not long after Hamilton created quaternions, mathematicians working in other domains introduced even stranger algebras. The famous algebraic geometer Arthur Cayley (1821–1895) introduced matrices, which are square or rectangular arrays of numbers. These too are subject to the usual operations of algebra but, as in the case of quaternions, lack the commutative property of multiplication. Moreover, the product of two matrices can be zero even though neither factor is zero. Quaternions and matrices were but the forerunners of a host of new algebras with stranger and stranger properties. Hermann Günther Grassmann (1809–1877), created a variety of such algebras. These were even more general than Hamilton's quaternions. Unfortunately, Grassmann was a high-school teacher and so it took many years before his work received the attention it deserved. In any case, Grassmann's work added to the variety of new types of what are now called hypernumbers.

The creation of new algebras for special purposes did not in itself challenge the truth of ordinary arithmetic and its extensions in algebra and analysis. After all, the ordinary real and complex numbers were used for totally different purposes where their applicability seemed unquestionable. Nevertheless, the very fact that new algebras appeared on the scene made men doubt the truth of the familiar arithmetic and algebra, just as people who learn about the customs of a strange civilization begin to question their own.

The sharpest attack on the truth of arithmetic came from Hermann von Helmholtz (1821–1894), a superb physician, physicist, and mathematician. In his *Counting and Measuring* (1887) he considered the main problem in arithmetic to be the justification of the *automatic* application of arithmetic to physical phenomena. His conclusion was that only experience can tell us where the laws of arithmetic do apply. We cannot be sure a priori that they do apply in any given situation.

Helmholtz made many pertinent observations. The very concept of number is derived from experiences. Some kinds of experience suggest the usual types of number, whole numbers, fractions, and irrational numbers, and their properties. To these experiences the familiar numbers are applicable. We recognize that virtually equivalent objects exist and so we recognize that we may speak, for example, of two cows. However, these objects must not disappear or merge or divide. One raindrop added to another does not make two raindrops. Even the notion of equality cannot be applied automatically to experience. It would seem certain that if object a equals c and b equals c that a must equal b. But two pitches of sound may seem to equal a third and yet the ear might distinguish the first two. Here things equal to the same thing need not equal each other. Likewise colors a and b may seem the same as do colors b and c but a and c can be distinguished.

Many examples may be adduced to show that the naïve application of arithmetic would lead to nonsense. Thus if one mixes two equal volumes of water, one at 40° Fahrenheit and the other at 50°, one does not get two volumes at 90°. If two simple sounds, one of 100 cycles per second and another of 200 cycles per second, are superposed one does not get a sound whose frequency is 300 cycles per second. In fact the composite sound has a frequency of 100 cycles per second. If two resistances of magnitudes R_1 and R_2 are connected in parallel in an electrical circuit, their combined effective resistance is $R_1R_2/(R_1+R_2)$. Further, as Henri Lebesgue facetiously pointed out, if one puts a lion and a rabbit in a cage, one will not find two animals in the cage later on.

We learn in chemistry that when one mixes hydrogen and oxygen he obtains water. But if someone takes two volumes of hydrogen and one volume of oxygen, he obtains, not three, but two volumes of water vapor. Likewise, one volume of nitrogen and three volumes of hydrogen yield two volumes of ammonia. We happen to know the physical explanation of these surprising arithmetic facts. By Avogadro's hypothesis, equal volumes of any gas, under the same conditions of temperature and pressure, contain the same number of *particles*. If, then, a given volume of oxygen contains 10 molecules, the same volume of hydrogen will also contain 10 molecules. Then there are 20 molecules in two volumes of hydrogen. Now it happens that the molecules of oxygen and hydrogen are diatomic; that is, each contains two atoms. Each of these 20 diatomic hydrogen *molecules* combines with one atom of the 10 molecules of oxygen to form 20 molecules of water or two volumes of water, but not three. Thus arithmetic fails to describe correctly the result of combining gases by volumes.

Ordinary arithmetic also fails to describe the combination of some liquids by volume. If a quart of gin is mixed with a quart of vermouth one does not get two quarts of the mixture but a quantity slightly less. One quart of alcohol and one quart of water yield about 1.8 quarts of vodka. This is true of most mixtures of alcoholic liquids. Three tablespoons of water and one tablespoon of salt do not make four tablespoons. Some chemical mixtures not only do not add up by volume but explode.

Not only do the properties of the whole numbers fail to apply to many physical situations, but there are practical situations in which a different arithmetic of fractions must be applied. Let us consider baseball, certainly a subject of interest to millions of Americans.

Suppose a player goes to bat 3 times in one game and 4 times in another. How many times in all did he go to bat? There is no difficulty here. He went to bat a total of 7 times. Suppose he hit the ball successfully, that is, got to first base or farther, twice in the first game and 3 times in the second. How many hits did he make in both games? Again there is no difficulty. The total number of hits is $2 + 3$, or 5. However, what the audience, and the player himself, is usually most interested in is the batting average, that is, the ratio of the number of hits to the number of times at bat. In the first game this ratio was 2/3; in the second game the ratio was 3/4. And now suppose the player or a baseball fan wishes to use these two ratios to compute the batting average for both games. One would think that all he would have to do was to add the two fractions by the usual method of adding fractions. That is,

$$\frac{2}{3} + \frac{3}{4} = \frac{17}{12}.$$

Of course this result is absurd. The player could not get 17 hits in 12 times at bat. Evidently, the ordinary method of adding fractions does not serve to give the batting average for both games by addition of the batting averages for the separate games.

How can we obtain the correct batting average for the two games from those of the separate games? The answer is to use a new method of adding fractions. We know that the combined average is 5/7 and the separate batting averages are 2/3 and 3/4. We see that, if we add the numerators and add the denominators of the separate fractions and then form the new fraction, we get the correct answer. That is,

$$\frac{2}{3} + \frac{3}{4} = \frac{5}{7}$$

provided this plus sign means adding the numerators and adding the denominators.

This method of adding fractions is useful in other situations. A salesman who wishes to keep a record of his effectiveness may note that he made a sale in 3 out of 5 calls one day and 4 out of 7 calls the next day. To obtain the ratio of his successful calls to his total number of calls he would have to add the 3/5 and 4/7 in precisely the manner in which the batting averages are added. His record for the two days of work is 7 successes in a total of 12 calls and this 7/12 is 3/5 + 4/7, *provided* plus means adding the numerators and adding the denominators.

Still another application is even more common. Suppose an automobile covers 50 miles in 2 hours and 100 miles in 3 hours. What is its average speed for both trips? One could say that the automobile covered 150 miles in 5 hours so that its average speed is 30 miles per hour. However, it is often useful to be able to compute the average speed for the total amount of travel from the average speeds for the individual trips. The average speed for the first trip is 50/2 and the average for the second trip is 100/3. If one adds the numerators and adds the denominators of these two fractions, he obtains the correct average for the total trip.

Ordinarily 4/6 = 2/3. However, in adding two fractions in the arithmetic under discussion, say, 2/3 + 3/5, it would not do to replace 2/3 by 4/6, for the answer in one case is 5/8 and in the other is 7/11, and these two answers are not equal. Further, in normal arithmetic, fractions such as 5/1 and 7/1 behave exactly as do the integers 5 and 7. However, if we add 5/1 and 7/1 as fractions in the new arithmetic we do not obtain 12/1 but 12/2.

These examples of what one might call baseball arithmetic do show that we can introduce operations different from the familiar ones and so create an applicable arithmetic. There are in fact many other arithmetics in mathematics. However, the sound mathematician does not create an arithmetic just to indulge whims. Each arithmetic is designed to represent some class of phenomena of the physical world. By defining the operations to fit what takes place in that class of phenomena, just as the above addition of fractions fits the combination of two batting averages, one can use the arithmetic to study conveniently what happens physically. Only experience can tell us where ordinary arithmetic applies to any given physical phenomena. Thus one cannot speak of arithmetic as a body of truths that necessarily apply to physical phenomena. Of course, since algebra and analysis are extensions of arithmetic, these branches, too, are not bodies of truth.

Thus the sad conclusion which mathematicians were obliged to draw is that there is no truth in mathematics, that is, truth in the sense of laws about the real world. The axioms of the basic structures of arithmetic and geometry are suggested by experience, and the structures as a consequence have a limited applicability. Just where they are applicable can be determined only by experience. The Greeks' attempt to guarantee the truth of mathematics by starting with self-evident truths and by using only deductive proof proved futile.

To many thoughtful mathematicians the fact that mathematics is not a body of truths was too repugnant to swallow. It seemed as though God had sought to confound them with several geometries and several algebras just as He had confounded the people of Babel with different languages. Hence they refused to accept the new creations.

William R. Hamilton, certainly one of the outstanding mathematicians, expressed his objection to non-Euclidean geometry in 1837:

> No candid and intelligent person can doubt the truth of the chief properties of *Parallel Lines,* as set forth by Euclid in his *Elements,* two thousand years ago; though he may well desire to see them treated in a clearer and better method. The doctrine involves no obscurity nor confusion of thought, and leaves in the mind no reasonable ground for doubt, although ingenuity may usefully be exercised in improving the plan of the argument.

In his presidential address of 1883 to the British Association for the Advancement of Science, Arthur Cayley affirmed:

> My own view is that Euclid's twelfth axiom [usually called the fifth or parallel axiom] in Playfair's form of it does not need demonstration, but is part of our notion of space, of the physical space of our experience—which one becomes acquainted with by experience, but which is the representation lying at the foundation of all external experience. . . . *Not* that the propositions of geometry are only approximately true, but that they remain absolutely true in regard to that Euclidean

space which has been so long regarded as being the physical space of our experience.

Felix Klein (1849–1925), one of the truly great mathematicians of recent times, expressed about the same view. Though Cayley and Klein had themselves worked in non-Euclidean geometries, they regarded them as novelties that result when artificial new distance functions are introduced in Euclidean geometry. They refused to grant that non-Euclidean geometry is as basic and as applicable as Euclidean. Of course their position, in pre-relativity days, was tenable.

Bertrand Russell too still believed in the truth of mathematics, though he limited the truths somewhat. In the 1890s he took up the question of what properties of space are necessary and can be presupposed before experience. That is, experience would be meaningless if any of these a priori properties were denied. In his *Essay on the Foundations of Geometry* (1897), he agreed that Euclidean geometry is not a priori knowledge. He concluded, rather, that projective geometry,* a basic qualitative geometry, is a priori for all geometry, an understandable conclusion in view of the importance of that subject around 1900. He then added to projective geometry, as a priori, axioms common to Euclidean and all the non-Euclidean geometries. These axioms, which concern the homogeneity of space, finite dimensionality, and a concept of distance, make measurement possible. Russell also pointed out that qualitative considerations must precede quantitative ones, and this contention reinforces the case for the priority of projective geometry.

As for the metrical geometries, that is, the Euclidean and the several non-Euclidean, the fact that they can be derived from projective geometry by the introduction of a specific concept of distance Russell regarded as a technical achievement having no philosophical significance. In any case their specific theorems are not a priori truths. With respect to the several basic metric geometries, he departed from Cayley and Klein and regarded all these geometries as being on an equal logical footing. Since the only metric spaces that possess the above a priori properties are the Euclidean, hyperbolic, and single and double elliptic, Russell concluded that these are the only possible metrical geometries and of course Euclidean is the only physically applicable one. The others are of philosophical importance in showing that there can be

---

* Projective geometry studies the properties common to figures that result from projecting a figure from one plane to another. Thus if a circle is held in front of a flashlight, its shadow can be seen on a screen or a wall. The shape of the shadow changes as the circle is more or less inclined to the vertical. Yet the circle and the various shapes have common geometrical properties.

other geometries. With hindsight it is now possible to see that Russell replaced the Euclidean bias by a projective bias. Russell admitted many years later that his *Essay* was a work of his youth and could no longer be defended. As we shall see later, however, he and others formulated a new basis for establishing truth in mathematics (Chapter X).

The persistence of mathematicians in searching for some basic truths is understandable. To accept the fact that mathematics is not a collection of diamonds but of synthetic stones, after the centuries of brilliant successes in describing and predicting physical phenomena, would be hard for anyone and especially for those who might be blinded by pride in their own creations. Gradually, however, mathematicians granted that the axioms and theorems of mathematics were not necessary truths about the physical world. Some areas of experience suggest particular sets of axioms and to these areas the axioms and their logical consequences apply accurately enough to be taken as a useful description. But if any area is enlarged the applicability may be lost. As far as the study of the physical world is concerned, mathematics offers nothing but theories or models. And new mathematical theories may replace older ones when experience or experiment shows that a new theory provides closer correspondence than an older one. The relationship of mathematics to the physical world was well expressed by Einstein in 1921:

> Insofar as the propositions of mathematics give an account of reality they are not certain; and insofar as they are certain they do not describe reality. . . . But it is, on the other hand, certain that mathematics in general and geometry in particular owe their existence to our need to learn something about the properties of real objects.

Mathematicians had given up God and so it behooved them to accept man. And this is what they did. They continued to develop mathematics and to search for laws of nature, knowing that what they produced was not the design of God but the work of man. Their past successes helped them to retain confidence in what they were doing, and, fortunately, hosts of new successes greeted their efforts. What preserved the life of mathematics was the powerful medicine it had itself concocted— the enormous achievements in celestial mechanics, acoustics, hydrodynamics, optics, electromagnetic theory, and engineering—and the incredible accuracy of its predictions. There had to be some essential— perhaps magical—power in a subject which, though it had fought under the invincible banner of truth, has actually achieved its victories through some inner mysterious strength (Chapter XV). And so mathematical creation and application to science continued at an even faster pace.

The recognition that mathematics is not a body of truths has had shattering repercussions. Let us note first the effect on science. From Galileo's time onward scientists recognized that the fundamental principles of science, as opposed to mathematics, must come from experience, although for at least two centuries they believed that the principles they did find were imbedded in the design of nature. But by the early 19th century they realized that scientific theories are not truths. The realization that even mathematics derives its principles from experience and that they could no longer assert their truth made scientists recognize that insofar as they use axioms and theorems of mathematics their theories are all the more vulnerable. Nature's laws are man's creation. We, not God, are the lawgivers of the universe. A law of nature is man's description and not God's prescription.

The reverberations of this disaster have reached almost all areas of our culture. The attainment of apparent truths in mathematics and mathematical physics had encouraged the expectation that truths could be acquired in all other fields of knowledge. In his *Discourse on Method* of 1637 Descartes had voiced this expectation:

> The long chains of simple and easy reasoning by means of which geometers are accustomed to reach the conclusions of their most difficult demonstrations, led me to imagine that all things, to the knowledge of which man is competent, are mutually connected in the same way, and there is nothing so far removed from us as to be beyond our reach or so hidden that we cannot discover it, provided only we abstain from accepting the false for the true, and always preserve in our thoughts the order necessary for the deduction of one truth from another.

Descartes wrote these words when the successes of mathematical inquiry were still few. By the middle of the 18th century the successes were so numerous and so deep that intellectual leaders were confident they could secure truths in all fields by application of reason and mathematics. D'Alembert spoke for his age:

> . . . a certain exaltation of ideas which the spectacle of the universe produces in us . . . [has] brought about a lively fermentation of minds. Spreading through nature in all directions like a river which has burst its dams, this fermentation has swept with a sort of violence everything along with it which has stood in its way. . . . Thus from the principles of the secular sciences to the foundations of religious revelation, from metaphysics to matters of taste, from music to morals, from the scholastic disputes of theologians to matters of trade, from the laws of princes to those of people, from natural law to the arbitrary law of nations . . . everything has been discussed and analyzed, or at least mentioned.

This confidence that truths would be discovered in all fields was shattered by the recognition that there is no truth in mathematics. The hope and perhaps even the belief that truths can be obtained in politics, ethics, religion, economics, and many other fields may still persist in human minds, but the best support for the hope has been lost. Mathematics offered to the world proof that man can acquire truths and then destroyed the proof. It was non-Euclidean geometry and quaternions, both triumphs of reason, that paved the way for this intellectual disaster.

As William James put it, "The intellectual life of man consists almost wholly in his substitution of a conceptual order for the perceptual order in which his experience originally comes." But the conceptual order is not a veridical account of the perceptual.

With the loss of truth, man lost his intellectual center, his frame of reference, the established authority for all thought. The "pride of human reason" suffered a fall which brought down with it the house of truth. The lesson of history is that our firmest convictions are not to be asserted dogmatically; in fact they should be most suspect; they mark not our conquests but our limitations and our bounds.

The history of the belief in the truth of mathematics can be summarized in the words of Wordsworth's "Intimations of Immortality." In 1750 mathematicians could say of their creations,

> But trailing clouds of glory do we come
> From God, who is our home.

By 1850 they were obliged to admit ruefully,

> But yet I know where'er I go,
> That there hath passed away a glory from the earth.

But the history should not be too discouraging. Evariste Galois said of mathematics, "[This] science is the work of the human mind, which is destined rather to study than to know, to seek the truth rather than to find it." Perhaps it is in the nature of truth that it wishes to be elusive. Or as the Roman philosopher Lucius Seneca put it, "Nature does not at once disclose all Her mysteries."

V

The Illogical Development
of a Logical Subject

> We will grieve not, rather find
> strength in what remains behind.
> WORDSWORTH

For over two thousand years mathematicians believed that they were highly successful in uncovering the mathematical design of nature. But now they were obliged to recognize that mathematical laws were not truths. During these two millennia mathematicians also believed that they were adhering to the Greek plan for arriving at truths, namely, to apply deductive reasoning to mathematical axioms, thus ensuring conclusions as reliable as the axioms. Because the mathematical laws of science were remarkably accurate the few disagreements that did take place about the correctness of some mathematical arguments were brushed aside. Even the most acute mathematicians were sure that any blemishes in the reasoning could readily be removed. However, in the 19th century the equanimity of mathematicians with respect to their reasoning was shattered.

What happened to open the eyes of the mathematicians? How did they come to recognize that they were not reasoning soundly? A few men had already become disturbed in the early decades of the 19th century by the attacks on the soundness of the calculus which were not being satisfactorily countered. But it was primarily the very same creations that forced mathematicians to abandon their claim to truth — non-Euclidean geometry and quaternions—that drove home to most mathematicians the sad state of the logic.

The work on non-Euclidean geometry, wherein constant reference to analogous theorems and proofs of Euclidean geometry was natural, resulted in the surprising revelation that Euclidean geometry, which had been hailed for two thousand years by the experts as the model of rigorous proof, was from a logical standpoint seriously defective. The creation of new algebras, commencing with quaternions (Chapter IV), disturbed mathematicians enough to make them want to reexamine the logical underpinnings of the arithmetic and algebra of the ordinary real and complex numbers, if for no other reason than to reassure themselves that the properties of these numbers were securely established. And what they found in this area too was startling. What they had thought to be highly logical subjects had actually been developed totally illogically.

The most fertile source of insight is hindsight, and it is with the hindsight afforded and sharpened by the new creations that mathematicians finally saw what generations of their predecessors had failed to see or had seen and glossed over in their impetuously eager drive to obtain truths. The mathematicians were certainly not going to abandon their subject. Beyond its continuing remarkable effectiveness in science, mathematics proper was a body of knowledge in and for itself which many mathematicians, following Plato, regarded as an extra-sensible reality. Hence the mathematicians decided that the least they could do was to reexamine the logical structure of mathematics and to supply or reconstruct the defective portions.

We know that deductive mathematics began with the Greeks and that the first seemingly sound structure was Euclid's *Elements*. Euclid started with definitions and axioms and deduced theorems. Let us look at some of Euclid's definitions.

> Definition 1. A point is that which has no parts.
>
> Definition 2. A line [curve in modern terminology] is breadthless length.
>
> Definition 4. A straight line is a line that lies evenly with the points on itself.

Now Aristotle had pointed out that a definition must describe the concept being defined in terms of other concepts already known. Since one must start somewhere, there must be, he asserted, undefined concepts to begin with. Though there are many indications that Euclid, who lived and worked in Alexandria about 300 B.C., knew the works of the classical Greeks and of Aristotle in particular, he nevertheless defined all his concepts.

There have been two explanations of this flaw. Either Euclid did not agree that there must be undefined terms or, as some defenders of Euclid state, he realized that there must be undefined terms but intended his initial definitions to give only some intuitive ideas of what the defined terms meant, so that one could know whether the axioms which were to follow were indeed correct assertions. In the latter case he should not have included the definitions in his text proper. Whatever Euclid's intentions were, practically all the mathematicians who followed him for two thousand years failed to note the need for undefined terms. Pascal in his *Treatise on the Geometrical Spirit* (1658) called attention to this need but his reminder was ignored.

What about Euclid's axioms? Presumably following Aristotle, he stated five common notions that apply in all reasoning and five postulates that apply only to geometry. One common notion states that things equal to the same thing are equal to each other. Euclid applied the word "things" to lengths, areas, volumes, and whole numbers. Certainly the word "things" was too vague. More misguided, however, was his common notion: Things which coincide are equal. He used this axiom to prove that two triangles are congruent by placing one triangle on top of another and showing, by using some given facts, that the triangles must coincide. However, to place one triangle on another he had to move it and in so doing he assumed that it did not change properties during the motion. The axiom really says that our space is homogeneous; that is, the properties of figures are the same wherever they may be placed. This may be a reasonable assumption but it is an additional assumption nevertheless. Also, the concept of motion was not treated in the definitions.

Moreover, Euclid used many axioms he never stated. As Gauss noted, Euclid spoke of points lying *between* other points and lines lying between other lines but never treated the notion of betweenness and its properties. Evidently he had geometrical figures in mind and took over into his reasoning properties the actual figures possessed but which were not incorporated in the axioms. Figures may be an aid to thinking and to memory but they cannot be the basis of reasoning. Another axiom used without explicit mention—this was noted by Leibniz—involves what is technically called continuity. Euclid used the fact that a line joining A (Fig. 5.1) on one side of a line l to B on the other side has a point in common with l. This is certainly evident from a figure. But no axiom about lines assures us of the existence of this common point. Can one even speak of two sides of the line l? This, too, requires an axiom.

Beyond failings in its definitions and axioms, the *Elements* also contains many inadequate proofs. Some theorems are incorrectly proved. Others prove only a special case or a special configuration of the theorem asserted. These latter failings are, however, relatively minor in that they can be more readily remedied. Euclid had supposedly given

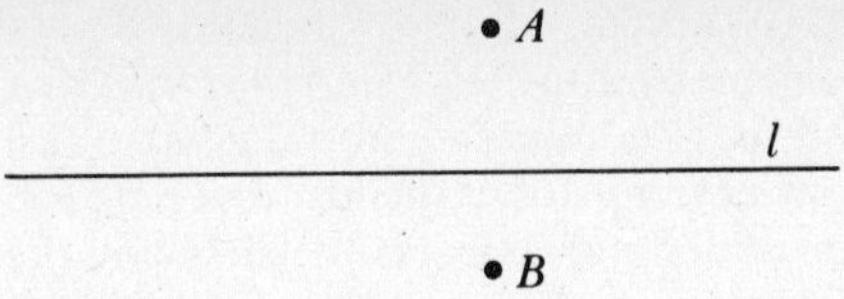

• *A*

_____ *l*

• *B*

Figure 5.1

accurate proofs of loosely drawn figures. However, if one judges Euclid's work as a whole one must say that in fact he gave loose proofs of accurately drawn figures. In short, Euclid's presentation was woefully defective.

Despite all these defects in the *Elements*, the best mathematicians, scientists, and philosophers before about 1800 regarded it as the ideal of rigorous proof. Pascal said in his *Pensées*, "The geometrical spirit excels in all those subjects that are capable of a perfect analysis. It starts with axioms and draws inferences whose truth can be demonstrated by universal logical rules." Isaac Barrow, Newton's teacher and predecessor at Cambridge University, listed eight reasons for the certainty of geometry: the clarity of its concepts; its unambiguous definitions; our intuitive assurance of the universal truth of its common notions; the clear possibility and easy imaginability of its postulates; the small number of its axioms; the clear conceivability of the mode by which magnitudes are generated; the easy order of the demonstrations; and the avoidance of things not known. Many more such testimonials could be adduced. As late as 1873 Henry John Stephen Smith, a noted number theorist, said: "Geometry is nothing if it be not rigorous. . . . The methods of Euclid are, by almost universal consent, unexceptionable in point of rigour."

However, the work on non-Euclidean geometry revealed so many defects that the logical perfection of Euclidean geometry could no longer be admired. Non-Euclidean geometry was the reef on which the logic of Euclidean geometry foundered. A piece of ground that had confidently been thought to be solid proved to be marshland.

Euclidean geometry is, of course, only a part of mathematics. Since 1700 by far the larger part has been the mathematics of number. Let us see how the logical development of numbers fared. The Egyptians and Babylonians worked with whole numbers, fractions, and even irrational numbers such as $\sqrt{2}$ and $\sqrt{3}$. Irrationals they approximated in practical applications. However, since the mathematics of these peoples and even of the Greeks before the 4th century B.C. was intuitively or empirically based, there could be no praise or criticism of logical structure.

The first logical treatment of the whole numbers known to us is in

Books VII, VIII, and IX of Euclid's *Elements*. Here Euclid offered definitions such as, a *unit* is that by virtue of which each of the things that exist is called one, and a *number* is a multitude composed of units. Clearly these are inadequate, to say nothing about the fact that here too he did not recognize the need for undefined concepts. In deducing the properties of whole numbers Euclid used the common notions mentioned above. Unfortunately some of his proofs are faulty. Nevertheless, the Greeks and their successors did believe that the theory of whole numbers had secured a satisfactory logical foundation. They also permitted themselves to speak without more ado about ratios of whole numbers, which later generations called fractions, though the concept of ratio was not defined.

The Greeks did encounter one to them insuperable difficulty in the logical development of numbers. As we know, the Pythagoreans of the 5th century B.C. were the first to stress the importance of the whole numbers and ratios of whole numbers in the study of nature and in fact they insisted that the whole numbers were the "measure" of all things. They were startled and disturbed to discover that some ratios, for example, the ratio of the hypotenuse of an isosceles right triangle to an arm, cannot be expressed by whole numbers. They called ratios which are expressed by whole numbers commensurable ratios and those that are not so expressible, incommensurable ratios. Thus what we express as the irrational number $\sqrt{2}$ was an incommensurable ratio. The discovery of incommensurable ratios is attributed to Hippasus of Metapontum (5th century B.C.) An old story relates that the Pythagoreans were at sea at the time and Hippasus was thrown overboard for having produced an element in the universe which denied the Pythagorean doctrine that all phenomena in the universe can be reduced to whole numbers or their ratios.

The proof that $\sqrt{2}$ is incommensurable with 1, or irrational, was given by the Pythagoreans and according to Aristotle the method was a *reductio ad absurdum,* that is, the indirect method. It shows that if the hypotenuse of an isosceles right triangle is commensurable with an arm then the same number must be both odd and even, which of course is untenable. The proof runs as follows: Suppose the ratio of hypotenuse to arm is expressible as a/b, where a and b are whole numbers, and if a and b have a common factor suppose that this is cancelled. If $a/b = \sqrt{2}$, then $a^2 = 2b^2$. Since a^2 is even, a must be even, for the square of any odd number is odd.* Now the ratio a/b is in its lowest terms and therefore since a is even b must be odd. Since a is even, let $a = 2c$. Then $a^2 = 4c^2$

---

* Any odd whole number can be expressed as $2n + 1$ for some n. Then $(2n + 1)^2 = 4n^2 + 4n + 1$, and this is necessarily odd.

and since $a^2 = 2b^2$, $4c^2 = 2b^2$, or $2c^2 = b^2$, and so b^2 is even. Then b is even, for if it were odd its square would be odd. But b is also odd, and so there is a contradiction.

The Pythagoreans and the classical Greeks generally would not accept irrational numbers because an understanding of the concept eluded them. The Pythagorean proof tells us that $\sqrt{2}$ is not a ratio of whole numbers but it does not tell us what an irrational is. The Babylonians, as we have noted, did work with such numbers. But they undoubtedly did not know that their decimal (sexagesimal) approximations could never be made exact. We may hail their blithe spirits but mathematicians they never were. The classical Greeks were of a different intellectual breed and could not be content with approximation.

The discovery of irrationals posed a problem which is central to Greek mathematics. Plato in the *Laws* called for a knowledge of incommensurables. The resolution of the problem, which was due to Eudoxus, for a time a pupil of Plato, was to think of all magnitudes geometrically. Thus lengths, angles, areas, and volumes, some of which would be irrational if expressed numerically, were treated geometrically. For example, Euclid states the Pythagorean theorem in the form, the square *on* the hypotenuse of a right triangle is the sum of the squares *on* the arms. And by the sum of the squares he meant that the two areas geometrically combined equalled the area of the square on the hypotenuse. The resort to geometry is understandable. When 1 and $\sqrt{2}$ are treated as lengths, that is, line segments, the distinction between 1 and $\sqrt{2}$ is obliterated.

The problem presented by irrational numbers was greater than just that of representing lengths, areas, and volumes numerically. Since the roots of quadratic equations, e.g., $x^2 - 2 = 0$, can very well be irrational numbers, the classical Greeks solved these equations geometrically so that the roots appeared as line segments and again the need to use irrational numbers was avoided. This development is referred to as geometrical algebra. Euclid's *Elements* is consequently a treatise on algebra as well as geometry.

The conversion of all of mathematics except the theory of whole numbers into geometry had several major consequences. For one thing, it forced a sharp separation between number and geometry, for only geometry could handle incommensurable ratios. From Euclid's time onward these two branches of mathematics had to be sharply distinguished. Also, because geometry embraced the bulk of mathematics it became the basis for almost all "rigorous" mathematics until at least 1600. We still speak of x^2 as x square and x^3 as x cube instead of x second or x third because for general magnitudes x^2 and x^3 once had only geometrical meaning.

The geometric representation of numbers and of operations with numbers was, of course, not practical. One might be satisfied logically to think of $\sqrt{2} \cdot \sqrt{3}$ as the area of a rectangle but if one needed to know the product numerically this would not suffice. For science and engineering, geometrical figures are not nearly so useful as a numerical answer which can be calculated to as many decimal places as needed. Applied science and engineering must be quantitative. When a ship at sea needs to know its location, it must know it numerically, in terms of degrees of latitude and longitude. To construct efficiently buildings, bridges, ships, and dams, one must know the quantitative measures of lengths, areas, and volumes to be employed, so that the parts will fit nicely; in fact this quantitative knowledge must be obtained in advance of the construction. But the classical Greeks, who regarded exact reasoning as paramount in importance and deprecated applications to commerce, navigation, construction, and calendar-reckoning, were content with their geometrical solution of the difficulty with irrational numbers.

Classical Greek civilization was succeeded about 300 B.C. by Alexandrian Greek civilization (Chapter I). This was a fusion of classical Greek civilization and the Egyptian and Babylonian civilizations. From the standpoint of the logical development it produced a curious blend of deductive and empirical mathematics. The chief mathematicians, Archimedes and Apollonius, pursued in the main the axiomatic, deductive geometry of Euclid's *Elements*. Even in his treatises on mechanics, Archimedes started with axioms and proved theorems. However, the Alexandrians, influenced by the more practically minded Egyptians and Babylonians, put mathematics to use. And so we find in Alexandria that formulas which would permit quantitative measures of lengths, areas, and volumes were derived. Thus the Alexandrian-Egyptian engineer Heron (1st century A.D.) in his *Metrica* gives as the formula for the area of a triangle:

$$\sqrt{s(s-a)(s-b)(s-c)},$$

wherein a, b, and c are the lengths of the sides and s is half the perimeter. The values given by such formulas are often irrational numbers. This particular formula is remarkable. Whereas the classical Greeks regarded the product of more than three numbers as meaningless because the product had no geometrical significance, Heron had no such qualms. In the many pure and applied sciences that the Alexandrian Greeks developed, calendar-reckoning, the measurement of time, the mathematics of navigation, optics, geography, pneumatics, and hydrostatics (Chapter I), irrational numbers were used freely.

The supreme achievement of the Alexandrians was the creation of a quantitative astronomy by Hipparchus and Ptolemy, a geocentric astronomy which enabled man to predict the motions of the planets, the sun, and the earth's moon (Chapter I). To develop this quantitative astronomy Hipparchus and Ptolemy created trigonometry, the branch of mathematics that enables one to calculate some parts of a triangle on the basis of information about other parts. Because Ptolemy approached trigonometry differently from the modern method, he was obliged to calculate the lengths of chords of a circle. Though he used deductive geometry to establish his basic results as to how some chords relate to others in size, he then proceeded to use arithmetic and bits of algebra to calculate the lengths of the chords with which he was finally concerned. Most of these lengths are irrational. Ptolemy was content to obtain rational approximations but in the course of his work he did not hesitate to use irrationals.

However, the arithmetic and algebra used so freely by the Alexandrian Greeks and taken over from the Egyptians and the Babylonians, had no logical foundation. Ptolemy and the other Alexandrian Greeks generally adopted the Egyptian and Babylonian attitude. Irrational numbers such as π, $\sqrt{2}$, $\sqrt{3}$, and the like were used uncritically and approximated where necessary. For example, the most famous use of irrationals was Archimedes' calculation that π lies between 3 1/17 and 3 10/71. Whether or not he knew that π was irrational, he calculated any number of square roots that he surely knew to be irrational in order to obtain the approximations.

As noteworthy for present purposes as the free use of irrationals was the resumption, independent of geometry, of Egyptian and Babylonian algebra. Outstanding in the revival were Heron and Diophantus (3rd century, A.D.), another Greek of Alexandria. Both treated arithmetical and algebraic problems in and for themselves and did not depend upon geometry either for motivation or to bolster the logic. Heron formulated and solved algebraic problems by purely arithmetical procedures. For example, he treated this problem: Given a square such that the sum of its area and perimeter is 896 feet, find the side. To solve the pertinent quadratic equation Heron completed the square by adding 4 to both sides and took the square root. He did not prove but merely described what operations to perform. There are many such problems in Heron's work.

In his *Geometrica* Heron spoke of adding an area, a circumference, and a diameter. In using such words he meant, of course, that he wanted to add the numerical values. Likewise when he said he multiplied a square by a square he meant that he was finding the product of the numerical values. Heron also translated much of the Greek geometrical algebra into arithmetical and algebraic processes. Some of his and his successors' problems are exactly those which appeared in Babylonian and Egyptian texts of 2000 B.C. This Greek algebraic work was written in verbal form. No symbolism was used. Nor were any proofs of the procedures given. From the time of Heron onward, problems leading to equations were also a common form of puzzle.

The highest point of Alexandrian Greek algebra was reached with Diophantus. We know almost nothing about his origins or life. His work, which towers above his contemporaries', unfortunately came too late to be highly influential in his times because a destructive tide (Chapter II) was already engulfing the civilization. Diophantus wrote several books which are entirely lost, but we do have six parts of his greatest work, the *Arithmetica,* which, Diophantus said, comprised thirteen parts. The *Arithmetica,* like the Egyptian Rhind papyrus, is a collection of separate problems. The dedication says that it was written as a series of exercises to help one of his students learn the subject.

One of Diophantus's big steps was to introduce some symbolism in algebra. Since we do not have manuscripts written by him but only much later ones, dating from the 13th century on, we do not know his precise forms, but he did use symbols corresponding to our x, powers of x up to x^6, and our $1/x$. The appearance of such symbolism is of course remarkable, but the use of powers higher than three is all the more extraordinary because, as we have noted, for the classical Greeks a product of more than three factors had no geometrical significance. On a purely arithmetical basis, however, such products do have a meaning, and this is precisely the basis Diophantus adopted.

Diophantus wrote out his solutions in a continuous text, as we write prose. His execution of the operations is entirely arithmetical; that is, there is no appeal to geometry to illustrate or substantiate his assertions. Thus, $(x - 1)(x - 2)$ is carried out algebraically as we do it. He also used algebraic identities such as $a^2 - b^2 = (a - b)(a + b)$ and more complicated ones. Strictly, he made steps which used identities, but these identities themselves do not appear.

Another unusual feature of Diophantus's algebra is his solution of indeterminate equations, for example, one equation in two unknowns. Such equations had been considered before, in the Pythagorean work on integral solutions of $x^2 + y^2 = z^2$ and in other writings. Diophantus, however, pursued the subject extensively and was the founder of the branch of algebra now called, in fact, Diophantine analysis.

Though Diophantus is notable for his use of algebra, he accepted only positive rational roots and ignored all others. Even when a quadratic equation in one unknown had two positive rational roots, he gave only one, the larger one. When an equation clearly led to two negative roots, or irrational or imaginary roots, he rejected the equation and said it was not solvable. In the case of irrational roots, he retraced his steps and showed how by altering the equation he could get a new one which had rational roots. Here Diophantus differed from Heron and Archimedes. Heron was an engineer and the quantities he sought could be irrational. Hence, he accepted them, though, of course, he approximated to obtain a useful value. Archimedes also sought exact answers, and when they were irrational he obtained inequalities to bound the irrational.

We do not know how Diophantus arrived at his methods. Since he made no appeal to geometry, it is not likely that he translated Euclid's methods for solving quadratics. Indeterminate problems, moreover, do not occur in Euclid and as a class are new with Diophantus. Because we lack information on the continuity of thought in the late Alexandrian period, we cannot find many traces of Diophantus's work in his Greek predecessors. His methods are, in fact, closer to Babylonian, and there are vague indications of Babylonian influences. However, unlike the Babylonians, he used symbolism and undertook the solution of indeterminate equations. His work as a whole is a monument in algebra.

The works of Heron and Diophantus, and of Achimedes and Ptolemy as far as arithmetic and algebra are concerned, read like the procedural texts of the Egyptians and Babylonians which tell us how to do things. The deductive orderly proof of the geometry of Euclid, Apollonius, and Archimedes was gone. The problems were treated inductively in that they showed methods for solving concrete problems presumably applicable to general classes whose extent was not specified. The various types of numbers, whole numbers, fractions, and irrationals (except for Euclid's imperfect work with the whole numbers) were certainly not defined. Nor was there any axiomatic basis on which a deductive structure could be erected.

Thus, the Greeks bequeathed two sharply different and dissimilarly developed branches of mathematics. On the one hand, there was the deductive, systematic, albeit somewhat faulty geometry, and on the other, the empirical arithmetic and its extension to algebra. In view of the fact that the classical Greeks required mathematical results to be derived deductively from an explicit axiomatic basis, the emergence of an independent arithmetic and algebra with no logical structure of its own proffered what became one of the great anomalies of the history of mathematics.

The Hindus and Arabs, who carried the ball for mathematics after the final destruction of the Alexandrian Greek civilization by the Arabs, violated still further the classical Greek concept of mathematics. They of course used whole numbers and fractions, but they also used irrational numbers unhesitatingly. In fact, they introduced new and correct rules for adding, subtracting, multiplying, and dividing irrational numbers. Since these rules did not have a logical foundation, how could they have been devised and why were they correct? The answer is that the Hindus and Arabs reasoned by analogy. Thus the rule $\sqrt{ab} = \sqrt{a}\sqrt{b}$ was justified for all numbers a and b because evidently $\sqrt{36} = \sqrt{4}\sqrt{9}$ was correct. In fact, the Hindus said that radicals could be handled like integers.

The Hindus were far more naïve than the Greeks in that they failed to see the logical difficulties involved in the concept of irrational numbers. Their interest in calculation caused them to overlook distinctions that were fundamental in Greek thought. However, in casually applying to irrationals procedures like those used for rationals, they helped mathematics to make progress. Moreover, their entire arithmetic was completely independent of geometry.

The Hindus added to the logical woes of mathematicians by introducing negative numbers to represent debts. In such uses positive numbers represented assets. The first known use was by Brahmagupta about A.D. 628, who merely stated rules for the four operations with negative numbers. No definitions, axioms, or theorems appeared. Bhaskara, the leading Hindu mathematician of the 12th century, pointed out that the square root of a positive number is twofold, positive and negative. He brought up the matter of square root of a *negative* number but said that there is no square root because the square of it would be a negative number and a negative number can not be a square.

Negative numbers were not accepted by all the Hindus. Even Bhaskara said, while giving 50 and −5 as two solutions of a problem, "The second value is in this case not to be taken, for it is inadequate; people do not approve of negative solutions." Nevertheless, negative numbers were used increasingly after they were introduced.

The Hindus also made some progress in algebra. They used abbreviations of words and a few symbols to describe operations and unknowns. The symbolism, though not extensive, was enough for us to regard Hindu algebra as superior in this respect to Diophantus's. Of solutions only the steps were given. No reasons or proofs accompanied them. Generally, negative and irrational roots of quadratic equations were recognized.

The Hindus used algebra even more freely than we have thus far indicated. For example, we learn in trigonometry that if A is any angle then $\sin^2 A + \cos^2 A = 1$. For Ptolemy, one of the creators and the systematic expositor of trigonometry, this equation was phrased as a geometrical statement involving a relation between the chords of a circle. Though, as we noted, Ptolemy did use arithmetic freely to calculate unknown lengths in terms of known ones, his basic mathematics and arguments were geometrical. The Hindus expressed the trigonometric relationships much as stated above. Moreover, to calculate cos A from sin A, they could and did use the above equation and simple algebra. That is, Hindu trigonometry relied far more on algebra than on geometry to express and derive relations among the sines and cosines of angles.

As our survey indicates the Hindus were interested in and contributed to the arithmetical and computational activities of mathematics rather than to the deductive pattern. Their name for mathematics is *ganita*, which means the science of calculation. There is much good procedure and technical facility but no evidence that they considered proof at all. They had rules but apparently no logical scruples. Moreover, no general methods or new viewpoints were arrived at in any area of mathematics.

It is fairly certain that the Hindus did not appreciate the significance of their own contributions. The few good ideas they had, such as separate symbols for the numbers from 1 to 9, the conversion from positional notation in base 60 to base 10, negative numbers, and the recognition of 0 as a number, were introduced casually with no apparent realization that they were valuable innovations. They were not sensitive to mathematical values. The fine ideas which they themselves advanced, they commingled with the crudest ideas of the Egyptians and Babylonians. The Arab historian al-Bîrûnî (973–1048) said of Hindus, "I can only compare their mathematical and astronomical literature . . . to a mixture of pearl shells and sour dates, or of pearl and dung, or of costly crystals and common pebbles. Both kinds of things are equal in their eyes, since they cannot raise themselves to the methods of a strictly scientific deduction." Since their special gift was arithmetical and they had added to arithmetic and algebra, the effect of the Hindu work was to enlarge the portion of mathematics which rested on empirical and intuitive foundations.

Whereas the Hindus practically ignored deductive geometry, the Arabs made critical studies of the Greek geometrical works and appreciated fully the role of deductive proof in establishing this branch of mathematics. However, in the field of arithmetic and algebra, which played a larger role in Arabic mathematics, the Arabs proceeded much like the Hindus. They were content to treat these subjects on the same empirical, concrete, and intuitive basis as their Hindu predecessors. Some Arabs did give geometrical arguments to support their solutions of quadratic equations but the primary approach and methodology of solution, unlike that of the classical Greeks, was algebraic. In the case of the solution of cubic equations, such as $x^3 + 3x^2 + 7x + 5 = 0$, they gave only geometric constructions because the algebraic process was yet to be discovered. But these constructions could not be carried out with straightedge and compass and the arguments were not strictly deductive. During all the centuries in which they were active in mathematics, the Arabs in their own contributions manfully resisted the lures of exact reasoning.

The most interesting feature of the mathematics of the Hindus and Arabs is their self-contradictory concept of the subject. That the Egyptians and Babylonians were content to accept their few arithmetic and geometric rules on an empirical basis is not surprising. This is a natural basis for almost all human knowledge. But the Hindus and Arabs were aware of the totally new concept of mathematical proof promulgated by the Greeks. Yet in arithmetic and algebra they did not concern themselves at all with the notion of deductive proof. The Hindu attitude is somewhat rationalizable. Though they did indeed possess some knowledge of the Greek works, they paid little attention to them and followed primarily the Alexandrian Greek treatment of arithmetic and algebra. But the Arabs were fully aware of Greek geometry and, as we have observed, even made critical studies of it. Moreover, in both civilizations, conditions for the pursuit of pure science were favorable for a number of centuries so that the pressure to produce practically useful results need not have caused mathematicians to sacrifice proof for immediate utility. How could both peoples treat the two areas of mathematics so differently from the classical and many of the Alexandrian Greeks?

There are numerous possible answers. Both civilizations were on the whole uncritical, despite the Arabic commentaries on deductive geometry. Hence they may have been content to take mathematics as they found it: geometry was to be deductive but arithmetic and algebra could be empirical or heuristic. A second possibility is that both of these peoples, the Arabs especially, appreciated the widely different standards for geometry as opposed to arithmetic and algebra but did not see how to supply a logical foundation for arithmetic. One fact which seems to vouch for such an explanation is that the Arabs did at least support their solution of quadratic and cubic equations by giving the geometric basis.

There are other possible explanations. Both the Hindus and Arabs favored arithmetic, algebra, and the algebraic formulation of trigonometric relationships. This predisposition may bespeak a different mentality or may reflect the response to the needs of the civilizations. Both of these civilizations were practically oriented and, as we have had occasion to note in connection with the Alexandrian Greeks, practical needs do call for quantitative results that are supplied by arithmetic and algebra. One bit of evidence which favors the thesis that there may be different mental traits comes from the reaction of the Europeans to the mathematical heritage handed down by the Hindus and Arabs. As we shall see, the Europeans were far more troubled by the disparate states of arithmetic and geometry. In any case, Hindu and Arab venturesomeness brought arithmetic and algebra to the fore once again and, insofar as utility is concerned, placed it almost on a par with geometry.

When the Europeans of the late medieval period and the Renaissance acquired the existing knowledge of mathematics partly through the Arabs and partly through the direct acquisition of Greek manuscripts, they attempted to face the dilemma presented by the two kinds of mathematics. Real mathematics seemed definitely to be the deductive geometry of the Greeks. On the other hand, the Europeans could not deny the utility and efficacy of the arithmetic and algebra that had been developed since ancient times but had no logical foundation.

The first problem they faced was what to do about irrational numbers. The Italian mathematician Luca Pacioli (c.1445–c.1514), the prominent German monk and professor of mathematics at Jena, Michael Stifel (1486?–1567), the physician, scholar, and rogue Jerome Cardan (1501–1576), and the military engineer Simon Stevin (1548–1620) used irrational numbers in the tradition of the Hindus and Arabs and introduced more and more types. Thus Stifel worked with irrationals of the form $\sqrt[m]{a} + \sqrt[n]{b}$. Jerome Cardan worked with irrationals involving cube roots. The extent to which irrationals were used is exemplified by François Vieta's (1540–1603) expression for π. By considering regular polygons of 4, 8, 16, $\cdots$ sides inscribed in a circle of unit radius, Vieta found that

$$\frac{2}{\pi} = \sqrt{\frac{1}{2}} \cdot \sqrt{\frac{1}{2} + \frac{1}{2}\sqrt{\frac{1}{2}}} \cdot \sqrt{\frac{1}{2} + \frac{1}{2}\sqrt{\frac{1}{2} + \frac{1}{2}\sqrt{\frac{1}{2}}}} \cdots$$

Irrational numbers were used freely in one of the new creations of the Renaissance—logarithms. Logarithms for positive numbers were invented in the late 16th century by John Napier (1550–1617) for the very purpose to which they have since been put, namely, to speed up arithmetic processes. Though the logarithms of most positive numbers are irrational—and Napier's method of calculating them used irrationals freely—all mathematicians welcomed the saving of labor that they afforded.

Calculations with irrationals were carried on freely, but the problem of whether irrationals were really numbers troubled some of the very same people who worked with them. Thus Stifel in his major work, the

Arithmetica integra (1544), which dealt with arithmetic and algebra, echoed Euclid's view of magnitudes (the geometrical theory of Eudoxus) as different from numbers, but then, in line with the new developments, considered expressing irrationals in the decimal notation. What troubled him was that an irrational expressed as a decimal requires an infinite number of digits. On the one hand he argued,

> Since, in proving geometrical figures, when rational numbers fail us irrational numbers take their place and prove exactly those things which rational numbers could not prove, . . . we are moved and compelled to assert that they truly are numbers, compelled that is, by the results which follow from their use—results which we perceive to be real, certain, and constant. On the other hand, other considerations compel us to deny that irrational numbers are numbers at all. To wit, when we seek to subject them to numeration [decimal form] . . . we find that they flee away perpetually so that not one of them can be apprehended precisely in itself. . . . Now that cannot be called a true number which is of such a nature that it lacks precision. . . . Therefore, just as an infinite number is not a number, so an irrational number is not a true number, but lies hidden in a kind of cloud of infinity.

Then Stifel added that real numbers are either whole numbers or fractions, and, obviously, irrationals are neither and so are not real numbers. A century later, Pascal and Barrow said that irrational numbers are mere symbols which have no existence independent of continuous geometrical magnitude, and the logic of operations with irrationals must be justified by the Euclidean theory of magnitudes, but this theory was inadequate for the purpose.

There were, on the other hand, positive assertions that the irrational numbers are legitimate entities. Stevin recognized irrationals as numbers and approximated them more and more closely by rationals and John Wallis (1616–1703) in his *Algebra* (1685) also accepted irrationals as numbers in the full sense. However, neither Stevin nor Wallis supplied any logical foundations.

Moreover, when Descartes in his *Geometry* (1637) and Fermat in his manuscript of 1629 created analytic geometry, neither had any clear notion about irrational numbers. Nevertheless, both presupposed a one-to-one correspondence between all the positive real numbers and the points on a line, or that the distance of any point on a line from some origin can be expressed as a number. Since many of these numbers were irrational, both men implicitly accepted irrationals, even though these had no logical foundation.

The Europeans also had to face negative numbers. These numbers became known in Europe through Arab texts, but most mathematicians of the 16th and 17th centuries did not accept them as numbers or, if they did, would not accept them as roots of equations. In the 15th century Nicolas Chuquet (1445?–1500?) and in the 16th Stifel both spoke of negative numbers as absurd numbers. Cardan gave negative numbers as roots of equations but considered them impossible solutions, mere symbols. He called them fictitious, whereas he called positive roots real. Vieta discarded negative numbers entirely. Descartes accepted them to an extent. He called negative roots of equations false on the ground that they claim to represent numbers less than nothing. However, he had shown that given an equation one can obtain another whose roots are larger than the original one by any given quantity. Thus an equation with negative roots could be transformed into one with positive roots. Hence, he said, we can turn false roots into real roots and so reluctantly accepted negative numbers. However, he never did make his peace with them. Pascal regarded the subtraction of 4 from 0 as pure nonsense. He said in his *Pensées*, "I have known those who could not understand that to take four from zero there remains zero."

An interesting argument against negative numbers was given by Antoine Arnauld (1612–1694), theologian, mathematician, and close friend of Pascal. Arnauld questioned that $-1 : 1 = 1 : -1$ because, he said, -1 is less than $+1$; hence, how could a smaller be to a greater as a greater is to a smaller? Leibniz agreed that there is an objection here but argued that one can calculate with such proportions because their *form* is correct just as one calculates with imaginary quantities (which as we shall shortly see had already been introduced). Nevertheless, he equivocated, one should call imaginary (non-existent) all those quantities which have no logarithm. On this basis -1 does not exist because positive logarithms belong to numbers greater than 1 and negative logarithms [!] belong to numbers between 0 and 1. There is then no logarithm available for negative numbers. In fact, if there were a number for $\log(-1)$, then, according to the laws of logarithms, $\log \sqrt{-1}$ would have to be half of that number. But certainly $\sqrt{-1}$ has no logarithm.

One of the first algebraists who occasionally placed a negative number by itself on one side of an equation was Thomas Harriot (1560–1621). But he did not accept negative roots and even "proved" in his posthumously published *Artis analyticae praxis* (Analytical Arts Applied, 1631) that such roots are impossible. Raphael Bombelli (16th century) gave rather clear definitions for negative numbers, though he could not justify the rules of operation because the necessary foundation even for the positive numbers was not available.* Stevin used posi-

---

* As W. H. Auden put it:

> Minus times minus is plus,
> The reason for this we need not discuss.

tive and negative coefficients in equations and accepted negative roots. In his *Invention nouvelle en algèbre* (New Invention in Algebra, 1629) Albert Girard (1595–1632) placed negative numbers on a par with positive numbers and gave both roots of a quadratic equation even when both were negative. Both Girard and Harriot used the minus sign for the operation of subtraction and for negative numbers, though separate symbols should be used because a negative number is an independent concept whereas subtraction is an operation.

On the whole, not many of the 16th- and 17th-century mathematicians felt at ease with or accepted negative numbers at all, and of course they did not recognize them as true roots of equations. There were also curious beliefs about them. Though Wallis was advanced for his times and accepted negative numbers, he thought they were larger than ∞ as well as less than 0. In his *Arithmetica infinitorum* (Arithmetic of Infinitesimals, 1655) he argued that, since the ratio $a/0$, when a is positive, is infinite, then when the denominator is changed to a negative number, as in a/b with b negative, the ratio should be greater than $a/0$ because the denominator is smaller. Thus the ratio must be greater than ∞.

Some of the more advanced thinkers, Bombelli and Stevin, proposed a representation which certainly aided in the ultimate acceptance of the whole real number system. Bombelli supposed that there is a one-to-one correspondence between real numbers and lengths on a line (with a chosen unit), and he defined for the lengths the four basic operations. He regarded the real numbers and their operations as defined by these lengths and the corresponding geometric operations. Thus, the real number system was rationalized on a geometric basis. Stevin, too, regarded real numbers as lengths and believed that with this interpretation the difficulties with irrationals also are resolved. Of course, in this view the real numbers were still tied to geometry.

Without having overcome their difficulties with irrational and negative numbers, the Europeans added to their burdens by blundering into what we now call complex numbers. By extending the use of the arithmetical operation of square root to whatever numbers appeared, for example, in solving quadratic equations, they obtained these new numbers. Thus Cardan in Chapter 37 of his *Ars magna* (The Great Art, 1545) set up and solved the problem of dividing 10 into two parts whose product is 40. This seemingly absurd problem has a solution for, as d'Alembert remarked, "Algebra is generous; it often gives more than one demands of it." If x is one part, the equation for x is $x(10-x)=40$. Cardan obtained the roots $5+\sqrt{-15}$ and $5-\sqrt{-15}$ and then says there are "sophistic quantities which though ingenious are useless." "Putting aside the mental tortures involved," multiply $5+\sqrt{-15}$ and $5-\sqrt{-15}$; the product is $25-(-15)$ or 40. He then stated, "So progresses arithmetic subtlety, the end of which, as is said, is as refined as it is useless."

Cardan became further involved with complex numbers in the algebraic method of solving cubic equations which he presented in his book. Though he sought and obtained only the real roots, his formula also gives the complex roots when these are present. Peculiarly, when all the roots are real, his formula yields complex numbers from which the real roots should be derivable. Hence, he should have attached great importance to complex numbers but, since he did not know how to find the cube root of complex numbers and thereby derive the real roots, he left this difficulty unresolved. He found the real roots in another way.

Bombelli, too, considered complex numbers as the solutions of cubic equations and he formulated in practically modern form the four operations with complex numbers, but he still regarded them as useless and "sophistic." Albert Girard did recognize complex numbers as at least formal solutions of equations. In his *New Invention in Algebra* he stated, "One could say: What are these impossible solutions [complex roots]? I answer: For three things, for the certitude of the general rules, that there are no other solutions, and for their utility." However, Girard's advanced views were not influential.

Descartes also rejected complex roots and coined the term imaginary. He said in his *Geometry*, "Neither the true nor the false [negative] roots are always real; sometimes they are imaginary." He argued that whereas negative roots can at least be made "real" by transforming the equation in which they occur into another equation whose roots are positive, this cannot be done for complex roots. These, therefore, are not real but imaginary; they are not numbers.

Not even Newton regarded complex roots as significant, most likely because in his day they lacked physical meaning. In fact, he said in his *Universal Arithmetic* (2nd ed., 1728), "But it is just that the Roots of Equations should be often impossible [complex], lest they should exhibit the cases of Problems that are impossible as if they were possible." That is, problems which do not have a physically or geometrically meaningful solution should have complex roots.

The lack of clarity about complex numbers is illustrated by Leibniz's oft-quoted statement, "The Divine Spirit found a sublime outlet in that wonder of analysis, that portent of the ideal world, that amphibian between being and not-being, which we call the imaginary root of negative unity." Though Leibniz worked with complex numbers formally, he possessed no understanding of their nature. To justify the use he and John Bernoulli made of them in the calculus, Leibniz said no harm came of it.

Despite the lack of any clear understanding during the 16th and

17th centuries, the operational procedures with real and complex numbers were improved and extended. In his *Algebra* (1685) Wallis showed how to represent geometrically the complex roots of a quadratic equation with real coefficients. Wallis said, in effect, that complex numbers are no more absurd than negative numbers and, since the latter can be represented on a directed line, it should be possible to represent complex numbers in a plane. He did give an incomplete representation of complex numbers and also a geometrical construction for the roots of $ax^2 + bx + c = 0$ when the roots are real or complex. Though Wallis's work was correct, it was ignored because mathematicians were not receptive to the use of complex numbers.

Though other problems in the logic of mathematics arose in the 17th century, which we shall deal with in the next chapter, here we shall pursue the difficulties mathematicians encountered in the 18th century in seeking to understand and justify their work with irrational, negative, and complex numbers and with algebra. As for (positive) irrational numbers, though nothing had been done to define them and establish their properties, they were intuitively more acceptable because their properties were those of whole numbers and fractions; hence, mathematicians used them freely and raised no new challenges about their meaning or properties. Some, Euler among them, believed that the logical basis was at hand in the Eudoxian theory of magnitudes, which is expounded in Book V of Euclid's *Elements*. Eudoxus did give a theory of proportion for magnitudes, a theory tied to geometry, but by no means a theory of irrational numbers. However, the consciences of these 18th-century men, if not their logic, were clear about irrationals.

Negative numbers troubled mathematicians far more than irrational numbers did, perhaps because negatives had no readily available geometrical meaning and the rules of operation were stranger. Though negative numbers were used rather freely from about 1650 onward, since the concept and its logical foundation were not clear mathematicians continued to fudge the justification or protest their use. In his article "Negative" in the famous *Encyclopédie,* one of the great intellectuals of the Age of Reason, Jean Le Rond d'Alembert (1717–1783) stated that "a problem leading to a negative solution means that some part of the hypothesis was false but assumed to be true." In his paper on negative quantities, he added, "arriving at a negative solution means that the opposition of the number [the corresponding positive] is the desired solution.

Leonhard Euler, the greatest of 18th-century mathematicians, wrote one of the outstanding algebra texts of all times. In his *Complete Introduction to Algebra* (1770), he justified the operation of subtracting $-b$ as equivalent to adding b because "to cancel a debt signifies the same as giving a gift." He also argued that $(-1)(-1) = +1$ because the product must be $+1$ or -1 and, since he had already shown that $1(-1) = -1$, then $(-1)(-1)$ must be $+1$. The best texts of the 18th century continued to confuse the minus sign used to denote subtraction and the same sign as used in, say, -2 to denote a negative number.

Many objections to negative numbers were voiced throughout the 18th century. The English mathematician, Baron Francis Masères (1731–1824), a Fellow of Clare College in Cambridge and a member of the Royal Society, wrote respectable papers in mathematics and put out a substantial treatise on the theory of life insurance. In 1759 he published his *Dissertation on the Use of the Negative Sign in Algebra*. He showed how to avoid negative numbers (except to indicate, without actually performing, the subtraction of a larger quantity from a smaller one), and especially negative roots, by carefully segregating the types of quadratic equations. Those with negative roots are considered separately and of course the negative roots are rejected. He does the same with cubics. Then he says of negative roots:

> . . . they serve only, as far as I am able to judge, to puzzle the whole doctrine of equations, and to render obscure and mysterious things that are in their own nature exceeding plain and simple. . . . It were to be wished therefore that negative roots had never been admitted into algebra or were again discarded from it: for if this were done, there is good reason to imagine, the objections which many learned and ingenious men now make to algebraic computations, as being obscure and perplexed with almost unintelligible notions, would be thereby removed; it being certain that Algebra, or universal arithmetic, is, in its own nature, a science no less simple, clear, and capable of demonstration, then geometry.

The debates about the meanings and use of complex numbers were even more strident. The quandaries were compounded by the fact that some mathematicians introduced logarithms of negative numbers (as well as logarithms of complex numbers) which were complex numbers.

Beginning in 1712, Leibniz, Euler, and John Bernoulli debated hotly through letters and papers the meaning of complex numbers and especially logarithms of negative and complex numbers. Leibniz and Bernoulli used Descartes's term "imaginary" for complex numbers and by imaginary meant that such numbers (and negative numbers) did not exist, though both miraculously managed to make good use of these non-existent numbers in the calculus.

Leibniz, as we have noted, gave various arguments to the effect that there are no logarithms for negative numbers. John Bernoulli's position was that $\log a = \log (-a)$ and supported this with several arguments. One, using the theorem on logarithms for positive numbers, was that

$$\log (-a) = \tfrac{1}{2}\log(-a)^2 = \tfrac{1}{2}\log a^2 = \log a.$$

Another, using the calculus, drew the same conclusion. Many letters were exchanged over the years by Leibniz and John Bernoulli. Most of what was written was nonsense.

Euler had the real answer. He published his result in a paper which appeared in 1751, "Investigations on the Imaginary Roots of Equations." His final result, correct but obtained by an incorrect argument, applies to all complex numbers, which include the real, for if y is 0 in $x + iy$, the number is real. His result is:

$$\log(x + iy) = \log (\rho e^{i\phi}) = \log\rho + i(\phi \pm 2n\pi).*$$

However, Euler's paper was not understood by his contemporaries.

Euler had communicated his results to d'Alembert in a letter of April 15, 1747. He even pointed out that there were infinitely many logarithms of a positive real number but that only one was a real number; this is the one we usually use in computing with real numbers. D'Alembert was not convinced by the correspondence or the paper, and in his essay "On Logarithms of Negative Quantities," he advanced all sorts of metaphysical, analytical, and geometrical arguments to deny the existence of such logarithms. He succeeded in shrouding the subject in more mystery. He also tried to cover up his differences with the masterful Euler by saying that they were just a matter of words.

All of the men involved in this controversy were inconsistent in their own thinking. During the first half of the 18th century it was believed that some operations on complex numbers, for example, a complex power of a complex number, might call for a totally new kind of number. But it was d'Alembert himself, in *Reflections on the General Cause of Winds* (1747), who proved that all operations on complex numbers would lead only to complex numbers. His proof did have to be mended by Euler and Lagrange; yet in this matter d'Alembert had made a significant step. Perhaps d'Alembert recognized the confused state of his own ideas on complex numbers, because in the *Encyclopédie*, for which he wrote the mathematical articles, he said nothing about them.

Apparently Euler, too, was still not clear about complex numbers. In his *Algebra* of 1770, the best algebra text of the 18th century, he said,

---

* Euler uses here what is called the polar form of a complex number. $\rho = \sqrt{x^2 + y^2}$ and ϕ is the angle that the line segment from the origin to $x + iy$ makes with the x-axis. If y is 0, then ϕ is 0.

The square roots of negative numbers are neither zero, nor less than zero, nor greater than zero. Then it is clear that the square roots of negative numbers cannot be included among the possible numbers [real numbers]. Consequently we must say that these are impossible numbers. And this circumstance leads us to the concept of such numbers, which by their nature are impossible, and ordinarily are called imaginary or fancied numbers, because they exist only in the imagination.

He also made mistakes with complex numbers. In his *Algebra* he writes $\sqrt{-1} \cdot \sqrt{-4} = \sqrt{4} = 2$ because $\sqrt{a}\sqrt{b} = \sqrt{ab}$.

Though he calls complex numbers impossible numbers, Euler says they are useful. And his idea of their use is that they tell us which problems do or do not have an answer. Thus, if we are asked to separate 12 into two parts whose product is 40 (shades of Cardan) we would find that the parts are $6 + \sqrt{-4}$ and $6 - \sqrt{-4}$. Thereby, he says, we recognize that the problem cannot be solved.

Despite the many objections to complex numbers, throughout the 18th century complex numbers were used effectively by applying rules that applied to real numbers, and so mathematicians acquired some confidence in them. Where used in intermediate stages of mathematical arguments, the final results proved to be correct and this fact had telling effect. Yet doubts as to the validity of the arguments and often even of the results continued to plague mathematicians.

The general attitude toward acceptance of the several troublesome types of numbers, irrational, negative, and complex, was expressed by d'Alembert apropos of negative numbers in the *Encyclopédie*. The article is not at all clear. D'Alembert concluded that the "algebraic rules of operation with negative numbers are generally admitted by the whole world and acknowledged as exact, whatever idea we may have about these quantities."

During the centuries that the Europeans struggled to understand the various types of numbers, another major logical problem was thrust to the fore, that of supplying the logic of algebra. The first work that organized significantly new results was Cardan's *The Great Art* which showed how to solve cubic equations such as $x^3 + 3x^2 + 6x = 10$ and quartic equations such as $x^4 + 3x^3 + 6x^2 + 7x + 5 = 0$. Within about one hundred years, a large number of other results, mathematical induction, the binomial theorem, and approximate methods of finding roots of low and high degree equations, were added to the body of algebra, the chief contributors being Vieta, Harriot, Girard, Fermat, Descartes, and Newton. But there were no proofs of these results. It is true that Cardan and the later algebraists Bombelli and Vieta gave geometrical arguments to support their methods for solving cubic and quartic equations but, since they ignored negative and complex roots, these arguments were certainly not proofs. Moreover, the introduction of higher degree equations such as the quartic and quintic meant that geometry which was limited to three dimensions could not be the basis of proof. Results given by the other authors were most often merely assertions suggested by concrete examples.

A step in the right direction was made by Vieta. From the time of the Egyptians and the Babylonians right up to Vieta's work, mathematicians solved linear, quadratic, cubic, and quartic equations with numerical coefficients only. Thus, equations such as $3x^2 + 5x + 6 = 0$ and $4x^2 + 7x + 8 = 0$ were regarded as different from each other though it was clear that the same method of solution applied to both. Moreover, to avoid negative numbers, an equation such as $x^2 - 7x + 8 = 0$ was for a long time, treated in the form $x^2 + 8 = 7x$. Hence, there were many types of equations of the same degree and each was treated separately. Vieta's contribution, a major one, was to introduce literal coefficients.

Vieta was trained as a lawyer. Generally, he pursued mathematics as a hobby and printed and circulated his work at his own expense. Though a number of men had used letters for sporadic and incidental purposes, Vieta is the first man to do so purposefully and systematically. The chief new use was not just to represent an unknown or powers of an unknown but as general coefficients. Thus all second-degree equations could be treated in one swoop by writing (in our notation) $ax^2 + bx + c = 0$, wherein a, b, and c, the literal coefficients, could stand for any numbers, whereas x stands for an unknown quantity or quantities whose values are to be found.

Vieta called his new algebra *logistica speciosa* (calculation with types) as opposed to *logistica numerosa* (calculation with numbers). He was fully aware of the fact that when he studied the general second-degree equation $ax^2 + bx + c = 0$, he was studying an entire class of expressions. In making the distinction between *logistica numerosa* and *logistica speciosa* in his *In artem analyticam isagoge* (Introduction to the Analytic Art, 1591), Vieta drew the line between arithmetic and algebra. Algebra, he said, is a method of operating on species or forms of things and this is the *logistica speciosa*. Arithmetic and equations with numerical coefficients deal with numbers and this is the *numerosa*. Thus, in this one step by Vieta, algebra became a study of general types of forms and equations, since what is done for the general case covers an infinity of special cases.

The merit of Vieta's use of letters to denote a class of numbers was that, if one could prove that a method of solving $ax^2 + bx + c = 0$ was correct, the method would justify the solution of the infinite number of concrete equations such as $3x^2 + 7x + 5 = 0$. One can say of Vieta's contribution that it made possible generality of proof in algebra. However, if one is to perform operations on a, b, and c, where these letters stand for any real or any complex number, one must know that these operations are correct for all real numbers and all complex numbers. But since these operations had not been justified logically—even the very definitions of the types of numbers had not been formulated—the justifications for operating with a general a, b, and c were certainly not within reach. Vieta himself rejected negative and complex numbers, so that the generality he envisioned in his *logistica speciosa* was a limited one.

Vieta's thinking is certainly inexplicable if not irrational. On the one hand he made the highly significant contribution of employing literal coefficients and he fully realized that this step made possible general proofs. But his failure to recognize negative numbers and his refusal to let the literal coefficients stand for them, too, shows the severe limitations of the best of human minds. The rules for operations with negative numbers had been in existence for about eight hundred years and the rules gave correct results. Vieta could not have objected to rules for that was pretty much all that algebra had to offer in his time. But negative numbers lacked the intuitive and physical meanings that the positive numbers possessed. It seems clear that not logic but intuition was determining what mathematicians would accept. It was not until 1657 that John Hudde (1633–1704) allowed these literal coefficients to stand for negative and positive numbers. Thereafter, most mathematicians did so freely.

In Vieta's time, the late 16th century, algebra was a minor appendage to geometry. The solution of single equations involving one unknown or of two equations involving two unknowns, motivated by simple practical problems arising in geometry or in commerce, had been effected. But the power of algebra was not recognized until the 17th century. The masterful step was made by Descartes and Pierre de Fermat (1601–1665), namely, the creation of coordinate geometry (which should be called algebraic geometry). The basic idea here is, of course, that curves can be represented by equations. For example, $x^2 + y^2 = 25$ represents a circle of radius 5. With algebraic representation, one could prove any number of properties of curves far more readily than by the purely geometric or synthetic methods of the classical Greeks.

However, in 1637 when Descartes published his *Geometry*, neither he nor Fermat in his work of 1629 (published posthumously) was prepared to accept negative numbers. Hence, the idea but not the full power of the algebraic approach to geometry was evident. But the successors of Descartes and Fermat introduced negative numbers into coordinate geometry, which became basic to major developments in both analysis and geometry.

The second innovation which pushed algebra to the fore was the use of algebraic formulas to represent functions. Galileo, as we know (Chapter II), had introduced the idea of describing the motions of objects by formulas. Thus, the height above ground of an object thrown up with a velocity of, say, 100 feet per second is given by the formula $h = 100t - 16t^2$. From this formula any number of facts about the motion can be deduced by algebraic means, for example, the maximum height it would attain, the time required to reach that height, and the time it would take to fall back to the ground. Indeed, the power of algebra was soon recognized to be so great that mathematicians began to use it extensively and it burgeoned into a commanding position over geometry.

The free use of algebra provoked a host of protests. Philosopher Thomas Hobbes, though only a minor figure in mathematics, nevertheless spoke for many mathematicians when he objected to the "whole herd of them who apply their algebra to geometry." Hobbes said these algebraists mistake the symbols for geometry and characterized John Wallis's book on the algebraic treatment of conics as a scurvy book and as a "scab of symbols." Many mathematicians, including Blaise Pascal and Isaac Barrow, objected to the use of algebra because it had no logical foundation; they insisted on geometric methods and proofs. Some contented themselves with the belief that one could establish the logical foundation for algebra by falling back on geometry but this, as we have noted, was a delusion.

Nevertheless, most mathematicians used algebra freely on a pragmatic basis. The value of algebra proper in treating all sorts of real problems and the superiority of algebra even in treating geometric problems were so evident that the mathematicians plunged deeply into its waters.

Unlike Descartes, who still regarded algebra as the servant of geometry, John Wallis and Newton recognized the full power of algebra. It was nevertheless with great reluctance that the mathematicians abandoned the geometric approach. According to Henry Pemberton, who edited the third edition of Newton's *Principles,* Newton not only constantly expressed great admiration for the geometers of Greece but censured himself for not following them more closely than he did. In a letter to David Gregory (1661–1708), a nephew of James Gregory (1638–1675), Newton remarked that "Algebra is the analysis of bunglers in mathematics." But his own *Universal Arithmetic* of 1707 did as much as any single work to establish the supremacy of algebra. Here he set up arithmetic and algebra as the basic mathematical science and allowed geometry only where it provided demonstrations. Nevertheless, on the whole this book is just a collection of rules. There are few proofs or even intuitive arguments for assertions about numbers and algebraic processes. Newton's position was that the letters in algebraic expressions stood for numbers and no one could doubt the certainty of arithmetic.

Leibniz, too, noted the growing dominance of algebra and fully appreciated its effectiveness. Yet, concerned about the lack of proof, he felt obliged to say, "Often the geometers could demonstrate in a few words what is very lengthy in the calculus . . . ; the use of algebra is assured, but it is not better." He characterized the work in algebra of his time as a "mélange of good fortune and chance." However, Leonhard Euler in his *Introduction to Infinitesimal Analysis* (1748) praised algebra openly and unreservedly as being far superior to the geometric methods of the Greeks. By 1750 the reluctance to use algebra had been overcome; by then algebra was a full-grown tree with many branches but no roots.

The development of the number system and algebra provides a striking contrast to that of geometry. Geometry was deductively organized by 300 B.C. The few defects could be and, as we shall see later, were readily corrected. However, arithmetic and algebra had no logical foundation whatsoever. It would seem that the absence of any logical basis should have troubled all mathematicians. How could the Europeans, well versed in Greek deductive geometry, have entertained and applied the various types of numbers and algebra, which had never been logically established?

There are several reasons. The basis for accepting the properties of the whole numbers and fractions had certainly been experience. As new types of numbers were added to the number system, the rules of operation already accepted for the positive integers and fractions on an empirical basis were applied to the new elements, with geometrical thinking as a handy guide. Letters, when introduced, were just representations of numbers and so could be treated as such. The more complicated algebraic techniques *seemed* justified either by geometrical arguments such as Cardan used or by sheer induction on specific cases. Of course, none of these procedures were logically satisfactory. Geometry even where called upon did not supply the logic for negative, irrational, and complex numbers. Certainly, the solution of fourth degree (quartic) equations could not be justified geometrically.

Secondly, at the outset, especially in the 16th and 17th centuries, algebra was not regarded as an independent branch of mathematics requiring a logical foundation of its own. It was regarded as a method of analyzing what were essentially geometric problems. The many who did take to algebra, notably Descartes, thought of it as a method of analysis. The titles of Cardan's book *The Great Art* and of Vieta's *In-*

troduction to the Analytic Art support the assertion that they used the word art in the sense in which one might use it today as opposed to a science. The term analytic geometry, which became current for Descartes's algebraic geometry, confirms this attitude toward algebra. As late as 1704, Edmund Halley in an article published in the Philosophical Transactions of the Royal Society spoke of algebra as the analytic art. But Descartes's analytic geometry was perhaps the crucial creation that impressed mathematicians with the power of algebra.

Finally, the use of negative and irrational numbers and of algebra in scientific work produced excellent agreement with observations and experiments. Whatever doubts mathematicians may have had while employing negative numbers, for example, in some scientific work, were swept away when the final mathematical result proved physically sound. Scientific application was the chief concern and any measure or device that promised effectiveness in that work was almost unscrupulously adopted. The needs of science prevailed over logical scruples. Doubts about the soundness of algebra were cast aside, much as greedy industrialists cast aside ethical principles, and the mathematicians proceeded blithely and confidently to employ the new algebra. Thereafter, mathematicians gradually converted algebra into an independent science which embraced and "established" results about number and geometry. In fact, Wallis affirmed that the procedures of algebra were no less legitimate than those of geometry.

By the end of the 17th century, number and algebra were recognized to be independent of geometry. Why then didn't the mathematicians undertake the logical development? Given the model for the deductive organization of geometry as embodied in Euclid's *Elements,* why didn't the mathematicians develop one for number and algebra? The answer is that geometrical concepts, axioms, and theorems are intuitively far more accessible than those of arithmetic and algebra. Pictures (in the case of geometry, diagrams) aided in suggesting the structure. But the concepts of irrational, negative, and complex numbers are far more subtle and even pictures, when they became available, did not suggest the logical organization of number qua number and of literal expressions built upon number. The problem of building a logical foundation for the number system and algebra was a difficult one, far more difficult than any 17th-century mathematician could possibly have appreciated. We shall have occasion to examine it later (Chapter VIII). And it is fortunate that mathematicians were so credulous and even naïve, rather than logically scrupulous. For free creation must precede formalization and logical foundation, and the greatest period of mathematical creativity was already under way.

VI

The Illogical Development:
The Morass of Analysis

> One must make a start in any line of research, and this
> beginning almost always has to be a very imperfect at-
> tempt, often unsuccessful. There are truths that are un-
> known in the way that there are countries the best road to
> which can only be learned after having tried them all.
> Some persons have to take the risk of getting off the track
> in order to show the right road to others. . . . We are al-
> most always condemned to experience errors in order to
> arrive at truth.   DENIS DIDEROT

On the non-existent logical foundations of arithmetic and algebra and
the somewhat insecure foundation of Euclidean geometry, mathematicians built analysis, whose core is the calculus. The calculus is the most
subtle subject in all of mathematics. In view of the defects we have already noted in comparatively simpler areas, one might guess that the
concepts and logical structure of the calculus would strain the intellectual capacities of mathematicians. This expectation was fulfilled.

The calculus utilizes the concept of function, which, loosely put, is a
relationship between variables. If a ball, say, is dropped from the roof
of a building, the distance it falls and the time of fall increase. Distance
and time are variables, and the function that relates distance and time,
if we neglect the resistance of air, is given by the formula $d = 16t^2$,
wherein t represents the time in seconds that the ball has fallen and d
represents the distance in feet fallen in time t.

The origins of any important idea can always be traced back decades
and even hundreds of years and this is also true of the concept of function. However, it gained explicit recognition in the 17th century. The
details of the history are not significant. More relevant is the fact that
though the concept of a function is rather straightforward, even the

simplest functions involve all types of real numbers. Thus in the case above, one could certainly ask for the value of d when $t = \sqrt{2}$. Or one could ask for the value of t when d is 50, say, and this value of t would be $\sqrt{50/16}$, which also is irrational. Now irrationals, we know, were not well understood in the 17th century. Hence, all of the logic that was missing in the treatment of number was also missing in the work with functions. However, since irrationals were used freely by 1650, this failing was glossed over.

The calculus uses functions but introduces two new and far more complex concepts, the derivative and the definite integral, and these require a logical foundation in addition to whatever might serve for number.

The two concepts were tackled by the greatest mathematicians of the 17th century, the most famous of whom were Kepler, Descartes, Bonaventura Cavalieri (1598–1647), Fermat (1601–1665), Blaise Pascal (1623–1662), James Gregory (1638–1675), Gilles Persone de Roberval (1602–1675), Christian Huygens (1629–1695), Isaac Barrow (1630–1677), John Wallis (1616–1703), and of course Isaac Newton (1642–1727) and Gottfried Wilhelm Leibniz (1646–1716). Each of these men made his own approach to the problems of defining and computing the derivative and definite integral. Some reasoned purely geometrically, others purely algebraically, and still others used a mixture of the two methods. Our concern is to see how well or how poorly the men adhered to the standards of mathematical reasoning. For this purpose a few typical examples may suffice. In fact, many of the methods were very limited and do not warrant attention here.

The nature of the derivative is perhaps best understood by thinking in terms of velocity, much as Newton did. If an object travels 200 feet in 4 seconds, one can say that its *average* velocity is 50 feet per second and, if the object travels at a constant velocity, the average velocity is also the velocity at each instant during the 4 seconds. However, most motions do not take place at a constant velocity. An object which falls to earth, a projectile shot out from a gun, and a planet moving around the sun move at continuously changing velocities. For many purposes it is important to know the velocity at some instants during the motion. Thus, the velocity at the instant a projectile strikes a man is crucial. If the velocity is 0 feet per second, the projectile falls to the ground. If the velocity is 1000 feet per second, the man falls to the ground. Now an instant, by its very meaning, is zero lapse of time. And in zero time an object travels zero distance. Hence, if one were to compute the instantaneous velocity as one computes average velocity, by dividing the distance traveled by the time, one would have 0/0 and this is meaningless.

The way out of this impasse, which the 17th-century mathematicians glimpsed but did not grasp, may be put thus. Suppose an object is falling to earth and one wishes to calculate its velocity precisely 4 seconds after it began to fall. Now if we take any interval, as opposed to an instant, of time during which it falls and if we divide the distance it falls during that interval by the time, we can get the average velocity during that interval. Let us then calculate the average velocity during the ½ second that begins at the fourth second, during the ¼ second that begins at the fourth second, during the ⅛ second that begins at the fourth second, and so on. Surely the smaller the interval of time, the closer the average velocity will be to the velocity at exactly 4 seconds. Presumably, all we have to do then is to note what the various average velocities amount to and to see what number they are *approaching;* that number should be the desired velocity at exactly 4 seconds. This scheme seems reasonable enough but as we shall see there are inherent difficulties. At any rate, the velocity at the fourth second, if calculable, is called the *derivative* of $d = 16t^2$ at $t = 4$.

We can see the difficulties more readily if we now translate the verbal account above into symbolic language. The mathematical formulation, essentially the one finally adopted, is due to Fermat. We shall calculate the velocity at the fourth second of a ball whose fall is described by the function

(1) $$d = 16t^2.$$

When $t = 4$, $d = 16 \cdot 4^2$ or 256. Now let h be any increment of time. In the time $4 + h$, the ball will fall 256 feet plus some incremental distance k. Then

$$256 + k = 16(4 + h)^2 = 16(16 + 8h + \mathrm{h}^2)$$

or

$$256 + k = 256 + 128h + h^2.$$

Then by subtracting 256 from both sides

$$k = 128h + h^2$$

and the average velocity in h seconds is

(2) $$\frac{k}{h} = \frac{128h + h^2}{h}.$$

Fermat was fortunate in the case of this simple function and others he considered in that he could divide the numerator and denominator of the right side by h and obtain

(3) $$k/h = 128 + h.$$

He then let h be 0 and obtained as the velocity at the fourth second of fall

(4) $\dot{d} = 128.$

(The notation $\dot{d}$ is Newton's.) Thus $\dot{d}$ is the derivative of $d = 16t^2$ at $t = 4$.

The objection to this process is that one starts with an h which is not zero and performs operations such as dividing numerator and denominator by h, which are correct only when h is not zero. Thus (3) is correct only when h is not 0. One cannot then let h be 0 to draw a conclusion. Moreover, in the case of the simple function $d = 16t^2$, (2) simplifies to (3). But for more complicated functions, we are obliged to deal with the form (2) and here, when $h = 0$, k/h is 0/0, which is meaningless.

Fermat never did justify what he did and in fact, though he must be credited with being one of the progenitors of the calculus, he did not carry this work very far. He was careful not to assert any general theorems when he advanced an idea he could not justify completely. He was satisfied that he had a correct process because he could give it a geometrical interpretation, and he believed that ultimately a suitable geometrical proof could be obtained.

The second concept of the calculus that perplexed the creators of the calculus, the definite integral, is involved, for example, in calculating the areas of figures bounded in whole or in part by *curves*, the volumes of figures bounded by *surfaces*, and centers of gravity of variously shaped bodies. To see the difficulty involved let us consider the problem of finding the area bounded in part by a curve.

Let us consider finding the area *DEFG* (Fig. 6.1) which is bounded by the arc *FG* of the curve whose equation is $y = x^2$, by *DE*, and by the ver-

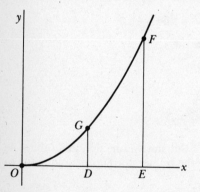

Figure 6.1

Figure 6.2

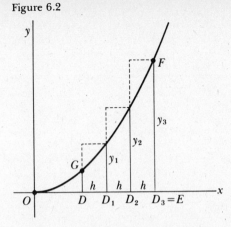

tical line segments DG and EF. Here, too, the quantity we want must be approached through better and better approximations, and this much the 17th-century mathematicians undertook. We subdivide the interval DE into three equal parts, each of length h, and denote the points of subdivision by D_1, D_2, and D_3, where D_3 is the point E (Fig. 6.2). Let y_1, y_2, and y_3 be the ordinates at the points of subdivision. Now y_1h, y_2h, and y_3h are the areas of the three rectangles shown, and the sum

$$(5) \qquad\qquad y_1h + y_2h + y_3h$$

is the sum of the three rectangular areas and thus an approximation to the area $DEFG$.

We can obtain a better approximation to the area $DEFG$ by using smaller rectangles and more of them. Suppose that we subdivide the interval DE into six parts. Figure 6.3 shows in particular what happens to the middle rectangle of Figure 6.2. This rectangle is replaced by two, and because we use the y-value at each point of subdivision as the height of a rectangle, the shaded area in Figure 6.3 is no longer a part of the sum of the areas of the six rectangles which now approximates the area $DEFG$. Therefore, the sum

$$(6) \qquad\qquad y_1h + y_2h + y_3h + y_4h + y_5h + y_6h$$

is a better approximation to the area $DEFG$ than the sum (5).

We can make a more general statement concerning this process of approximation. Suppose that we divide the interval DE into n parts. There would then be n rectangles, each of width h. The ordinates at the points of subdivision are $y_1, y_2, \cdots y_n$, where the dots indicate that all

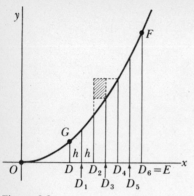

Figure 6.3

intervening y-values at points of subdivision are included. The sum of the areas of the n rectangles is then

(7) $$y_1 h + y_2 h + \cdots + y_n h,$$

and the dots again indicate that all intervening rectangles are included. In view of what we said just above about the effect of subdividing DE into smaller intervals, the approximation to the area $DEFG$ given by the sum (7) improves as n increases. Of course, as n gets larger, h gets smaller because $h = DE/n$. We see so far how figures formed by line segments—rectangles in the present case—can be used to provide better and better approximations to an area bounded by a curve.

It seems intuitively clear that the larger the number of rectangles the better the sum of their areas would approximate the desired curvilinear area. However, if one stopped at 50 or 100 rectangles the sum would still not be the area sought. The thought that occurred to the 17th-century mathematicians who fashioned this approach was to let n become infinite. However, just what infinite meant was not clear. Was it a number and if so, how did one calculate with this number? When Fermat arrived at expressions for the sum of n rectangles such as (7) and necessarily found that these contained terms such as $1/n$ and $1/n^2$, he dismissed them on the ground that these are negligible when n is infinite. Here, too, as in the case of the derivative, Fermat believed that precise proofs could be made, most likely by resorting to the method of exhaustion, introduced by Eudoxus (a very limited and rather complicated geometrical method and used so skillfully by Archimedes).

Of the early works to obtain areas and volumes by means of the definite integral, perhaps Bonaventura Cavalieri's work deserves attention because it influenced many contemporary and later men and because it is typical of the vague thinking of the times. Cavalieri regarded an area such as the one shown in Figure 6.1 as a sum of an infinite number of units which he called indivisibles and which presumably could be lines. However, Cavalieri was not clear about just what his indivisibles were, except to suggest that, if one divided up the area into smaller and smaller parts such as the rectangles shown in Figure 6.3, one would arrive at the indivisibles. In one of his books, *Exercitationes geometricae sex* (Six Geometrical Exercises, 1647), he "explained" that the area in question is made up of indivisibles just as a necklace is made up of beads, a cloth is of threads, and a book is of pages. With this concept he did manage to compare two areas or two volumes and obtain the correct relationship between the two.

Cavalieri's critics were not satisfied. One contemporary, Paul Guldin (1557–1643), accused him of having confounded the geometry of the Greeks instead of understanding it. And one recent historian said of Cavalieri's work that if there were a prize for obscurity he would win it without contest. Since Cavalieri could not explain how an infinite number of elements, his indivisibles, could make up an object of finite extent, he evaded an answer by rejecting any precise interpretation of his indivisibles. Sometimes he resorted to speaking of an infinite sum of lines without explicitly explaining the nature of this infinity. At other times he claimed his method was just a pragmatic device to avoid the complicated Greek method of exhaustion. Further, argued Cavalieri, contemporary geometers had been freer with concepts than he had been—witness Kepler in his *Stereometria doliorum vinariorum* (Measuring Volumes of Wine Casks, 1616). These geometers, he continued, had been content in their calculation of areas to imitate Archimedes' method but had failed to give the complete proofs which the great Greek had used to make his work rigorous. Nevertheless, these geometers were satisfied with their calculations because the results were useful. Adopting the same point of view, Cavalieri said that his procedures could lead to new inventions and that his method did not at all oblige one to consider a geometrical structure as composed of an infinite number of elements; it had no other object than to establish correct ratios between areas or volumes. But these ratios preserved their sense and value whatever opinion one might have about the composition of the figures. As a last resort he relegated the conceptual questions to philosophy, and therefore as unimportant. "Rigor," he said, "is the concern of philosophy not of geometry."

Pascal defended Cavalieri. In his *Letters of Dettonville* (1658), he affirmed that the geometry of indivisibles and classical Greek geometry were in agreement. "What is demonstrated by the true rules of indivisibles could be demonstrated also with the rigor and the manner of the ancients." They differed only in terminology. Further, the method of indivisibles had to be accepted by any mathematician who pretended to rank among geometers. Actually Pascal, too, had ambivalent feelings about rigor. At times he argued that the proper "finesse" rather than geometrical logic is what is needed to do the correct thing just as the religious appreciation of grace is above reason. The paradoxes of geometry that appeared in the calculus are like the apparent absurdities of Christianity, and the indivisible in geometry has the same worth as our justice has in comparison to God's.

Moreover, the correctness of ideas is often determined by the heart (Chapter II). He said in his *Pensées,* "We know the truth not only by reason but also by the heart. It is from this last source that we know the first principles and it is in vain that reason which has no part in it attempts to combat it. . . . And it is on our knowledge of the heart and instinct that reason necessarily rests and that it founds on them all its discourse." Of course, Pascal contributed nothing to the clarification of Cavalieri's method.

The biggest contributors to the creation of the calculus were Newton and Leibniz. Newton did very little with the integral concept, but he did use the derivative intensively. His method of obtaining the derivative was essentially that of Fermat. However, he was no clearer about the logical justification of this basic concept. He wrote three papers on the calculus, and he issued three editions of his masterpiece, *The Mathematical Principles of Natural Philosophy.* In the first paper (1669), in which he presents his method of finding the derivative, he remarked that the method is "shortly explained rather than accurately demonstrated." Here he used the fact that h and k are indivisibles. In the second paper (1671), he professed to do better because, he said, he had changed his point of view toward variables and instead of regarding them as changing by discrete amounts, in which case the h's are ultimate, indivisible units, he now regarded them as changing continuously. He said that he had removed the harshness of the doctrine of the indivisibles which he employed in his first paper. However, the process of calculating a fluxion, his term for the derivative, is essentially the same and the logic no better than in the first paper.

In his "Quadrature of Curves," the third paper on the calculus (1676), Newton repeated that he had abandoned the infinitely small quantity (ultimate indivisibles). He then criticized the dropping of terms involving what we have denoted by h in (3) above, for, he said, "in mathematics the minutest errors are not to be neglected." He then proceeded to give a new explanation of what he meant by a fluxion. "Fluxions are, as near as we please, as the increments of fluents [variables] generated in times, equal and as small as possible, and to speak accurately, they are in the prime ratio of nascent increments. . . ." Of course such vague phraseology didn't help much. As to his method of calculating a fluxion, Newton's third paper is logically as crude as the first one. He dropped all terms involving powers of h higher than the first, such as the h^2 in (2), and thereby obtained the derivative.

In his great work, *The Mathematical Principles of Natural Philosophy* (1st edition, 1687), Newton made several statements concerning his fluxions. He rejected ultimate indivisible quantities in favor of "evanescent divisible quantities," quantities which can be diminished without end. In the first and third editions of the *Principles,* Newton said:

> Ultimate ratios in which quantities vanish are not, strictly speaking, ratios of ultimate quantities, but limits to which the ratios of these quantities decreasing without limit, approach, and which, though they can come nearer than any given difference whatever, they can neither pass over nor attain before the quantities have diminished indefinitely.

Though by no means clear, this is the clearest statement Newton ever gave as to the meaning of his fluxions. Newton did touch here on the key word "limits," but he did not pursue the concept.

Undoubtedly he realized that his explanation of a fluxion was not satisfactory and so, perhaps in desperation, he resorted to physical meaning. He said in the *Principles:*

> Perhaps it may be objected that there is no ultimate ratio of evanescent quantities because the ratio before the quantities have vanished is not ultimate; and when they have vanished is none. But by the same argument, it might as well be maintained, that there is no ultimate velocity of a body arriving at a certain place, when its motion is ended: because the velocity before the body arrives at its ultimate place is not its ultimate velocity; when it has arrived is none. But the answer is easy; for by the ultimate velocity is meant that by which the body is moved, neither before it arrives at its last place and the motion ceases, nor after, but at the very instant it arrives; that is, the velocity with which the body arrives at its last place, and with which the motion ceases. And, in like manner, by the ultimate ratio of evanescent quantities is to be understood the ratio of the quantities not before they vanish, nor afterwards, but with which they vanish.

Since the results of his mathematical work were physically true, Newton spent very little time on the logical foundations of the calculus. In the *Principles,* he used geometric methods and he gave theorems on limits in geometric form. He admitted much later that he used analysis to find the theorems in the *Principles,* but he formulated the proofs geometrically to make the arguments as secure as those of the ancients. Of

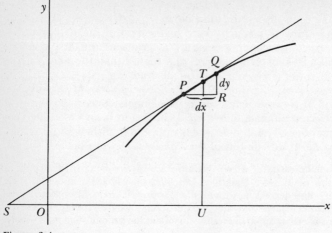

Figure 6.4

course, the geometric proofs were not at all rigorous. Newton had faith in Euclidean geometry but no factual evidence that it could support the calculus.

Leibniz's approach to the calculus was somewhat different. He argued that as our h and k decreased (he wrote dx and dy) they reached values which are "vanishingly small" or "infinitely small." At this stage, h and k were not zero but yet were smaller than any given number. Hence any powers of h such as h^2 or h^3 could surely be neglected, and the resulting ratio of k/h was indeed the quantity sought, that is, the derivative, which he denoted by the quotient dy/dx.

Geometrically, Leibniz described h and k as follows. When P and Q are "infinitely near" points on a curve, then dx is the difference in their abscissas and dy is the difference in their ordinates (Fig. 6.4). Moreover, the tangent line at T then coincides with the arc PQ. Hence dy divided by dx is the slope of the tangent. The triangle PQR, called the characteristic triangle, was not original with Leibniz. It had been used earlier by Pascal and Barrow, whose works Leibniz had studied. Leibniz also maintained that triangle PQR is similar to triangle STU and used that fact to prove some results about dy/dx.

Leibniz also worked extensively with the integral concept, and he arrived independently at the idea of summing rectangles as in (7) above. However, the passage from the sum of a finite number of rectangles to the infinite number was vague. He argued that, when h becomes "infinitely small," then the infinite sum is attained. This he indicated by $\int y dx$.

He did manage to evaluate such integrals and in fact discovered independently what is now called the fundamental theorem of the calculus, which shows that such sums can be obtained by reversing the process of finding the derivative (antidifferentiation). After about twelve years of struggling with his version of the calculus, he published his first paper on the subject in the *Acta eruditorum* of 1684. Perhaps the most appropriate comment on this paper was made by his friends, the Bernoulli brothers James and John. They described it as "an enigma rather than an explication."

Newton's and Leibniz's ideas were not clear, and both received criticism. Whereas Newton did not respond to critics, Leibniz did. His attempts to explain particularly his notion of infinitesimals (infinitely small numbers) are so numerous that many pages can be devoted to them. In an article in the *Acta* of 1689, he said that infinitesimals are not real but fictitious numbers. However, these fictitious or ideal numbers, he asserted, are governed by the same laws as ordinary numbers.

Alternatively, in this article he appealed to geometry to argue that a higher differential (infinitesimal) such as $(dx)^2$ is to a lower one, dx, as a point is to a line and that dx is to x as a point is to the earth or as the radius of the earth is to that of the heavens. The ratio of two infinitesimals he thought of as a quotient of unassignables or indefinitely small quantities but a ratio which could be expressed in terms of definite quantities; for example, geometrically this ratio of dy to dx is the ratio of ordinate to subtangent (TU to SU in Fig. 6.4).

Leibniz's work was criticized by Bernhard Nieuwentijdt (1654–1718), and Leibniz responded in an article in the *Acta* of 1695. There he attacked "over-precise" critics and said that excessive scrupulousness should not cause us to reject the fruits of invention. Leibniz then said that his method differed from Archimedes' only in the expressions used but that his own expressions were better adapted to the art of discovery. The terms "infinite" and "infinitesimal" merely signified quantities that one can take as large or as small as one wishes in order to show that the error incurred is less than any number that can be assigned; in other words, that there is no error. One can use these ultimate things—that is, "actually infinite" and "infinitely small" quantities—as a tool, much as algebraists use imaginary roots with great profit. (We should recall here the status of imaginary numbers in Leibniz's time.)

In a letter to Wallis of 1699, Leibniz gave a somewhat different explanation:

> It is useful to consider quantities infinitely small such that when their ratio is sought, they may not be considered zero but which are rejected as often as they occur with quantities incomparably greater. Thus if we have $x + dx$, dx is rejected. But it is different if we seek the difference

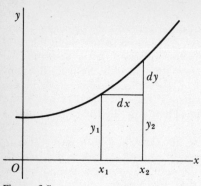

Figure 6.5

> between $x + dx$ and x. Similarly we cannot have $x\ dx$ and $dx \cdot dx$ standing
> together. Hence if we are to differentiate [find the derivative of] xy we
> write $(x + dx)\ (y + dy) - xy = x\ dy + y\ dx + dx\ dy$. But here $dx\ dy$ is to be
> rejected as incomparably less than $x\ dy + y\ dx$. Thus, in any particular
> case, the error is less than any finite quantity.

Thus far Leibniz argued that his calculus used only legitimate mathematical concepts. But since he could not satisfy his critics, he enun-
ciated a philosophical principle known as the law of continuity, which
was practically the same as one already stated by Kepler. As Leibniz
stated it early in his work on the calculus in a letter of March 19, 1678,
to Herman Conring, the principle affirmed that, "If a variable at all
stages enjoys a certain property, its limit will enjoy the same property."

In 1687 in a letter to Pierre Bayle, Leibniz expressed this principle
more fully: "In any supposed transition, ending in any terminus, it is
permissible to institute a general reasoning, in which the final terminus
may also be included." He then applied the principle to the calculation
of dy/dx for the parabola $y = x^2$. After obtaining $dy/dx = 2x + dx$, he said,
"Now, since by our postulate it is permissible to include under the one
general reasoning the case also in which (Fig. 6.5) the ordinate x_2y_2 is
moved up nearer and nearer to the fixed ordinate x_1y_1 until it ultimately coincides with it; it is evident that in this case dx becomes equal
to 0 and should be neglected. . . ." Leibniz did not say what meaning
should be given to the dx and dy that appear at the left side of the equation when dx is 0.

Of course, he said, things which are absolutely equal have a difference which is absolutely nothing.

> Yet a state of transition may be imagined, or one of evanescence, in
> which indeed there has not yet arisen exact equality or rest, . . . but in

which it is passing into such a state, that the difference is less than any assignable quantity; also that in this state there will still remain some difference, some velocity, some angle, but in each case one that is infinitely small. . . .

For the present, whether such a state of instantaneous transition from inequality or equality . . . can be sustained in a rigorous or metaphysical sense, or whether infinite extensions successively greater and greater, or infinitely small ones successively less and less, are legitimate considerations, is a matter that I own to be possibly open to question. . . .

It will be sufficient if, when we speak of infinitely great (or more strictly unlimited) or of infinitely small quantities (i.e., the very least of those within our knowledge), it is understood that we mean quantities that are indefinitely great or indefinitely small, i.e., as great as you please, or as small as you please, so that the error that any one may assign may be less than a certain assigned quantity.

On these suppositions, all the rules of our algorithm, as set out in the *Acta eruditorum* for October 1684, can be proved without much trouble.

Leibniz then discussed these rules but added nothing in the way of clarification.

Leibniz's principle of continuity certainly was not and is not today a mathematical axiom. Nevertheless he emphasized it and gave many arguments based on this principle. For example, in a letter to Wallis in 1698, Leibniz defended his use of the characteristic triangle (Fig. 6.4) as a form without magnitude, where form remains after the magnitudes have been reduced to zero, and challengingly asked, "Who does not admit a form without magnitude?" Similarly, in a letter to Guido Grandi in 1713, he said the infinitely small is not a simple and absolute zero but a relative zero, that is, an evanescent quantity which yet retains the character of that which is disappearing. However, Leibniz also said at other times that he did not believe in magnitudes truly infinite or truly infinitesimal.

Until the end of his life in 1716 Leibniz continued to make explanations of what his infinitely small quantities (infinitesimals, differentials) and his infinitely large quantities were. These explanations were no better than what we have seen above. He had no clear concepts or logical justifications of his calculus.

That Newton and Leibniz should be so crude in their reasoning is rather surprising. Before they tackled the calculus, other great mathematicians had already made considerable progress, and both men had read the works of many of their predecessors. In fact, Newton's teacher, Isaac Barrow, already had a number of fundamental results in geometrical form. When Newton said, "If I have seen farther than others it is because I have stood on the shoulders of giants," it was not mere modesty but fact. As for Leibniz, he was one of the great intellects. We have already cited his contributions to many fields (Chapter III). He ranks with Aristotle in intellectual strength and breadth. Of course, the calculus involved new and very subtle ideas, and the best of creative minds do not necessarily grasp even their own creations firmly.

Not able fully to clarify concepts and justify operations, both men relied upon the fecundity of their methods and the coherence of their results and pushed ahead with vigor but without rigor. Leibniz, less concerned about rigor though more responsive to critics than Newton, felt that the ultimate justification of his procedures lay in their effectiveness. He stressed the procedural or algorithmic value of what he had created. Somehow he had confidence that if he formulated clearly the rules of operation and that if these were properly applied, reasonable and correct results would be obtained, however vague the meanings of the concepts involved might be. Like Descartes, he was a man of vision who thought in broad terms. He saw the long-term implications of new ideas and did not hesitate to declare that a new science was coming to light.

The foundations of the calculus remained unclear. The proponents of Newton's work continued to speak of prime and ultimate ratios while the followers of Leibniz used infinitesimals, the infinitely small, non-zero quantities. The existence of these dissimilar approaches added to the difficulty of erecting the proper logical foundations. Moreover, many of the English mathematicians, perhaps because they were in the main still tied to Greek geometry, were more concerned with rigor and so distrusted both approaches to the calculus. Other British mathematicians chose to study Newton instead of mathematics and so made no progress toward rigorization. Thus the 17th century ended with the calculus, as well as arithmetic and algebra, in a muddled state.

Despite the muddle, uneasiness, and some opposition, the great 18th-century mathematicians not only vastly extended the calculus but derived entirely new subjects from it: infinite series, ordinary and partial differential equations, differential geometry, the calculus of variations, and the theory of functions of a complex variable, subjects which are at the heart of mathematics today and are collectively referred to as analysis. Even the doubters and critics used the various types of numbers and the algebraic and calculus processes freely in these extensions as though there were no longer any problem of logical foundations.

The extension of the calculus to new branches introduced new concepts and methodologies which compounded the problem of rigorizing the subject. The treatment of infinite series may serve to illustrate the additional complications. Let us note first just what problems infinite series presented to mathematicians.

The function $1/(1+x)$ can be written as $(1+x)^{-1}$ and, by applying the binomial theorem to the latter form, one finds that

$$(8) \qquad \frac{1}{1+x} = (1+x)^{-1} = 1 - x + x^2 - x^3 + x^4 \cdots,$$

wherein the dots indicate that the terms continue indefinitely and follow the pattern indicated by the few terms already shown. Now the original intent in introducing infinite series into the calculus was to use them in place of the functions for operations, such as differentiation (finding derivatives) and antidifferentiation because technically it is easier to work with the simpler terms of the series. Moreover, series for functions such as sin x were used to calculate the values of the functions. In all these uses, it is important to know that the series is the equivalent of the function. Now functions have numerical values when values are assigned to x. The first question then one must raise about the series is what value does it yield for a given value of x. In other words, what do we mean by, and how do we obtain, the sum of an infinite series? The second question is whether the series represents the function for all values of x, or at least for all values for which the function has meaning.

In his first paper on the calculus (1669), Newton proudly introduced the use of infinite series to expedite the processes of the calculus. Thus, to integrate (antidifferentiate) $y = 1/(1+x^2)$, he used the binomial theorem to obtain

$$y = 1 - x^2 + x^4 - x^6 + x^8 - \cdots,$$

and integrated term by term. He noted that if, instead, one writes the same function as $y = 1/(x^2 + 1)$, one obtains by means of the binomial theorem

$$y = \frac{1}{x^2} - \frac{1}{x^4} + \frac{1}{x^6} - \frac{1}{x^8} + \cdots.$$

He then remarked that when x is small enough, the first expansion is to be used, but when x is large, the second one is to be used. Thus he was somewhat aware that what we now call convergence is important, but he had no precise notion about this.

The justification given by Newton for his use of infinite series exemplifies the logic of the time. He said in this paper of 1669:

> Whatever the common Analysis [algebra] performs by Means of Equations of a finite Number of Terms (provided that can be done) this [new analysis] can always perform the same by Means of infinite Equa-

tions [series] so that I have not made any question of giving this the name of Analysis likewise. For the reasonings in this are no less certain than in the other; nor the equations less exact; albeit we Mortals whose reasoning powers are confined within narrow limits, can neither express, nor so conceive all the Terms of these Equations, as to know exactly from thence the quantities we want.

Thus for Newton infinite series were just a part of algebra, a higher algebra which dealt with an infinite instead of a finite number of terms.

As Newton, Leibniz, the several Bernoullis, Euler, d'Alembert, Lagrange, and other 18th-century men struggled with the strange problem of infinite series and employed them in analysis, they perpetrated all sorts of blunders, made false proofs, and drew incorrect conclusions; they even gave arguments that now with hindsight we are obliged to call ludicrous. A brief examination of some of the arguments may show the bewilderment and the confusion in their handling of infinite series.

When $x = 1$, the series (8) which represents $1/(1 + x)$,

$$(8) \qquad \frac{1}{1+x} = 1 - x + x^2 - x^3 + x^4 \cdots$$

becomes

$$1 - 1 + 1 - 1 + 1 \cdots.$$

The question of what the sum of this series is engendered endless disputation. It seemed clear that by writing the series as

$$(1 - 1) + (1 - 1) + (1 - 1) \cdots$$

the sum should be 0. It seemed equally clear, however, that by writing it as

$$1 - (1 - 1) - (1 - 1) \cdots$$

the sum should be 1. But it is also true that if S denotes the sum of the series, then

$$S = 1 - (1 - 1 + 1 - 1 + \cdots)$$

or

$$S = 1 - S.$$

Hence $S = \frac{1}{2}$. This last-mentioned result was supported by another argument. The series is geometric with common ratio -1, and the sum of an infinite geometric series whose first term is a and whose common ratio is r is $a/(1 - r)$. In the present case then the sum is $1/[1 - (-1)]$ or $\frac{1}{2}$.

Guido Grandi (1671–1742), in his little book *Quadratura circuli et hyperbolae* (*Quadrature of Circles and Hyperbolas*, 1703), obtained the third result, ½, by another method. He set $x = 1$ in equation (8) and obtained

$$\frac{1}{2} = 1 - 1 + 1 - 1 \cdots .$$

Grandi therefore maintained that 1/2 was the sum of the series. He also argued contrariwise that the sum was 0, and so he had proved that the world could be created out of nothing.

In a letter to Christian Wolf published in the *Acta eruditorum* of 1713, Leibniz treated the same series. He agreed with Grandi's result but thought that it should be possible to obtain it without resorting to the original function. Instead Leibniz argued that if one takes the first term, the sum of the first two, the sum of the first three, and so forth, one obtains 1, 0, 1, 0, 1, $\cdots$. Thus 1 and 0 are equally probable; one should therefore take as the sum the arithmetic mean, namely 1/2, which is also the most probable value. This argument was accepted by James, John, and Daniel Bernoulli and Lagrange. Leibniz conceded that his argument was more metaphysical than mathematical but went on to say that there is more metaphysical truth in mathematics than is generally recognized.

In a letter of 1745 and a paper of 1754/55, Euler took up the summation of series. A series in which by continually adding terms we approach closer and closer to a fixed number is said to be convergent and the fixed number is its sum. This according to Euler will happen when the terms continually decrease. A series whose terms do not decrease and may even increase is divergent and since this type, he said, comes from known explicit functions, one can take the value of the function as the sum of the series.

Euler's theory led to additional problems. He took up the expansion:

$$\frac{1}{(1+x)^2} = (1+x)^{-2} = 1 - 2x + 3x^2 - 4x^3 + \cdots .$$

For $x = -1$, he obtained

$$\infty = 1 + 2 + 3 + 4 + \cdots .$$

This sum seemed reasonable. However, Euler then considered the series for $1/(1-x)$, namely,

$$\frac{1}{(1-x)} = 1 + x + x^2 + x^3 + \cdots ,$$

and let $x = 2$. Then

$$-1 = 1 + 2 + 4 + 8 + \cdots .$$

Since the sum of the right-hand side of this series should exceed the sum of the preceding one, Euler concluded that -1 is larger than infinity. Some of Euler's contemporaries argued that negative numbers larger than infinity are different from those less than zero. Euler objected and argued that ∞ separates positive and negative numbers just as zero does.

Euler's views on convergence and divergence were not sound. Even in his day, series were known whose terms continually decrease but which do not have a sum in his sense. Also he himself worked with series that do not come from explicit functions. Hence his "theory" was incomplete. Further, in a letter (now lost) of 1743, Nicholas Bernoulli (1687–1759) must have pointed out to Euler that the same series may come from different expressions and so, according to Euler's definition, one would have to give the sums of these series different values. But Euler replied (in a letter to Goldbach in 1745) that Bernoulli gave no examples and he himself did not believe that the same series could come from two truly different algebraic expressions. However, Jean-Charles Callet (1744–1799) did give an example of the same series coming from two different functions, which Lagrange tried to brush aside by an argument that was later seen to be fallacious.

Euler's treatment of infinite series was inadequate for additional reasons. Series are differentiated and integrated, and the fact that the differentiation and integration of a series also yield the derivative and antiderivative of the function must be justified. Nevertheless, Euler declared, "Whenever an infinite series is obtained as the development of some closed expression [formula for a function], it may be used in mathematical operations as the equivalent of that expression, even for values of the variable for which the series diverges." Thus, he said, we can preserve the utility of divergent series and defend their use from all objections.

Other 18th-century mathematicians also recognized that a distinction must be made between what we now call convergent and divergent series, though they were not at all clear as to what the distinction should be. The source of the difficulty was of course that they were dealing with a new concept, and like all pioneers they had to struggle to clear the forest. Certainly, the initial thought of Newton, adopted by Leibniz, Euler, and Lagrange—that series are just long polynomials and so belong in the domain of algebra—could not serve to rigorize the work with series.

The formal view dominated 18th-century work on infinite series. Mathematicians even resented any limitations on their procedures, such as the need to think about convergence. Their work produced useful results and they were satisfied with this pragmatic sanction.

They did exceed the bounds of what they could justify, but they were on the whole prudent in their use of divergent series.

Though the logic of the number system and of algebra was in no better shape than that of the calculus, mathematicians concentrated their attacks on the calculus and attempted to remedy the looseness there. The reason for this is undoubtedly that the various types of numbers appeared familiar and more natural by 1700, whereas concepts of the calculus, still strange and mysterious, seemed less acceptable. In addition, while no contradictions arose from the use of numbers, contradictions did arise from the use of the calculus and its extensions to infinite series and the other branches of analysis.

Newton's approach to the calculus was potentially easier to rigorize than Leibniz's, though Leibniz's methodology was more fluid and more convenient for application. The English still thought they could secure rigor for both approaches by tying them to Euclidean geometry. But they also confused Newton's moments (his indivisible increments) and his use of continuous variables. The Continentals followed Leibniz and tried to rigorize his concept of differentials (infinitesimals). The books written to explain and justify Newton's and Leibniz's approaches to the calculus are too numerous and too misguided to warrant examination.*

While these efforts were being made to rigorize the calculus, some thinkers were attacking its soundness. The strongest attack was made by the philosopher Bishop George Berkeley (1685–1753), who feared the growing threat to religion of the mathematically inspired philosophy of mechanism and determinism. In 1734 he published *The Analyst Or a Discourse Addressed to an Infidel Mathematician.* [The infidel was Edmund Halley.] *Wherein It Is Examined Whether the Object, Principles, and Inferences of the Modern Analysis Are More Distinctly Conceived, or More Evidently Deduced, than Religious Mysteries and Points of Faith. "First Cast the Beam Out of Thine Own Eye; and Then Shalt Thou See Clearly To Cast Out the Mote Out of Thy Brother's Eye."* Berkeley rightly complained that the mathematicians were proceeding mysteriously and incomprehensibly: they did not give the logic or reasons for their steps. Berkeley criticized many arguments of Newton, and, in particular, he pointed out that Newton in his paper "Quadrature of Curves" (using x for the increment which we have denoted by h) performed some algebraic steps, and then dropped terms involving h because h was now 0. (Compare equation (4) above.) This, Berkeley said, was a defiance of the law of contradiction. Such reasoning would not be allowed in theology. He

---

* An account of these books can be found in Florian Cajori: *A History of the Conceptions of Limits and Fluxions in Great Britain from Newton to Woodhouse,* The Open Court Publishing Co., Chicago, 1915. Also Carl Boyer: *The Concepts of the Calculus,* reprint by Dover Publications, 1949; original edition, Columbia University Press, 1939.

said the first fluxions (first derivatives) appeared to exceed the capacity of man to understand because they were beyond the domain of the finite.

> And if the first [fluxions] are incomprehensible, what should we say of the second and third, [derivatives of derivatives] etc.? He who can conceive the beginning of a beginning, or the end of an end . . . may be perhaps sharpsighted enough to conceive these things. But most men will, I believe, find it impossible to understand them in any sense whatever. . . . He who can digest a second or third fluxion . . . need not, methinks, be squeamish about any point in Divinity.

Speaking of the vanishing of h and k, Berkeley said, "Certainly when we suppose the increments to vanish, we must suppose that their proportions, their expressions, and everything else derived from the supposition of their existence, to vanish with them." As for the derivatives propounded by Newton as the ratio of the evanescent quantities h and k, "They are neither finite quantities, nor quantities infinitely small, nor yet nothing. May we not call them the ghosts of departed quantities."

Berkeley was equally critical of Leibniz's approach. In an earlier work, *A Treatise Concerning the Principles of Human Knowledge* (1710, rev. ed. 1734), he attacked Leibniz's concepts.

> Some there are of great note who, not content with holding that finite lines may be divided into an infinite number of parts, do yet farther maintain that each of those infinitesimals is itself subdivisible into an infinity of other parts or infinitesimals of a second order $[(dx)^2]$ and so on *ad infinitum*. These, I say, assert there are infinitesimals of infinitesimals of infinitesimals, without ever coming to an end! . . . Others there be who hold all orders of infinitesimals below the first to be nothing at all. . . .

He continued his attack on Leibniz's ideas in his *Analyst*.

> Leibnitz and his followers in their *calculus differentialis* making no manner of scruple, first to suppose, and secondly to reject, quantities infinitely small; with what clearness in the apprehension and justness in the reasoning, any thinking man, who is not prejudiced in favor of these things, may easily discern.

The ratio of the differentials, Berkeley said, should geometrically determine the slope of the secant and not the slope of the tangent. One undoes this error by neglecting higher differentials. Thus "by virtue of a twofold mistake you arrive, though not at a science, yet at the truth," because errors were compensating for each other. He also picked on the second differential, Leibniz's $d(dx)$, which he said is the difference of a quantity dx that is itself the least discernible quantity.

With respect to both approaches Berkeley asked rhetorically "whether the mathematicians of the present age act like men of science in taking so much more pains to apply their principles than to understand them." "In every other science," he said, "men prove their conclusions by their principles, and not their principles by their conclusions."

Berkeley ended *The Analyst* with some queries, among which are these:

> Whether mathematicians who are so delicate in religious points, are strictly scrupulous in their own science? Whether they do not submit to authority, take things upon trust, and believe points inconceivable? Whether they have not their mysteries, and what is more, their repugnances and contradictions?

Dozens of mathematicians replied to Berkeley's criticisms and each attempted unsuccessfully to rigorize the calculus. The most important of these efforts was made by Euler. Euler rejected geometry entirely as a basis for the calculus and tried to work purely formally with functions, that is, to argue from their algebraic (analytic) representations. He denied Leibniz's concept of an infinitesimal, that is, a quantity which is less than any assignable magnitude and yet not zero. In his *Institutiones calculi differentialis* (Principles of Differential Calculus, 1755), an 18th-century classic on the calculus, he argued that:

> There is no doubt that every quantity can be diminished to such an extent that it vanishes completely and disappears. But an infinitely small quantity is nothing other than a vanishing quantity and therefore the thing itself equals 0. It is in harmony also with that definition of infinitely small things, by which the things are said to be less than each assignable quantity; it certainly would not be able not to be nothing; for unless it is equal to 0, an equal quantity can be assigned to itself, which is contrary to the hypothesis.

Since infinitesimals such as dx (in Leibniz's notation) are zero, so are $(dx)^2$, $(dx)^3$, etc., though, Euler said, because it is customary, one can speak of these as of higher order than dx. Then Leibniz's dy/dx which for Leibniz was a ratio of infinitesimals in his sense, was for Euler actually $0/0$. However, $0/0$, he maintained, could have many values. Euler's argument was that $n \cdot 0 = 0$ for *any* number n. Hence, if we divide by 0, we have $n = 0/0$. The usual process of finding the derivative determines the value of $0/0$ for the particular function involved. He illustrated this for the function $y = x^2$. He gave x the increment h (he used ω). At this stage presumably h is not zero. (Compare this with (1) through (4) above.) Consequently

$$\frac{k}{h} = 2x + h.$$

Where Leibniz allowed h to become infinitesimal but not zero, Euler said h was zero, and so in this case the ratio k/h, which is $0/0$, is found to equal $2x$.

Euler emphasized that these differentials, the ultimate values of h and k, were absolute zeros and that nothing could be inferred from them other than their mutual ratio, which was in the end evaluated as a finite quantity. There is more of this nature in the third chapter of the *Institutiones*. There he encouraged the reader by remarking that this notion does not hide so great a mystery as is commonly thought although in the mind of many it rendered the calculus suspect. Of course, Euler's justification of the process of finding the derivative was no sounder than that of Newton and Leibniz.

What Euler did contribute in his formalistic, incorrect approach to the calculus was to free it from geometry and base it on arithmetic and algebra. This step at least prepared the way for the ultimate justification of the calculus on the basis of number.

The most ambitious of the subsequent 18th-century attempts to build the foundations of the calculus was made by Lagrange. Like Berkeley and others, he believed that the correct results obtained by the calculus were due to the fact that errors were offsetting each other. He worked out his reconstruction in his *Théorie des fonctions analytiques* (1797; 2nd edition, 1813). The subtitle of his book is revealing: "Containing the principal theorems of the differential calculus without the use of the infinitely small, of vanishing quantities, of limits and fluxions, and reduced to the art of *algebraic analysis of finite quantities*" (italics added).

Lagrange criticized Newton's approach by pointing out that when Newton considered the limiting ratio of arc to chord, he considered chord and arc equal not before vanishing or after vanishing but *when* they vanished. Lagrange correctly pointed out:

> That method has the great inconvenience of considering quantities in the state in which they cease, so to speak, to be quantities; for though we can always well conceive the ratios of two quantities, as long as they remain finite, that ratio offers to the mind no clear and precise idea, as soon as its terms both become nothing at the same time.

He was equally dissatisfied with the infinitely small quantities of Leibniz and with the absolute zeros of Euler, both of "which although correct in reality are not sufficiently clear to serve as foundation of a science whose certitude should rest on its own evidence."

Lagrange wished to give the calculus all the rigor of the demonstrations of the ancients and he proposed to do this by reducing the calculus to algebra. Specifically, Lagrange proposed to use infinite series—which were regarded as part of algebra but whose logic was still more confused than that of the calculus—to found the logic of the calculus rigorously. With "modesty" he remarked that it was surprising that his [Lagrange's] method had not occurred to Newton.

We need not pursue the details of Lagrange's foundation of the calculus. Beyond using series in a manner totally unjustified, he performed a mass of algebraic steps which only made it more difficult for the reader to see that the proper definition of the derivative was lacking. Indeed, what he did was to obtain it in as crude a manner as any of his predecessors had. Lagrange believed that he had dispensed with the limit concept and had founded the calculus on algebra. Despite all his errors, Lagrange's foundation was accepted by a number of his outstanding successors.

The belief that the calculus was merely an extension of algebra was maintained by Sylvestre-François Lacroix (1765–1843) in an influential three-volume work of 1797–1800, wherein Lacroix followed Lagrange. In a briefer one-volume work, *An Elementary Treatise on the Differential and Integral Calculus* (1802), Lacroix used the theory of limits (insofar as that theory was understood at the time) but, Lacroix said, only to save space.

Some British mathematicians of the early 19th century decided to take over the superior Continental work on analysis. Charles Babbage (1792–1871), John Herschel (1792–1871), and George Peacock (1791–1858), who as undergraduates at Cambridge University founded the Analytical Society, translated Lacroix's briefer calculus. But the translators stated in their Preface:

> The work of Lacroix, of which a translation is now presented to the Public . . . may be considered as an abridgement of his great work in the Differential and Integral Calculus, although in the demonstration of the first principles, he has substituted the method of limits of d'Alembert, in the place of the most correct and natural method of Lagrange, which was adopted in the former. . . .

Peacock said the theory of limits was not acceptable because it separated the principles of the differential calculus from algebra. Herschel and Babbage agreed.

It was clear to the mathematical world of the late 18th century that proper foundations for the calculus were urgently needed, and at the suggestion of Lagrange the Mathematics section of the Berlin Academy of Sciences, of which he was director from 1766 to 1787, proposed in

1784 that a prize be awarded in 1786 for the best solution to the prob-lem of the infinite in mathematics. The announcement of the competi-tion read as follows:

> The utility derived from mathematics, the esteem it is held in, and the honorable name of "exact science" *par excellence* justly given it, are all the due of the clarity of its principles, the rigor of its proofs, and the precision of its theorems.
>
> In order to ensure the perpetuation of these valuable advantages in this elegant part of knowledge, there is needed *a clear and precise theory of what is called Infinite in Mathematics.*
>
> It is widely known that advanced Geometry [mathematics] regularly employs the *infinitely great* and the *infinitely small.* The geometers of an-tiquity, however, and even the ancient analysts, took pains to avoid anything approaching the infinite, whereas certain eminent modern analysts admit that the phrase *infinite magnitude* is a contradiction in terms.
>
> The Academy, therefore, desires an explanation of how it is that so many correct theorems have been deduced from a contradictory sup-position, together with enunciation of a sure, a clear, in short a truly mathematical principle that may properly be substituted for that of the *infinite* without, however, rendering investigations carried out by its means overly difficult or overly lengthy. It is required that the sub-ject be treated in all possible generality and with all possible rigor, clar-ity, and simplicity.

The competition was open to all comers with the exception of regu-lar members of the Academy. Twenty-three papers in all were submit-ted. The official finding on the outcome of the contest reads as follows:

> The Academy has received many essays on this subject. Their au-thors have all overlooked explaining *how so many correct theorems have been deduced from a contradictory supposition,* such as that of an infinite quantity. They have all, more or less, disregarded the qualities of *clar-ity, simplicity,* and above all *rigor* that were required. Most of them have not even perceived that the principle sought should not be restricted to the infinitesimal calculus, but should be extended to algebra and to geometry treated in the manner of the ancients.
>
> The feeling of the Academy, therefore, is that its query has not met with a full response.
>
> However, it has found that the entrant who comes closest to its in-tentions is the author of the French essay that has for its motto: "The Infinite is the abyss in which our thoughts are engulfed." The Acad-emy has, therefore, voted him the prize.

The winner was the Swiss mathematician, Simon L'Huillier. The Academy published his "Elementary Exposition of the Higher Calculus" in that same year, 1786. There is no doubt that the judgment made by the

Mathematics section of the Academy was essentially right. None of the other papers (except one by Carnot (Chapter VII)) made even so much as an attempt to explain how it was that infinitesimal analysis established theorems that are correct after starting from suppositions that are false. L'Huillier's undoubtedly stood out in its quality, although its fundamental idea was not in the slightest degree original. According to L'Huillier, his essay represented "the development of ideas . . . that M. d'Alembert had only sketched and, as it were, proposed in the article 'Différentiel' in the *Encyclopédie* and in his *Mélanges*." In the opening chapter of the Exposition, L'Huillier improved somewhat the theory of limits itself. He introduced for the first time in a printed text the symbol for a limit in the form lim. Thus he denoted the derivative dP/dx as lim $\Delta P/\Delta x$ (our k/h), but L'Huillier's contribution to the theory of limits was minuscule.

Though almost every mathematician of the 18th century made some effort or at least a pronouncement on the logic of the calculus and though one or two were on the right track, all of their efforts were unavailing. Any delicate questions were either ignored or overlooked. The distinction between a very large number and an infinite number was hardly made. It seemed clear that a theorem which held for any finite n must hold for n infinite. Likewise the difference quotient k/h (see (3)) was replaced by a derivative, and the sum of a finite number of terms, such as occur in (7), and an integral were hardly distinguished. They passed from one to the other freely. Their work could be summed up in Voltaire's description of the calculus as "the art of numbering and measuring exactly a Thing whose Existence cannot be conceived." The net effect of the century's efforts to rigorize the calculus, particularly those of giants such as Euler and Lagrange, was to confound and mislead their contemporaries and successors. They were, on the whole, so blatantly wrong that one could despair of mathematicians' ever clarifying the logic involved.

Mathematicians trusted the symbols far more than logic. Because infinite series had the same symbolic form for all values of x, the distinction between values of x for which series converged and values for which they diverged did not seem to demand attention. And even though it was recognized that some series, such as $1 + 2 + 3 \cdots$, had an infinite sum, mathematicians preferred to try to give meaning to the sum, rather than to question the applicability of a summation. Of course, they were fully aware of the need for some proofs. We have seen that Euler did try to justify his use of divergent series, and both he and Lagrange, among others, did attempt a foundation for the calculus. But the few efforts to achieve rigor—significant because they show that standards of rigor vary with the times—did not succeed in validating the work of the century and men almost willfully took the position that what cannot be cured must be endured.

One curious feature of the arguments used by 18th-century thinkers was their recourse to the term metaphysics. It was used to imply that there was a body of truths lying outside the domain of mathematics proper which could, if necessary, be called upon to justify their work, though just what these truths were was not clear. The appeal to metaphysics was meant to give credence to arguments that reason failed to support. Thus, Leibniz asserted that metaphysics is of more use in mathematics than we realize. His argument for taking $\frac{1}{2}$ to be the sum of the series $1 - 1 + 1 - \cdots$ and his principle of continuity, neither of which had more to recommend it than Leibniz's own assertion, were "justified" as metaphysical as though this "justification" placed them beyond dispute. Euler, too, appealed to metaphysics and argued that we must acquiesce to it in analysis. When they could provide no better argument for an assertion, 17th- and 18th-century mathematicians were wont to say the reason was metaphysical.

And so the 18th century ended with the logic of the calculus and of the branches of analysis built on the calculus in a totally confused state. In fact, one could say that the state of the foundations was worse in 1800 than in 1700. Giants, notably Euler and Lagrange, had given incorrect logical foundations. Because these men were authorities, many of their colleagues accepted and repeated uncritically what they proposed and even built more analysis on their foundations. Other lesser lights were not quite satisfied with what the masters had advanced, but they were confident that a totally clear foundation could be secured merely by clarification or minor emendation. Of course they were being led down false trails.

VII

The Illogical Development:
The Predicament *circa* 1800

> Ah, Why, ye Gods, should two and two make four?
>
> ALEXANDER POPE

By 1800, mathematics was in a highly paradoxical situation. Its successes in representing and predicting physical phenomena were beyond all expectation superlative. On the other hand, as many 18th-century men had already pointed out, the massive structure had no logical foundation, and there was therefore no assurance that the mathematics was correct. This paradoxical state continued in the first half of the 19th century. While many mathematicians ploughed ahead into new areas of physical science and achieved still greater successes, the logical foundation was not tackled. Rather, criticisms of negative and complex numbers, algebra, and the calculus and its extensions to what is called analysis continued.

Let us examine the predicament in the early 19th century. We can pass quickly over the few objections that were still made to the use of irrational numbers. These, as mentioned earlier, could be thought of as points on a line; intuitively they were not much more difficult to accept than whole numbers and fractions, and they obeyed the same laws as whole numbers and fractions. As to their utility, there could be no question. Hence, though there was no logical basis for irrational numbers, they were accepted. The troublesome and intuitively unacceptable elements were the negative numbers and the complex numbers. They were attacked and rejected in the 19th century with the same virulence as in previous centuries.

William Frend (1757–1841), father-in-law of Augustus De Morgan and a fellow of Jesus College, Cambridge, declared outright in the Preface to his *Principles of Algebra* (1796):

[A number] submits to be taken away from a number greater than itself but to attempt to take it away from a number less than itself is ridiculous. Yet this is attempted by algebraists who talk of a number less than nothing; of multiplying a negative number into a negative number and thus producing a positive number; of a number being imaginary. Hence they talk of two roots to every equation of the second order [degree], and the learner is to try which will succeed in a given equation; they talk of solving an equation, which requires two impossible roots to make it soluble; they can find out some impossible numbers, which, being multiplied together, produce unity. This is all jargon, at which common sense recoils; but, from its having been once adopted, like many other figments, it finds the most strenuous supporters among those who love to take things upon trust and hate the color of a serious thought.

In an essay included in a book by Baron Masères (Chapter V) published in 1800, Frend criticized the general rule that an equation has as many roots as its degree. He said this is true of only a few equations, and of course he cited those which necessarily have all positive roots. He then said of the mathematicians who accept the general rule, "they find themselves under a necessity of giving specious names to a parcel of quantities which they endeavor to make pass for roots of these equations, though in truth they are not so, in order to cover the falsehood of their general proposition, and give it, in words at least, an appearance of truth . . ."

Lazare N. M. Carnot (1753–1823), a well-known French geometer, was influential beyond his own original contributions by virtue of his *Reflections on the Metaphysics of Infinitesimal Calculus* (1797; 2nd edition revised, 1813), which was translated into many languages. He asserted flatly that the notion of something less than nothing was absurd. Negative numbers could be introduced into algebra as fictitious entities useful for calculation. However, they were certainly not quantities and could lead to erroneous conclusions.

The 18th-century controversies on the logarithms of negative and complex numbers had bewildered mathematicians so much that even in the 19th century they were impelled to question both negative and complex numbers. In 1801, Robert Woodhouse of Cambridge University published a paper "On the Necessary Truth of Certain Conclusions Obtained by Means of Imaginary Quantities." He said, "The paradoxes and contradictions mutually alleged against each other, by mathematicians engaged in the controversy concerning the application of logarithms to negative and impossible quantities, may be employed as arguments against the use of those quantities in investigation."

Cauchy, certainly one of the greatest mathematicians and the man who founded the theory of functions of a complex variable during the first few decades of the 19th century, refused to treat expressions such as $a + b\sqrt{-1}$ as numbers. In his famous *Cours d'analyse* (Course on Analysis, 1821), he said such expressions regarded as a totality have no meaning. However, they do say something about the real numbers a and b. For example, the equation

$$a + b\sqrt{-1} = c + d\sqrt{-1}$$

tells us that $a = c$ and $b = d$. "Every imaginary equation is only the symbolic representation of two equations between real quantities." As late as 1847 he gave a rather involved theory which would justify operations with complex numbers but avoid the use of $\sqrt{-1}$, a quantity which, he said, "we can repudiate completely and which we can abandon without regret because one does not know what this pretended sign signifies nor what sense one ought to attribute to it."

In 1831, Augustus De Morgan (1806–1871), a famous mathematical logician and a contributor to algebra, expressed his objections to both negative and complex numbers in his book *On the Study and Difficulties of Mathematics*. He said, incidentally, that his book contained nothing that could not be found in the best works then in use at Oxford and Cambridge.

> The imaginary expression $\sqrt{-a}$ and the negative expression $-b$ have this resemblance, that either of them occurring as the solution of a problem indicates some inconsistency or absurdity. As far as real meaning is concerned, both are equally imaginary, since $0 - a$ is as inconceivable as $\sqrt{-a}$.

Then De Morgan illustrates by means of a problem. A father is 56; his son is 29. When will the father be twice as old as the son? He solves $56 + x = 2(29 + x)$ and obtains $x = -2$. This result, he says, is absurd. But, he continues, if we change x to $-x$ and solve $56 - x = 2(29 - x)$, we get $x = 2$. He concludes that we phrased the original problem wrongly. The negative answer shows that we made a mistake in the first formulation of the equation.

Then turning to complex numbers, he stated:

> We have shown the symbol $\sqrt{-a}$ to be void of meaning, or rather self-contradictory and absurd. Nevertheless, by means of such symbols, a part of algebra is established which is of great utility. It depends upon the fact, which must be verified by experience that the common rules of algebra may be applied to these expressions [complex numbers] without leading to any false results. An appeal to experience of this nature appears to be contrary to the first principles laid down at the beginning of this work. We cannot deny that it is so in reality, but it

must be recollected that this is but a small and isolated part of an immense subject, to all other branches of which these principles apply in their fullest extent. [The principles he referred to are that mathematical truths follow necessarily by deductive reasoning from axioms.]

He then compared negative roots and complex roots:

There is, then, this distinct difference between the negative and the imaginary result. When the answer to a problem is negative, by changing the sign of x in the equation which produced that result, we may either discover an error in the method of forming that equation or show that the question of the problem is too limited and may be extended so as to admit of a satisfactory answer. When the answer to a problem is imaginary this is not the case.

Somewhat later in the book he said:

We are not advocates for stopping the progress of the student by entering fully into all the arguments for and against such questions, as the use of negative quantities, etc., which he could not understand, and which are inconclusive on both sides; but he might be made aware that a difficulty does exist, the nature of which might be pointed out to him, and he might then, by the consideration of a sufficient number of examples, treated separately, acquire confidence in the results to which the rules lead.

One of the great mathematicians, whose work in other areas we have had occasion to treat, William R. Hamilton, was no more inclined to accept negative and complex numbers. In a paper of 1837 he expressed his objections:

But it requires no peculiar scepticism to doubt, or even to disbelieve, the doctrine of Negatives and Imaginaries, when set forth (as it has commonly been) with principles like these: that a *greater magnitude may be subtracted from a less,* and that the remainder is *less than nothing;* that *two negative numbers,* or numbers denoting magnitudes each less than nothing, may be *multiplied* the one by the other, and that the product will be a *positive* number, or a number denoting a magnitude greater than nothing; and that although the *square* of a number, or the product obtained by multiplying that number by itself, is therefore *always positive,* whether the number be positive or negative, yet that numbers, called *imaginary,* can be found or conceived or determined, and operated on by all the rules of positive and negative numbers, as if they were subject to those rules, *although they have negative squares,* and must therefore be supposed to be themselves neither positive nor negative, nor yet null numbers, so that the magnitudes which they are supposed to denote can neither be greater than nothing, nor less than nothing, nor even equal to nothing. It must be hard to found a sci-ENCE on such grounds as these, though the forms of logic may build up from them a symmetrical system of expressions, and a practical art may be learned of rightly applying useful rules which seem to depend upon them.*

George Boole (1815–1864), who shares honors with De Morgan as a logician, in *An Investigation of the Laws of Thought* (1854), said of $\sqrt{-1}$ that it is an uninterpretable symbol. However, by using it in trigonometry, we pass from interpretable expressions through uninterpretable ones into interpretable ones.

What reconciled mathematicians somewhat to complex numbers was not logic but the geometrical representation of Wessel, Argand, and Gauss (Chapter IV). Nevertheless, in Gauss's work we can still find evidence of his reluctance to accept complex numbers. Gauss gave four proofs of the fundamental theorem of algebra, which states that every polynomial equation of the nth degree has exactly n roots. In the first three proofs (1799, 1815, and 1816), he dealt with polynomials whose coefficients are real numbers, and, in addition, he presupposed a one-to-one correspondence of points of the Cartesian plane and complex numbers, though he did not define the correspondence explicitly. There was no actual plotting of $x + iy$ but rather of x and y as coordinates of a point in the real plane. Moreover, the proofs did not really use complex function theory because he separated the real and imaginary parts of the functions involved. He was more explicit in a letter to Bessel of 1811, wherein he wrote that $a + ib$ was represented by the point (a,b) and that one could go from one point to another of the complex plane by many paths. There is no doubt, if one judges from the thinking exhibited in these three proofs and in other unpublished works, that Gauss was still concerned about the status of complex numbers and functions. In a letter of December 11, 1825, he said he could not tear himself away from "the true metaphysics of negative and imaginary quantities. The true sense of $\sqrt{-1}$ is always pressingly present in my mind but it would be difficult to grasp it in words."

However, by 1831, if Gauss still possessed any scruples about his own or other mathematicians' acceptance of complex numbers, he had overcome them, and he publicly described the geometric representation of complex numbers. In papers of that year Gauss was very explicit. He not only gave the representation of $a + bi$ as a point in the complex plane but described the geometrical addition and multiplication of complex numbers (Chapter IV). He then pointed out that the fractions, negative numbers, and real numbers were then well understood, whereas complex numbers had just been tolerated despite their great

---

* We shall see in the next chapter Hamilton's resolution of the problem presented by complex numbers.

value. To many they appeared to be just a play with symbols. But "Here [in this geometrical representation] the demonstration of an intuitive meaning of $\sqrt{-1}$ is completely grounded and more is not needed in order to admit these quantities into the domain of the objects of arithmetic." Thus, Gauss himself was content with an intuitive understanding. He also said that if the units 1, -1, and $\sqrt{-1}$ had not been given the names positive, negative, and imaginary units but direct, inverse, and lateral, people would not have gotten the impression that there is some dark mystery in these numbers. The geometrical representation, he said, puts the true metaphysics of imaginary numbers in a new light. He introduced the term "complex numbers"—as opposed to Descartes's term "imaginary numbers"—and used i for $\sqrt{-1}$. Gauss did not comment on what was equally significant, namely, that he and his contemporaries used real numbers freely despite the fact that these had no logical foundation.

In a paper of 1849, about which we shall say a little more later, Gauss used complex numbers much more freely because, he said, by now everyone is familiar with them. But this was not so. Long after the theory of complex functions of a complex variable was developed principally by Cauchy in the first third of the 19th century and employed in hydrodynamics, Cambridge University professors preserved an invincible repulsion to the objectionable $\sqrt{-1}$, and cumbrous devices were adopted to avoid its occurrence or use wherever possible.

In the first half of the 19th century, the logical foundation of algebra was also noticeable by its absence. The problem in this area was that letters were used to represent all types of numbers and were manipulated as though they possessed all the familiar and intuitively acceptable properties of the positive integers. The results of these manipulations were correct when any numbers—negative, irrational, or complex—were substituted for the letters. But because these types of numbers were not really understood nor were their properties logically established, this use of letters was certainly unjustified. It seemed as though the algebra of literal expressions possessed a logic of its own, which accounted for its effectiveness and correctness. Hence in the 1830s, mathematicians tackled the problem of justifying operations with literal or symbolic expressions.

This problem was first considered by George Peacock (1791–1858), professor of mathematics at Cambridge University. He made the distinction between arithmetical algebra and symbolic algebra. The former dealt with symbols representing the positive integers and so was on solid ground. Here, only operations leading to positive integers were permissible. Symbolic algebra, Peacock argued, adopts the rules of arithmetical algebra but removes the limitation to positive integers.

All the results deduced in arithmetical algebra, whose expressions are general in form but particular in value, are correct results likewise in symbolic algebra where they are general in value as well as in form. Thus $ma + na = (m + n)a$ holds in arithmetical algebra when a, m, and n are positive integers, and it therefore holds in symbolic algebra for all a, m, and n. Likewise the binomial expansion of $(a + b)^n$ when n is a positive integer, if it is exhibited in a general form without reference to a final term, holds for all n. Peacock's argument, known as the principle of the permanence of equivalent forms, was set forth in 1833 in his "Report on the Recent Progress and Present State of Certain Branches of Analysis" to the British Association for the Advancement of Science. He dogmatically affirmed:

> Whatever algebraical forms are equivalent when the symbols are general in form but specific in value [positive integers], will be equivalent likewise when the symbols are general in value as well as in form.

Peacock used this principle to justify in particular operations with complex numbers. He did try to protect his meaning by the phrase "when the symbols are general in form." Thus, one could not state properties which belong only to 0 and 1, because these numbers have special properties.

In the second edition of his *Treatise on Algebra* (1842–45, 1st edition 1830), Peacock *derived* his principle from axioms. He stated explicitly that algebra like geometry is a deductive science. Hence the processes of algebra have to be based on a complete statement of the body of laws or axioms which dictate the operations used in the processes. The symbols for the operations have, at least for the deductive science of algebra, no sense other than those given to them by the laws. Thus, addition means no more than any process which obeys the laws of addition. His laws were, for example, the associative and commutative laws of addition and multiplication, and the law that if $ac = bc$ and $c \neq 0$ then $a = b$. And so the principle of permanence of form was justified here by the adoption of axioms.

Throughout most of the 19th century the view of algebra affirmed by Peacock was accepted. It was endorsed with minor modifications by Duncan F. Gregory (1813–1844), Augustus De Morgan, and Hermann Hankel (1839–1873).

The principle was essentially arbitrary and begged the question of why the various types of numbers possess the same properties as the whole numbers. It actually sanctioned by fiat what was empirically correct but not logically established. Evidently, Peacock, Gregory, and De Morgan thought that they could make a science out of algebra independent of the properties of real and complex numbers. Of course, labelling some rule of thumb a principle does not improve its logical stature. But, as Bishop Berkeley remarked, "Ancient and rooted prejudices do often pass into principles; and those propositions which once obtain the force and credit of a *principle,* are not only themselves, but likewise whatever is deducible from them, thought privileged from all examination."

The principle of permanence of form treats algebra as a science of symbols and their laws of combination. This foundation was both vague and inelastic. Its advocates insisted on a parallelism between arithmetical and general algebra so rigid that, if maintained, it would destroy the generality of algebra, and they never seem to have realized that a formula true with one interpretation of the symbols might not be true for another. It so happens that the principle was vitiated by the creation of quaternions because these numbers—the first of what are now called hypernumbers—do not possess the commutative property of multiplication (Chapter IV). Hence, letters standing for a class of hypernumbers do not possess all the properties of real and complex numbers. Thus the principle was false. What Peacock and his followers failed to appreciate but what soon became evident after the introduction of quaternions is that there is not one algebra but many, and the algebra built on the real and complex numbers could be justified only by proving that the numbers for which the letters stood possessed all the properties attributed to the letters.

Beyond algebra, analysis in the early 1800s was still in a logical fog. Lagrange's foundation for the calculus (Chapter VI) did not satisfy all mathematicians, and some returned to the position taken by Berkeley and others that errors were offsetting each other. This position was taken by Lazare N. M. Carnot, who was also a great military leader of the French Revolution, in his *Reflections on the Metaphysics of Infinitesimal Calculus.* His metaphysics "explains" that errors do compensate for each other. After much discussion of the various approaches to the calculus that had been presented, Carnot concluded that, though all methods as well as the use of d'Alembert's limit concept were really equivalent to the Greek method of exhaustion, infinitesimals were more expeditious. Carnot did make a contribution to the clarification of the concepts of the calculus but it was not major. Moreover, in relating the ideas of Newton, Leibniz, and d'Alembert to the Greek method of exhaustion he introduced a false note. There was nothing in Greek geometry or algebra related to the derivative.

The blunders in analysis continued well into the 19th century. Examples are numerous but perhaps one or two may suffice. Fundamental to all of analysis are the concepts of a continuous function and the derivative of a function. Intuitively a continuous function is one which

Figure 7.1

can be represented by a curve drawn with an uninterrupted motion of a pencil (Fig. 7.1). The geometrical meaning of the derivative of such a function is the slope of the tangent at any point P of the curve. On an intuitive basis, it seemed clear that a continuous function must have a derivative at each point. However, some of the early 19th-century mathematicians were above using such intuitive evidence, and they set out to prove by logical arguments whatever they could.

Unfortunately, as Figure 7.2 shows, a continuous function that has what are now called corners, as at A, B, and C, does not possess derivatives at such points. Nevertheless, in 1806 André-Marie Ampère (1775–1836) "proved" that every function has a derivative at all points where it is continuous. Other or similar "proofs" were given by Lacroix in his famous three-volume work *Treatise on Differential and Integral Calculus* (2nd edition, 1810–19), and by almost all the leading texts of the 19th century. Joseph L. F. Bertrand (1822–1900) "proved" the differentiability in papers as late as 1875. All of these "proofs" were wrong. Some of these mathematicians can be excused on the ground that the concept of a function was for a long time not well defined, but by 1830 or so this failing was remedied.

When one considers that continuity and differentiability are the basic concepts in all of analysis and that analysis has been the major field of activity from about 1650 to the present, one can only be shocked to learn how vague and uncertain mathematicians were about these concepts. The mistakes were so gross that they would be inexcusable in an undergraduate mathematics student today; yet they were made by the most famous men—Fourier, Cauchy, Galois, Legendre, Gauss—and also by a multitude of lesser lights who were, nevertheless, leading mathematicians of their times.

Texts of the 19th century continued to use freely terms such as dif-

Figure 7.2

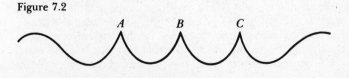

ferential and infinitesimal whose meanings were unclear or inconsistently defined to be both non-zero and zero. Students of the calculus remained perplexed, and the best they could do was to follow the advice of d'Alembert: "Persist and faith will come to you." Bertrand Russell, who attended Trinity College, Cambridge University from 1890 to 1894, wrote in *My Philosophical Development,* "Those who taught me the infinitesimal calculus did not know the valid proofs of its fundamental theorems and tried to persuade me to accept the official sophistries as an act of faith."

The logical difficulties which plagued the mathematicians during the 17th, 18th, and 19th centuries were most severe in the area of analysis, that is, in the calculus and in the domains such as infinite series and differential equations which were built on the calculus. However, in the early 19th century, geometry once again became a favorite field of study. Euclidean geometry was extended, and a new branch of geometry, projective geometry (concerned mainly with properties of figures which are retained when the figure is projected from one place to another as, for example, when a real two-dimensional scene is projected through the "eye" of a camera onto a film), was first properly envisioned by Jean-Victor Poncelet (1788–1867). As one might expect on the basis of earlier history, though Poncelet and others divined many theorems, they had endless troubles in proving them. By this time, algebraic methods of proving geometric results had become available thanks largely to the 17th-century work of Descartes and Fermat, but the geometers of the first part of the 19th century scorned algebraic methods as alien and impenetrable to the insights and values which geometry proper afforded.

To "establish" his results by purely geometric methods, Poncelet appealed to the principle of continuity. In his *Treatise on the Projective Properties of Figures* (1822), he phrased it thus: "If one figure is derived from another by a continuous change and the latter is as general as the former, then any property of the first figure can be asserted at once for the second figure." The determination of when both figures are general was not explained.

To "demonstrate" the soundness of the principle, Poncelet used a theorem of Euclidean geometry which states the equality of the products of the segments of intersecting chords of a circle ($ab = cd$ in Fig. 7.3). He noted that when the point of intersection moves outside the circle, one obtains the equality of the products of the secants and their external segments (Fig. 7.4). No proof was needed because the principle of continuity guaranteed the theorem. Further, when one secant becomes a tangent, the secant and its external segment become equal and their product continues to equal the product of the other secant

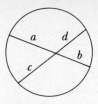

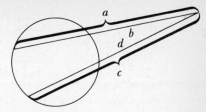

Figure 7.3   Figure 7.4

and its external segment ($ab = c^2$ in Fig. 7.5). The results Poncelet used
to illustrate the principle of continuity happen to be three separate,
well-established theorems and do satisfy or exemplify the principle. But
Poncelet, who coined the term "principle of continuity," advanced the
principle as an absolute truth and applied it boldly in his *Treatise* to
"prove" many new theorems of projective geometry.

The principle was not new with Poncelet. In a broad philosophical
sense the principle goes back to Leibniz. As a mathematical principle
Leibniz used it in the manner indicated earlier (Chapter VI) in connection with the calculus. The principle was used only occasionally until
Gaspard Monge (1746–1818) revived it to establish certain types of
theorems. He proved a general theorem first for a special position of a
figure and then maintained that the theorem was true generally, even
when some elements in the figure became *imaginary*. Thus to prove a
theorem about a line and a curve, he would prove it when the line cuts
the curve and then maintain that the result holds even when the line no
longer cuts the curve and the points of intersection are imaginary.

Some members of the Paris Academy of Science criticized the principle of continuity and regarded it as having only heuristic value. Cauchy
in particular criticized the principle and said:

> This principle is properly speaking only inductive, with the aid of
> which one extends theorems established under certain restrictions to
> cases where these restrictions no longer hold. Applied to curves of the

Figure 7.5

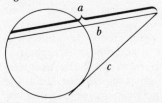

second degree it leads the author to exact results. Nevertheless we think that it cannot be admitted generally and applied freely to all sorts of questions in geometry or even in analysis. By according too much emphasis to it one could fall some times into obvious errors.

Unfortunately the examples Cauchy used to attack the soundness of this principle were ones wherein the principle did produce results that had been correctly proved by other means.

The critics also charged that the confidence which Poncelet and others had in the principle really rested on the fact that the principle could be justified on an algebraic basis. As a matter of fact some notes Poncelet made while he was a prisoner in Russia (he was a soldier in Napoleon's army) show that he did use algebra to *test* the soundness of the principle. Poncelet agreed that a proof could be based on algebra but he insisted that the principle did not depend on such a proof. However, it is quite certain that Poncelet relied on the algebraic method to see what should be the case and then affirmed the geometric results using the principle as a justification.

Despite some criticism, the principle of continuity was accepted during the 19th century as intuitively clear and therefore applicable as a method of proof. The geometers used it freely. However, from the standpoint of the logical development of mathematics, the principle of continuity was no more than a dogmatic ad hoc assertion intended to justify what the men of the time could not establish by purely deductive proofs. The principle was contrived and invoked to justify what visualization and intuition had adduced.

Poncelet's advocacy and application of the principle of continuity is but one example of the extraordinary lengths to which mathematicians went to justify what they could not establish by valid proof. But the logical state of geometry was sad in almost every area. As we know (Chapter V), it was chiefly the work on non-Euclidean geometry during the last part of the 18th and first part of the 19th century that revealed serious defects in the deductive structure of Euclidean geometry. Nevertheless, mathematicians did not hasten to repair these defects but in fact maintained the absolute certainty of the theorems. One may safely infer that the intuitive basis for the theorems and the corroboration which practical applications provided were so convincing that no one was deeply concerned about the defects.

The situation was somewhat different in the case of non-Euclidean geometry. In the early 1800s, a few men in addition to its creators—Lambert, Gauss, Lobatchevsky, and Bolyai—accepted the geometry they created as a proper branch even though its logical foundation was certainly not in as good a state as that of Euclidean geometry. However, especially after Gauss's and Riemann's work in this subject became known, not only these four men, but almost all their successors believed, though they had no proof, that the non-Euclidean geometry was consistent, that is, that its theorems would not contradict each other. They also recognized that Saccheri was wrong in believing that he had arrived at a contradiction.

However, the possibility remained open that contradictions within this geometry might still be discovered. Were this to happen, then the assumption of the parallel axiom of hyperbolic geometry would be invalid and, as Saccheri had believed, Euclid's parallel axiom would be a consequence of his other axioms. Thus, without any proof of consistency or evidence of the applicability of the new geometry, which might at least have served as a convincing argument, many mathematicians accepted what their predecessors had regarded as absurd. Their acceptance was an act of faith. The question of the consistency of non-Euclidean geometry remained open for another fifty years (Chapter VIII).

Evidently in the early 19th century no branch of mathematics was logically secure. The real number system, algebra, and Euclidean and the newer non-Euclidean and projective geometries had either inadequate or no foundations at all. Analysis, that is, the calculus and its extensions, lacked not only the logical foundations of the real numbers and algebra, which were used freely, but also the clarity of the concepts of calculus proper—the derivative, the integral, and infinite series. One could justifiably say that nothing in mathematics had been soundly established.

The attitude toward proof adopted by many mathematicians seems incredible in view of what mathematics had meant. In the 18th century, the obscurities in analysis were apparent and some mathematicians gave up on rigor in this area. Thus, Michel Rolle (1652–1719) taught that the calculus was a collection of ingenious fallacies. Others went further and, acting like the fox with the grapes, explicitly derided the rigor of the Greeks. Alexis-Claude Clairaut (1713–1765) said in his *Elements of Geometry* (1741):

> It is not surprising that Euclid goes to the trouble of demonstrating that two circles which cut one another do not have the same center, that the sum of the sides of a triangle which is enclosed within another is smaller than the sum of the sides of the enclosing triangle. This geometer had to convince obstinate sophists who glory in rejecting the most evident truths; so that geometry must, like logic, rely on formal reasoning in order to rebut the quibblers.

Clairaut then added, "But the tables have turned. All reasoning concerned with what common sense knows in advance, is today disregarded, and serves only to conceal the truth and to weary the reader."

The attitude of the 18th and early 19th centuries was expressed by J. Hoëne-Wronski (1775–1853), who was a great algorithmist but not concerned with rigor. A paper of his was criticized by a commission of the Paris Academy of Sciences as lacking in rigor, and Wronski replied that this was "pedantry which prefers means to the end."

Lacroix, in the second edition (1810–19) of his three-volume *Treatise on Differential and Integral Calculus,* said in the Preface to the first volume, "Such subtleties as the Greeks worried about we no longer need." The typical attitude of the times was to ask why one should go to the trouble of proving by abstruse reasoning things one never doubts in the first place or of demonstrating what is more evident by means of what is less evident.

Even later in the 19th century, Karl Gustav Jacob Jacobi (1804–1851), who left many points in his work on elliptic functions incomplete, said, "For Gaussian rigor we have no time." Many acted as though what defied proof needed no proof. For most men rigor was not a concern. Often what they said could be rigorized by the method of Archimedes, could not have been rigorized by a modern Archimedes. This is particularly true of the work on differentiation which had no parallel in Greek mathematics. What d'Alembert said in 1743, "Up to the present . . . more concern has been given to enlarging the building than to illuminating the entrance, to raising it higher than to giving proper strength to the foundations," applies to the work of the entire 18th and early 19th centuries.

By the middle of the 19th century, the regard for proof had fallen so low that some mathematicians did not even bother to execute full proofs where they might have been able to achieve them. Arthur Cayley (1821–1895), one of the superb algebraic geometers and the inventor of what is called matrix algebra (Chapter IV), stated a theorem on matrices known as the Cayley-Hamilton theorem. A matrix is a rectangular array of numbers and in the case of square matrices there are n numbers in each row and in each column. Cayley verified that his theorem was true for 2 by 2 matrices and in a paper of 1858 said, "I have not thought it necessary to undertake the labor of a formal proof of the theorem in the general case of a matrix of any degree [n by n]."

James Joseph Sylvester (1814–1897), an excellent British algebraist, spent the years from 1876 to 1884 as a professor at Johns Hopkins University. In his lectures he would say, "I haven't proved this, but I am as sure as I can be of anything that it must be so." He would then use the result to prove new theorems. Often at the end of the next lecture he would admit that what he had been so sure of was false. In 1889 he proved a theorem about 3 by 3 matrices and merely indicated a few additional points which have to be considered to prove the theorem for n by n matrices.

This history of illogical development, in view of the fine start made by Euclid in his treatment of geometry and the whole numbers, prompts the question, why did mathematicians have to struggle so much and so ineffectively to logicize the subsequent developments—irrational, negative and complex numbers, algebra, the calculus and its extensions? As we have noted (Chapter V), insofar as Euclidean geometry and the whole numbers are concerned, these are so readily and well understood intuitively that it was relatively easy to find fundamental principles or axioms from which to derive other properties, though at that Euclid's development was somewhat deficient. On the other hand, irrational, negative, and complex numbers, operations with letters, and the concepts of the calculus are far more difficult to grasp.

But there is a deeper reason. A subtle change in the nature of mathematics had been unconsciously made by the masters. Up to about 1500, the concepts of mathematics were immediate idealizations of or abstractions from experience. It is true that by that time negative and irrational numbers had made their appearance and had been accepted by the Hindus and the Arabs. However, though their contributions are not to be deprecated, they were content to be intuitive and empirical insofar as justification is concerned. When, in addition, complex numbers, an extensive algebra employing literal coefficients, and the notions of derivative and integral entered mathematics, the subject became dominated by concepts derived from the recesses of human minds. In particular, the notion of the derivative or an instantaneous rate of change, though of course it has some intuitive base in the physical phenomenon of velocity, is nevertheless far more an intellectual construct. It is entirely different in quality from the mathematical triangle. Likewise, efforts to understand infinitely large quantities, which the Greeks had studiously avoided, infinitely small ones, which the Greeks had skillfully circumvented, negative numbers, and complex numbers floundered because mathematicians failed to appreciate that these concepts were not grounded in immediate experience, but creations of the mind.

In other words, mathematicians were contributing concepts rather than abstracting ideas from the real world. For the genesis of their ideas they were turning from the sensory to the intellectual faculties. As these concepts proved to be more and more useful in application, they were at first grudgingly and later greedily seized upon. Familiarity in these cases did not breed contempt but uncriticalness and even naturalness. From 1700 on, more and more notions farther removed from nature and sprung full blown from human minds were to enter mathematics and be accepted with fewer qualms. Mathematicians, hoist by their own petard, were being compelled to view their subject from a level well above solid earth.

Because they failed to recognize this change in the character of the new concepts, they also failed to realize that a basis for axiomatic development other than self-evident truth was needed. Of course, the new concepts were far more subtle than the old ones and the proper axiomatic basis, as we now know, could not have been readily erected.

How then did mathematicians know where to head and, in view of their tradition of proof, how could they have dared merely to apply rules and assert the reliability of their conclusions? There is no question that solving physical problems supplied the goal. Once these were formulated mathematically, technical virtuosity took over and new methodologies and conclusions emerged. The physical meaning of the mathematics also guided the mathematical steps and often supplied partial arguments to fill in non-mathematical steps. The process was in principle no different from the proof of a theorem of geometry wherein some facts entirely obvious in a figure are used, even though no axiom or theorem supports them.

Beyond physical thinking, there is in all new mathematical work the role of sound intuition. The essential idea or method is always grasped intuitively long before any rational argument for the conclusion is devised. Great mathematicians, whatever license they may allow themselves, have a sure instinct for saving themselves from crushing disaster. The intuitions of great men are sounder than the deductive demonstrations of mediocrities.

Having grasped the essence of a physical problem in some mathematical formulation, the 18th-century mathematicians in particular were seduced by formulas. Apparently, formulas were so attractive to them that the mere derivation of one formula from another by a formal operation such as differentiation or integration was enough to justify it. The fascination of symbols overpowered and consumed their reason. The 18th century has been called the heroic age in mathematics because the mathematicians dared and achieved such magnificent scientific conquests with so little logical armament.

There is still the question of why the mathematicians were confident that their results were correct even though they knew, especially in the 18th century, that the concepts of the calculus were vaguely formulated and their proofs inadequate. Part of the answer is that many of the results were confirmed by experience and observation. Outstanding were the astronomical predictions (Chapter II). But another related factor induced the 17th- and 18th-century men to believe in their work. These men were convinced that God had designed the world mathematically and that mathematicians were discovering and revealing that design (Chapter II). Though what the men of the 17th and 18th centuries uncovered was partial, they thought it was part of underlying truth. The belief that they were discovering some of God's handiwork and that they would eventually arrive at the promised land of complete eternal truths sustained their spirit and courage while the fruitful scientific results were the manna which fed their minds and enabled them to persist.

The mathematicians had found only a portion of the sought-for treasures, but there were ample indications of more to be found. Need one quibble then if the mathematical laws that applied so accurately lacked the precision of mathematical proof? Religious conviction bolstered by scientific evidence substituted for weak or non-existent logical strength. They were so eager to secure God's truth that they continued to build without a secure foundation. They salved their consciences with success. Indeed, success was so intoxicating that most of the time theory and rigor were forgotten. An occasional recourse to philosophical or mystical doctrines cloaked some difficulties so that they were no longer visible. Logically the work of the 17th, 18th, and early 19th centuries was surely crude. But it was also masterfully creative. The errors and imprecision of this work have been stressed somewhat unfairly by late-19th- and 20th-century men in order to belittle its triumphs.

The mathematics of the 17th and 18th centuries was like that of a large business corporation which makes numerous deals and spectacular deliveries but because of mismanagement is fundamentally insolvent. Of course, the customers—the scientists who bought and used the mathematical merchandise—and the creditors—the public which invested unhesitatingly in the stock of mathematics—were unaware of the true financial state.

And so we find a highly paradoxical state of affairs. The logic of the now vastly expanded mathematics was never in a sorrier state. But the success of mathematics in representing and predicting the ways of nature was so impressive that, still more so than the Greeks, all the intellectuals of the 18th century proclaimed the mathematical design of nature and extolled mathematics as the superb and even sublime product of human reason. As Joseph Addison in his *Hymn* said of the heavenly bodies, in reason's ear they all rejoiced.

In retrospect, this glorification of mathematical reasoning seems incredible. To be sure, tatters of reasoning were employed. But especially in the 18th century when heated debates about the meaning and properties of complex numbers, logarithms of negative and complex numbers, the foundations of the calculus, the summation of series, and other issues we have not described filled the literature, the designation Age of Confusion seems more appropriate. By 1800, mathematicians were more certain of results than of their logical justification. From the standpoint of proof the results were beliefs. As we shall soon see, it was the work of the latter part of the 19th century that would justify far more the name Age of Reason (Chapter VIII).

While most mathematicians were content to pursue innovations without much concern for proof, a few of the leading mathematicians became alarmed by the illogical state of mathematics. The desperateness of the situation in analysis was underscored by the brilliant, precocious Norwegian Niels Henrick Abel (1802–1829) in a letter of 1826 to Professor Christoffer Hansteen. He complained about:

> the tremendous obscurity which one unquestionably finds in analysis. It lacks so completely all plan and system that it is peculiar that so many men have studied it. The worst of it is, it has never been treated stringently. There are very few theorems in advanced analysis which have been demonstrated in a logically tenable manner. Everywhere one finds this miserable way of concluding from the special to the general and it is extremely peculiar that such a procedure has led to so few of the so-called paradoxes.

Apropos of divergent series in particular, Abel wrote in January 1826 to his former teacher Berndt Holmböe:

> The divergent series are the invention of the devil, and it is a shame to base on them any demonstration whatsoever. By using them one may draw any conclusion he pleases and that is why these series have produced so many fallacies and so many paradoxes. . . . I have become prodigiously attentive to all this, for with the exception of the geometrical series, there does not exist in all of mathematics a single infinite series the sum of which has been determined rigorously. In other words, the things which are most important in mathematics are also those which have the least foundation. That most of these things are correct in spite of that is extraordinarily surprising. I am trying to find the reason for this; it is an exceedingly interesting question.

Among people at large, some are not content to drown their sorrows in alcohol. So among mathematicians, some were not content to drown their concern about the illogical state of mathematics in the inebriation of physical successes. Whatever solace these more courageous men may have derived from the belief that they were uncovering pieces of God's design was nullified by the late-18th-century abandonment of that belief (Chapter IV). Having lost that support they felt obliged to reexamine their work, and they faced the vagueness, the lack of proofs, the inadequacies in the existing proofs, the contradictions, and the sheer confusion about what was correct in what had been created. These men realized that mathematics had not been the paradigm of reason it was reputed to be. In place of reason, it was intuition, geometrical diagrams, physical arguments, ad hoc principles such as the principle of permanence of form, and the recourse to metaphysics that justified what had been accepted.

The ideal of a logical structure had certainly been made clear and proclaimed by the Greeks. And so the few mathematicians who undertook to achieve it in arithmetic, algebra, and analysis were buoyed up in their efforts by the belief that mathematicians had practically done so in at least one highly significant case, Euclidean geometry. They thought that if some had scaled Olympus once, others might scale it again. What these men did not foresee was that the task of supplying rigorous foundations for all of the existing mathematics would prove to be far more difficult and subtle than any mathematician of 1850 could possibly have envisioned. Nor did they foresee the additional troubles that were to ensue.

VIII

The Illogical Development:
At the Gates of Paradise

> One may say today that absolute rigor has been attained.
> HENRI POINCARÉ

The founders of what has been called the critical movement in mathematics realized that for over two thousand years mathematicians had wandered in a wilderness of intuitions, plausible arguments, inductive reasoning, and formal manipulation of symbolic expressions. They proposed to build the proper logical foundations of mathematics where none existed, to eliminate vague concepts and contradictions, and to improve the existing foundations in branches such as Euclidean geometry. This program was begun in the second decade of the 19th century. The movement was expanded and accelerated when the work on non-Euclidean geometry became more widely known, because this work revealed faults in the structure of Euclidean geometry. It became apparent that even that structure, the presumed stronghold and paradigm of rigorous proof, needed overhauling. Just a little later (1843), the creation of quaternions challenged the assurance with which real and complex numbers were manipulated. Of course, some mathematicians, cocksure of their own work, continued to bungle their reasoning and, when they obtained correct results, deceived themselves into believing that their proofs and textbook presentations were sound.

Though the critical thinkers acknowledged that the claim of mathematics to truth about the physical world had to be abandoned, they did appreciate the enormous achievements in celestial and terrestrial mechanics, acoustics, fluid dynamics, elasticity, optics, electricity, magnetism, and the many branches of engineering, and the incredible accuracy of predictions in these areas. Though mathematics had fought under the protection of the invincible banner of truth, it must have achieved its victories through some essential, perhaps mysterious strength. The extraordinary applicability of mathematics to nature was yet to be explained (Chapter XV), but no one could deny the fact itself or dared to throw away such an all-powerful tool. Certainly this power should not be imperiled by countenancing mazes of logical difficulties and contradictions. Moreover, although mathematicians had violated their own principles in disregarding logical soundness, they were not going to let their subject rest forever on a pragmatic basis. Their prestige was at stake. How could they otherwise distinguish their noble-minded activity from that of earth-grubbing engineers and artisans?

And so some mathematicians undertook to retravel the hasty progress through barely distinct trails and to hew sound, clear-cut paths to the objectives they had already attained. They decided to devote much of their energies to construction and in some areas reconstruction of the foundations of mathematics.

To put the house of mathematics in order called for strong measures. It was clear that there was no solid earth on which to base mathematics, for the seemingly firm ground of truth had proved to be deceptive. But perhaps the structure could be made stable by constructing a solid foundation of another kind. This would consist of complete, sharply worded axioms and definitions, and explicit proof of all results no matter how obvious they might seem to the intuition. Moreover, in place of reliance upon truth there was to be logical compatibility or consistency. The axioms and theorems were to be so completely dependent upon each other that the entire structure would be solid. No matter how it rested in relation to the earth it would hold together much as a skyscraper sways with the wind but remains firm from base to tip.

The mathematicians began by building the logic of the calculus. Since the calculus presupposes the real number system and algebra, neither of which had a logical foundation, the irrationality from a purely logical standpoint of this step may be seen by an analogy. A fifty-story office building becomes crowded with tenants, furniture, and other equipment and the owner suddenly realizes that the entire structure is shaky and must be rebuilt. He decides to do so by starting at the twentieth floor.

There is an explanation for the choice of starting point. We have already noted that by 1800 the various types of numbers had become so familiar that even though they had no logical basis there was not too much concern about the soundness of their properties. The sanctity of Euclidean geometry had also been questioned, but here, too, no difficulties in application had been encountered. In fact, two thousand years of reliable usage lent assurance to what logic had failed to demonstrate. The calculus, on the other hand, was the fount of analysis and in this huge area, loose proofs, paradoxes, and even contradictions had appeared and not all results had pragmatic sanction.

Early in the 19th century three men, the priest, philosopher, and mathematician Bernhard Bolzano (1781–1848), Niels Henrick Abel, and Augustin-Louis Cauchy (1789–1857), decided that the problem of rigorizing calculus must be tackled. Unfortunately, Bolzano worked in Prague and his writings did not become known for many decades. Abel died at the age of twenty-seven and so did not get far into the enterprise. Cauchy worked at the center of the mathematical world of his time and by 1820 was recognized as one of the great mathematicians. Hence it was Cauchy's role in starting the movement to rigorize mathematics that received most recognition and exerted the most influence.

Cauchy decided to build the logic of the calculus on number. Why build on number? The English, following Newton, had attempted to rigorize the calculus by using geometry and they failed. It was evident to Cauchy that geometry was not the proper base. Moreover, the Continentals, following Leibniz, had been using analytic methods. Also, though the work on non-Euclidean geometry had by 1820 not become widely known, perhaps enough was known to make mathematicians leery of geometry. On the other hand, in the realm of number no creation disturbed mathematicians until Hamilton produced quaternions in 1843, and even this did not challenge the correctness of the real number system.

Cauchy wisely also decided to found the calculus on the limit concept. As happens repeatedly in mathematics, this correct approach had already been recommended by several keen minds. John Wallis in his *Arithmetic of Infinitesimals* (1655) and the Scottish professor James Gregory (1638–1675) in his *True Quadrature of the Circle and Hyperbola* (1667) in the 17th century, and d'Alembert in the 18th were very certain that the limit concept was the proper one. D'Alembert's views are most significant because, by the time he wrote, Newton's, Leibniz's, and Euler's works were at hand. In his article "Limit" in the *Encyclopédie* (1751–65), d'Alembert was explicit:

> One says that a quantity is the limit of another quantity, when the second can approach the first more closely than any given quantity as small as one can suppose, though the quantity which approaches can never surpass the quantity it approaches. . . .
>
> The theory of limits is the base of the true metaphysics of the differential calculus. . . .

D'Alembert also wrote the *Encyclopédie* article entitled "Differential" in which he discussed the work of Barrow, Newton, Leibniz, Rolle, and others, and he said a differential (infinitesimal) is a quantity infinitely small or less than any assignable quantity. But he explained that he used such words to conform to current usage. This terminology, he said, was abbreviated and obscure, one hundred times more obscure than that which one wished to define. Limits were the correct language and approach. He criticized Newton's use of velocity to explain the derivative because there was no clear idea of velocity at an instant and it introduces a non-mathematical idea—motion. In his *Mélanges* (1767), d'Alembert repeated that "A quantity is either something or nothing; if it is something it has not vanished; if it is nothing it has vanished completely." He again suggested the limit concept. However, he did not work out the application of the concept to the calculus proper and his contemporaries failed to appreciate d'Alembert's suggestion.

Ideas on limits are also to be found in Carnot's *Reflections,* in L'Huillier's prize paper of 1786 which won the competition of the Berlin Academy, and in Carnot's paper which, though it did not win the prize, was at least respectable. It is very likely that Cauchy was influenced by these writings. In any case in the Introduction to his now famous work, *Cours d'analyse algébrique* (Course on Algebraic Analysis, 1821), Cauchy was very explicit that, "As to methods I have sought to give them all the rigor that one can demand in mathematics."

Despite the word "algebraic" in the title of his text, Cauchy objected to the current reliance upon the "generality of algebra." What he meant was that his contemporaries assumed what was true for real numbers was true for complex numbers, what was true for convergent series was true for divergent ones, and what was true for finite quantities was also true for infinitesimals. He therefore defined carefully and established the properties of the basic notions of the calculus—function, limit, continuity, the derivative, and the integral. He also distinguished between infinite series which have a sum in the sense he specified and those which do not, that is, between convergent and divergent series. The latter he outlawed.* In October 1826, Abel could write to his former teacher Holmböe that Cauchy "is at present the only one who knows how mathematics should be treated." Abel did add that Cauchy was a fool and a bigot but apparently he thought the devil should be given his due.

Though Cauchy undertook to rigorize analysis and stated in the 1829 revision of his *Cours* that he had brought the ultimate in rigor into analysis, the concepts he tackled are subtle and he made many mistakes. His definitions of function, limit, continuity, and derivative were

---

* We need not pursue the technical definitions and theorems Cauchy advanced. For our purposes the fact that the proper rigorization was finally undertaken by Cauchy is the significant one.

essentially correct but the language he used was nevertheless vague and imprecise. Like his contemporaries, he believed that continuity implied differentiability (Chapter VII) and so he stated many theorems in which the hypotheses called only for continuity, but where he used differentiability, and even after his attention was called to his mistake, he persisted. Cauchy defined the definite integral carefully and then proceeded to show that it has a precise value for every continuous function. His proof, however, was erroneous (because he did not recognize the need for uniform continuity). Though he distinguished clearly between convergent and divergent series, he proceeded to make assertions and proofs about convergent series that were false. For example, he asserted that the sum of an infinite series of continuous functions is continuous (which without the requirement of uniform convergence is false). He integrated an infinite series term-by-term and maintained that the integrated series represented the integral of the function represented by the original series. Here, too, he failed to recognize the need for uniform convergence. He gave a criterion for the convergence of a sequence, still known as the Cauchy condition, but failed to prove the sufficiency of this condition because the proof requires a knowledge of the real number system which neither Cauchy nor his contemporaries possessed. Cauchy believed that if a function of two variables possesses a limit at some point when each variable separately approaches that point, the function must approach a limit when both variables vary simultaneously and approach the point.

From the outset the work on the rigorization of analysis caused a considerable stir. After a scientific meeting of the Paris Academy of Sciences at which Cauchy presented his theory on the convergence of series, Laplace hastened home and remained in seclusion until he had examined the series in his *Celestial Mechanics*. Luckily he found every one to be convergent.

Paradoxically, Cauchy refused to be shackled by his own concern for rigor. Though he wrote three texts (1821, 1823, and 1829) which were aimed primarily at the establishment of rigor, he continued to write research papers in which he ignored it. He defined what continuity means but he never proved that any function he used was continuous. Though he had stressed the importance of convergence of series and improper integrals, he operated with series, Fourier transforms, and improper integrals as though there were no problems of convergence. He had defined the derivative as a limit but he also gave a purely formal approach such as Lagrange had given (Chapter VI). He admitted semi-convergent (oscillating) series such as $1 - 1 + 1 - 1 + \cdots$ and rearrangements of the terms in what are called conditionally convergent series (some series with positive and negative terms). He committed other "crimes" but he had a sure feeling for what was true even though he did not establish the truth by the standards of his own texts.

Numerous contributions to the rigorization of analysis were inspired by Cauchy's work. But the chief credit belongs to another master, Karl Weierstrass (1815–1897). With his work the rigorization of the fundamentals of analysis was completed. He began to present his foundations in his lectures of 1858–59 at the University of Berlin. The earliest record is the notes taken by H. A. Schwarz in the spring of 1861. Weierstrass's work finally freed analysis from all dependence upon motion, intuitive understanding, and geometric notions which were certainly suspect by Weierstrass's time.

Weierstrass was clear by 1861 that continuity does not imply differentiability. In view of the long held conviction to the contrary (Chapter VII), the world was shocked when Weierstrass presented to the Berlin Academy in 1872 (published for him by Paul du Bois-Reymond in 1875) an example of a function which is continuous for all real values of x but has no derivative at any value x. (There were earlier examples by Bolzano in 1830 in geometric form but not published, and by Charles Cellérier about 1830 but not published until 1890. Hence neither of these exerted any influence.)

In one respect it was fortunate that Weierstrass's example came late in the development of the calculus, for, as Emile Picard said in 1905, "If Newton and Leibniz had known that continuous functions need not necessarily have a derivative, the differential calculus would never have been created." Rigorous thinking can be an obstacle to creativity.

Cauchy and even Weierstrass at the outset of his work on the rigorization of analysis took for granted all the properties of the real and complex number systems. The first step in supplying the logical basis for real and complex numbers was made in 1837 by Hamilton, the inventor of quaternions. Hamilton knew that complex numbers could be used to represent vectors in one plane, and he sought (Chapter IV) three-unit numbers which could represent vectors in space. He therefore studied the properties of complex numbers with a view to generalizing them. One product of his work, contained in his paper "Algebraic Couples; with a Preliminary Essay on Time," was a logical basis for complex numbers which, however, assumed that the real numbers possessed the familiar properties. In place of complex numbers $(a + b\sqrt{-1})$, Hamilton introduced ordered couples (a,b) of real numbers and he defined the operations with these couples so that one obtained the same results as when operating with the complex numbers in the form $a + b\sqrt{-1}$. It is noteworthy that Hamilton was impelled to seek a new theory of complex numbers because, like all his predecessors, he could not reconcile himself to $\sqrt{-1}$ or even to negative numbers.

Later in the paper he said:

> The present Theory of Couples is published to make manifest that hidden meaning [of complex numbers], and to show, by this remarkable instance, that expressions which seem according to common views to be merely symbolical, and quite incapable of being interpreted, may pass into the world of thoughts, and acquire reality and significance.

He said further in this paper:

> In the Theory of Single Numbers, the symbol $\sqrt{-1}$ is *absurd* [Hamilton's italics] and denotes an Impossible Extraction or a merely Imaginary Number; but in the Theory of Couples, the same symbol $\sqrt{-1}$ is *significant* and denotes a Possible Extraction or a Real Couple, namely, (as we have just now seen) the principal square root of the couples $(-1, 0)$. In the latter theory, therefore, though not in the former, this sign $\sqrt{-1}$ may properly be employed; and we may write, if we choose, for any couple (a_1, a_2) whatever,
>
> $$(a_1, a_2) = a_1 + a_2 \sqrt{-1}$$
>
> . . . and interpreting the symbol $\sqrt{-1}$ in the same expression, as denoting the secondary unit or pure secondary couple $(0,1)$.

Thus, Hamilton removed what he called the "metaphysical stumbling-blocks" in the complex number system.

Hamilton presupposed in his theory of couples the properties of the real numbers. He did try in his essay of 1837 to give a logical development of the real number system. From the concept of time, he derived properties of the positive whole numbers and then extended this development to rational numbers (positive and negative whole numbers and fractions) and irrational numbers. But the development was logically poor and foundered especially in the treatment of irrational numbers. This work was not only unclear but incorrect. It was ignored by the mathematical community. Hamilton's own concern for the foundations of the real and complex numbers was limited. His objective was quaternions. Hence, like most men of his time when he worked in analysis he did not hesitate to use freely the properties of real and complex numbers.

Weierstrass was the first to realize that the rigorization of analysis could not be completed without a better understanding of the real number system, and he was the first to offer a rigorous definition and derivation of the properties of irrational numbers on the basis of the familiar properties of the rationals. He undertook some of this work in the 1840s but did not publish it at that time. It became known through his lectures delivered at the University of Berlin in the 1860s.

Several other men, notably Richard Dedekind and Georg Cantor, also taking for granted the properties of the rational numbers, correctly defined the irrational numbers and established their properties. Their work was published in the 1870s. Dedekind, much like Weierstrass, realized the need for a clear theory of irrationals while teaching the calculus. He wrote, in *Continuity and Irrational Numbers* (1872), that from 1858 on, "He felt more keenly than ever the lack of a rigorous foundation for arithmetic." Cantor recognized the need for a theory of irrational numbers in his work on theorems of analysis (Chapter IX). Thus, through the work of Weierstrass, Dedekind, and Cantor mathematics could finally prove that $\sqrt{2}\sqrt{3} = \sqrt{6}$.

The logic of the rational numbers was still missing. Dedekind realized this and, in *The Nature and Meaning of Numbers* (1888), he described the basic properties that one might use for an axiomatic approach to the rationals. Giuseppe Peano (1858–1932), utilizing Dedekind's ideas and some ideas in Hermann Grassmann's *Textbook on Arithmetic* (1861) succeeded in *Principles of Arithmetic* (1889) in producing a development of the rational numbers from axioms about the positive whole numbers. Thus, finally, the logical structure of the real and complex number systems was at hand.

As a by-product, the erection of the foundations of the number system also resolved the problem of the foundations of the familiar algebra. Why is it that, when letters are manipulated freely as though they stood for positive integers, the results apply equally well when any kind of real or complex numbers are substituted for the letters? The answer is that the other types of numbers possess the same formal properties as the positive integers. To put the matter somewhat loosely, not only is it true that $2 \cdot 3 = 3 \cdot 2$ but also $\sqrt{2}\,\sqrt{3} = \sqrt{3}\,\sqrt{2}$, so that if ab is replaced by ba, the replacement is correct whether a and b stand for positive integers or irrational numbers.

The sequence of events is noteworthy. Instead of starting with the whole numbers and fractions, then taking up the irrational numbers, the complex numbers, algebra, and the calculus, the mathematicians tackled these subjects in reverse order. They acted as though they were reluctant to tackle what could well be left alone as clearly understood and only when the need to logicize a subject was imperative did they undertake to do so. At any rate by about 1890, only six thousand years after the Egyptians and Babylonians began to work with whole numbers, fractions, and irrational numbers, the mathematicians could finally prove that $2 + 2 = 4$. It would appear that even the great mathematicians must be forced to consider rigor.

During the latter part of the 19th century, another outstanding problem was resolved. For roughly sixty years, from the time that Gauss expressed confidence that his non-Euclidean geometry was consistent, probably because he was certain it can be the geometry of the physical world, to about 1870 when Gauss's work on the subject and Riemann's qualifying lecture for *privatdozent* were published, most mathematicians had not taken non-Euclidean geometry seriously (Chapter IV). Its implications were too drastic to be faced. These mathematicians preferred to believe and hope that some day contradictions would be discovered in each of the several non-Euclidean geometries so that these creations could be dismissed as nonsense.

Fortunately, the question of the consistency of all the elementary non-Euclidean geometries was finally answered. The method warrants examination, especially in the light of what happened later. One of the several non-Euclidean geometries, the double elliptic geometry suggested by Riemann's 1854 paper (Chapter IV), differs from Euclidean geometry in essential respects. There are no parallel lines; any two lines meet in two points; the sum of the angles of a triangle is greater than 180°; and many other theorems differ from those in Euclidean geometry. Eugenio Beltrami (1835–1900) pointed out in 1868 that this double elliptic geometry of the plane applies to the surface of a sphere, provided that the lines in the double elliptic geometry are interpreted as great circles on the sphere (circles whose center is the center of the sphere such as circles of longitude).

This interpretation may not seem allowable. The creators of all the non-Euclidean geometries meant their straight lines to be the same as the lines of Euclidean geometry. However, we may recall (Chapter V) that Euclid's definitions of line and other concepts were superfluous. There must be undefined terms in any branch of mathematics, as Aristotle had stressed, and all that one can require of these lines is that they satisfy the axioms. But great circles on a sphere do satisfy the axioms of double elliptic geometry. As long as the axioms of the double elliptic geometry apply to the great circles on the sphere, the theorems of the geometry must also apply, because the theorems are logical consequences.

Granted the interpretation of line as great circle, the consistency of double elliptic geometry is established as follows. If there should be contradictory theorems in double elliptic geometry, then there would be contradictory theorems about the geometry of the surface of the sphere. Now the sphere is part of Euclidean geometry. Hence, *if Euclidean geometry is consistent,* then double elliptic geometry must also be so.

The case for the consistency of hyperbolic geometry (Chapter IV) cannot be made so simply. However, just as the consistency of double elliptic geometry was established by using the spherical surface as a model, so the consistency of hyperbolic geometry was established by using a somewhat more involved configuration of Euclidean geometry. We need not examine it. However, we should note that the fact that hyperbolic geometry is consistent means also that the Euclidean parallel axiom is independent of the other Euclidean axioms. For, if the Euclidean parallel axiom were not independent of the other Euclidean axioms—that is, if it were derivable from them—it would be a theorem of hyperbolic geometry since, aside from the parallel axiom, the other axioms of hyperbolic geometry are the same as those of Euclidean geometry. But this Euclidean "theorem" would contradict the parallel axiom of hyperbolic geometry and hyperbolic geometry would then be inconsistent. Thus, the age-long efforts to deduce the Euclidean parallel axiom from the other Euclidean axioms were doomed to failure.

The surprising fact that the non-Euclidean geometries, which were intended as geometries in which line had the usual meaning, could apply to figures totally different from the ones intended has weighty consequences. As we have explained, totally different interpretations are possible because undefined terms are necessarily present in any axiomatic development. These interpretations are called models. Thus, what we have seen is that a branch of mathematics created with one physical meaning intended may apply to an entirely different physical or mathematical situation.

The consistency of the non-Euclidean geometries was established on the assumption that Euclidean geometry is consistent. To the mathematicians of the 1870s and 1880s the consistency of Euclidean geometry was hardly open to question. Despite Gauss, Lobatchevsky, Bolyai, and Riemann, Euclidean geometry was still accepted as the necessary geometry of the physical world and it was inconceivable that there could be any contradictory properties in the geometry of the physical world. However, there was no logical proof that Euclidean geometry was consistent.

To many mathematicians who had almost scorned non-Euclidean geometry, these consistency proofs were welcomed for another reason. The proofs gave meaning to the non-Euclidean geometries but only as models within Euclidean geometry. Hence, one could accept them in this sense but need not as geometries that might apply to the physical world with the usual meaning of straight line. Of course this was contrary to the views of Gauss, Lobatchevsky, and Riemann.

Only one major problem of rigorization remained. The foundations of Euclidean geometry had been found to be defective. However, unlike the situation in analysis, the nature of geometry and its concepts were clear. Hence, it was a relatively simple task to fix on the undefined terms, sharpen the definitions, supply the missing axioms, and complete the proofs. This was done independently by Moritz Pasch (1843–1930), Giuseppe Veronese (1854–1917), and Mario Pieri (1860–1904). David Hilbert (1862–1943), who acknowledged Pasch's contributions, gave the version most commonly used today. Almost in the same breath, the foundations of the non-Euclidean geometry created by Lambert, Gauss, Lobatchevsky, and Bolyai and the foundations of other geometries created in the 19th century, notably projective geometry, were supplied.

By 1900, arithmetic, algebra, and analysis (on the basis of axioms for the whole numbers) and geometry (on the basis of axioms about points, lines, and other geometric concepts) had been rigorized. Many mathematicians favored going further and building up all of geometry on the basis of number, which could be done through analytic geometry. Geometry per se was still suspect. One lesson of non-Euclidean geometry, namely, that Euclidean geometry, which had been regarded as the model of rigor, had actually been defective, still rankled in the minds of mathematicians. However, the reduction of all of geometry to number was not actually carried out by 1900. Nevertheless, most mathematicians of that time spoke of the arithmetization of mathematics though it would have been more accurate to speak of the arithmetization of analysis. Thus, at the Second International Congress of Mathematicians, which was held in Paris in 1900, Poincaré asserted, "Today there remain in analysis only integers and finite and infinite systems of integers, interrelated by a net of relations of equality or inequality. Mathematics, as we say, has been arithmetized." Pascal had said, "Tout ce qui passe la Géométrie nous passe."* In 1900, mathematicians preferred to say, "Tout ce qui passe l'Arithmétique nous passe."

Movements initially limited in their objectives often tend, as they enlist more and more adherents, to embrace and even engulf many more issues than originally planned. The critical movement in the foundations of mathematics fastened also on logic, the principles of reasoning used in deducing one mathematical step from another.

The science of logic was founded by Aristotle in his *Organon* (Instrument [of reasoning], *c.*300 B.C.). He said explicitly that he noted the principles of reasoning used by the mathematicians, abstracted them, and recognized them to be principles applicable to all reasoning. Thus one of the fundamental principles is the law of excluded middle, which states that every meaningful statement is either true or false.

* All that transcends geometry transcends our comprehension.

This he may have abstracted from a mathematical statement such as every integer is odd or non-odd. In the main, Aristotle's logic consisted of the laws of syllogisms.

For over two thousand years, the intellectual world, of which mathematicians are a part, accepted Aristotle's logic. It is true that Descartes, who questioned all beliefs and doctrines, did raise the question of how we know that the principles of logic are correct. His answer was that God would not deceive us. The assurance we possess of the correctness of these principles was thereby justified to him.

Descartes and Leibniz thought of broadening the laws of logic into a universal science of reasoning that would be applicable to all fields of thought, a universal calculus of reasoning, and they had the idea of using symbolism, such as algebra does, to sharpen and facilitate the use of the laws of reasoning. Descartes said of mathematical method, "It is a more powerful instrument of knowledge than any other that has been bequeathed to us by human agency as being the source of all others."

Leibniz's plan for a universal logic, somewhat more specific than Descartes's, called for three main elements. The first was to be a *characteristica universalis*—a universal scientific language which could be partly or largely symbolic and apply to all truths derived by reasoning. The second component was to be an exhaustive collection of logical forms of reasoning—a *calculus ratiocinator*—which would permit any possible deductions from initial principles. The third—an *ars combinatoria*—was to be a collection of basic concepts in terms of which all other concepts could be defined, an alphabet of thought which would assign a symbol to every simple idea and permit the expression and treatment of more complicated concepts by combinations and operations with these symbols.

The fundamental principles would be, for example, the law of identity, namely, that A is A and A is not not-A. From such principles, all the truths of reason, including those of mathematics, would be derivable. There were, in addition, truths of fact but these were contingent on support by what he called the principle of sufficient reason, that they could not be otherwise. Leibniz was the founder of symbolic logic, but his work in this area was not known until 1901.

Neither Descartes nor Leibniz developed a symbolic calculus of reasoning. They wrote only fragments. Thus, until the 19th century, Aristotelian logic held sway. In 1797, in the second edition of his *Critique of Pure Reason,* Kant said logic was "a closed and complete body of doctrine." Though until about 1900 most mathematicians continued to reason in accordance with informal and verbally expressed Aristotelian principles, they also used others never entertained by Aristotle. They did not examine their own logical principles closely and were under the impression that they were using adequate deductive logic. Actually they were using intuitively reasonable but not explicitly logical principles.

While most mathematicians were concentrating on the rigorization of mathematics proper, a few took up the subject of the logic being used. The next major development was due to George Boole (1815–1864), a professor of mathematics at Queens College in Cork, Ireland.

The inspiration for Boole's work was undoubtedly the view of algebra espoused by Peacock, Gregory, and De Morgan (Chapter VII). Though their principle of permanence of form did not really justify the operations of algebra with literal coefficients representing real and complex numbers, they perhaps unintentionally espoused a new view of algebra as symbols and operations that could represent any objects. And Hamilton's work on quaternions (1843) did indeed show that other algebras were possible. Boole had himself written in 1844 a generalization of algebraic reasoning in what is called the calculus of operators. Hence, he was alerted to the idea that algebra need not treat just number and that the laws of algebra need not be those of real and complex numbers. He referred to this in the beginning of his *Mathematical Analysis of Logic* (1847) and then proposed an algebra of logic. His masterpiece was *An Investigation of the Laws of Thought* (1854). Boole's main idea, less ambitious than Leibniz's and more in tune with Leibniz's *calculus ratiocinator,* was that the existing laws of reasoning could be expressed in symbolic form and thereby one could sharpen and expedite the application of the existing logic. He said in this book:

> The design of the following treatise is to investigate the fundamental laws of those operations of the mind by which reasoning is performed, to give expression to them in the symbolic language of a calculus, and upon this foundation to establish the science of logic and construct its method.

Boole also had specific applications in mind, for example, to the laws of probability.

The advantages of symbolism are numerous. One might in the course of an argument unintentionally make the mistake of introducing meanings that are not intended or use incorrect deductive principles. Thus, in a discussion of light as an optical phenomenon, a reference to "seeing the light" or the "light weight" of an object might be confusing. But if one uses l for physical light, all further symbolic treatment of l can refer only to the optical phenomenon. Moreover, all proofs consist in transforming collections of symbols into new collections by rules for the transformation of symbols that replace the verbal laws of logic. The rules express the correct principles in a sharp and readily applied manner.

Just to get some appreciation of Boole's algebra of logic, let us note a few of his ideas. Suppose the symbols x and y stand for classes of objects, for example, the class of dogs and the class of red animals. Then xy stands for the class of objects that are in both x and y. In the case of dogs and red animals, xy would be the class of red dogs. For any x and y, it is true that $xy = yx$. If z now stands for white objects and if $x = y$, then $zx = zy$. It also follows from the very meaning of xy that $xx = x$.

The symbolism $x + y$ means the class of objects in x or in y or in both. (This is a later modification of Boole by William Stanley Jevons (1835–1882)). Thus, if x is the class of men and y the class of voters, $x + y$ is the class of men and voters (which would include women voters). One could then argue that, if z stands for people over age 35,

$$z(x + y) = zx + zy.$$

If x is a class, then $1 - x$ or $-x$ is the set of all objects not in x. Thus if 1 stands for all objects and x represents dogs, $1 - x$ or $-x$ stands for objects that are not dogs. Then $-(-x)$ means the set of dogs. The equation

$$x + (1 - x) = 1$$

says that everything is either a dog or not a dog. This is the law of excluded middle for classes. Boole showed how to carry out reasoning in various fields by means of such purely algebraic operations.

Boole also introduced what is called the logic of propositions, though the beginning of this use of logic can be traced back to the Stoics (4th century B.C.). In this interpretation, p would stand for, say, "John is a man," and to assert p is to say that "John is a man" is true. Then $1 - p$ (or $-p$) means that "John is a man" is not true. Also $-(-p)$ means it is not true that John is not a man, or John *is* a man. The law of excluded middle for propositions, which affirms that any proposition is either true or false, was expressed by Boole as $p + (-p) = 1$, where 1 represented truth. The product pq means both propositions p and q are true, whereas $p + q$ means either p or q or both are true.

Another innovation was made by De Morgan. In his chief work, *Formal Logic* (1847), De Morgan introduced the idea that logic must deal with relations in general. Aristotle's logic deals with the relation of the verb "to be." The classic example is "All men are mortal." As De Morgan put it, Aristotelian logic could not justify deducing from "A horse is an animal" that "A horse's head is an animal's head." One needs in addition the premise that all animals have heads. Aristotle did write on the logic of relations although vaguely and not extensively. Moreover, many of Aristotle's writings and the extensions made by medieval scholars were lost sight of by the 17th century.

The need for a logic of relations is readily seen. Thus, an argument based only on the relation "to be," namely,

A is a p;
B is a p;
Hence A and B are p's,

is readily seen to be false. For the argument

John is a brother;
Peter is a brother;
Hence John and Peter are brothers (of each other),

can obviously be incorrect. Likewise

An apple is sour;
Sour is a taste;
Hence an apple is a taste,

is also an incorrect conclusion. The failure to develop a logic of relations is the main defect of Aristotelian logic, a defect which Leibniz also noted.

Relations cannot usually be translated into a subject and predicate in which the predicate merely says that the subject is included in a class specified by the predicate. Hence, one must consider propositions which state relationships such as 2 is less than 3 or the point Q is between P and R. One must consider for such propositions what one means by their denial, converse, joint assertion, and other connections.

The logic of relations was expanded by Charles Sanders Peirce (1839–1914) in various papers from 1870 to 1893 and systematized by Ernst Schröder (1841–1902). Peirce introduced special symbolism to denote propositions expressing relationships. Thus l_{ij} expresses that i loves j. Actually, his algebra of relationships was complicated and not very useful. We shall see later how relations are treated in modern symbolic logic.

Another addition to the science of logic, briefly touched on by Boole, was effectively introduced by Peirce. He emphasized the notion of propositional functions. Just as mathematics deals with functions such as $y = 2x$ as opposed to statements about constants such as $10 = 2 \cdot 5$, so "John is a man" is a proposition, but "x is a man" is a propositional function wherein x is a variable. Propositional functions can include two or more variables as in x loves y. With Peirce's contribution reasoning could be extended to propositional functions.

Peirce also introduced what are called *quantifiers*. Ordinary language is ambiguous as to quantifiers. The two statements:

> An American led the War of Independence.
> An American believes in democracy.

use the term "an American" in two different senses. The first refers to a particular American, George Washington. The second refers to every American. Usually this ambiguity can be resolved by reference to the context in which the phrase is used. But this ambiguity cannot be tolerated in rigorous logical thinking. The statements must be clear in and for themselves. The resolution is the use of quantifiers. One might wish to assert that a propositional function is true for all individuals of some class, say, the people in the United States. If one then asserts that for all x, x is a man, he means that all people in the United States are men. The phrase "for all x" is a quantifier. On the other hand one might wish to say that there is at least one x such that x is a man in the United States. In this case, "There is at least one x such that" is the quantifier. These two are denoted respectively by the symbols Vx and $\exists x$.

The extension of logic to relations, to propositional functions, and to quantifiers embraces the types of reasoning used in mathematics and thereby makes logic more adequate.

The final 19th-century step in the direction of mathematizing logic was made by Gottlob Frege (1848–1925), a professor of mathematics at Jena. Frege wrote several major works, *Concept-Writing* (1879), *The Foundations of Arithmetic* (1884) and *The Fundamental Laws of Arithmetic* (vol. I, 1893; vol. II, 1903). He took over the ideas of a logic of propositions, propositions involving relations, propositional functions, and quantifiers. He also made several contributions of his own. He introduced the distinction between the mere statement of a proposition and the assertion that it is true. The assertion is denoted by placing the symbol $\vdash$ in front of the proposition. He distinguished between an object x and the set $\{x\}$ containing just x and between an object belonging to a set and the inclusion of one set in another.

Frege formalized a broader concept of implication called material implication, though its expression in verbal form can be traced back to Philo of Megara (*c.* 300 B.C.). Logic deals with reasoning about propositions and propositional functions and in this process implication is most important. Thus, if we know that John is wise and that wise men live long, we deduce the implication that John will live long.

Material implication is somewhat different from the one commonly used. When we assert, for example, that "If it rains, I shall go to the movies," there is not only some relation between the two propositions, but the implication that, if the antecedent "if it rains" holds, the consequent "I shall go to the movies" necessarily follows. However, the notion of material implication allows p and q, the antecedent and the consequent, to be any propositions. There need be no causal relationship or even any relationship of the propositions to each other. One can as well deal with "If x is an even number, I shall go to the movies." Moreover, material implication allows that even if it is false that x is an even number, then the consequent holds. Thus, "If x is not an even number, I shall go to the movies." Moreover, it allows the implication "If x is not an even number, I shall not go to the movies." The implication is false only in the case where x is an even number and I do not go to the movies.

Put more formally, if p and q are propositions and if p is true, the implication "p implies q" certainly means that q is true. However, material implication allows that even if p is false, then whether or not q is true or false, we still regard the implication p implies q as true. Only in the case where p is true and q is false is the implication false. This notion of implication is an extension of the usual meaning. The extension, however, does no harm because we use "p implies q" only when we know p is true. Moreover, material implication is even somewhat in keeping with ordinary usage. Consider the statement "If Harold is paid today, he will purchase food." Here the p is *Harold is paid today,* and the q is *he will purchase food.* Now he may still purchase food even if he is not paid today. Hence, we can include the case where p is false and q is true as a legitimate implication. Certainly the conclusion is not false. Similarly, "If Harold is not paid today, he will not buy food" is certainly not a false statement. As another, perhaps better example of the last case, we might consider "If wood were a metal, then wood would be malleable." We know that both statements are false, and yet the implication is true. Hence, we include this case of implication, wherein p is false and q is false, as a correct case of p implies q. The important use of the concept is to conclude q from the truth of p and from the implication p implies q. The extension to the cases where p is false is convenient in symbolic logic and the best available.

However, since the falsity of p implies q whether or not q is true or false, material implication allows a false proposition to imply any proposition. To this "failing" one could rebut that in a correct system of logic and mathematics, false propositions should not occur. Nevertheless there have been objections to the concept of material implication. Poincaré, for example, mocked it by citing the case of students who on examinations start with false equations and then draw true conclusions. However, despite efforts to improve upon the concept, material implication is now standard, at least in the symbolic logic used as a basis for mathematics.

Frege made one more contribution which assumed great importance later. Logic contains many principles of reasoning. These may be compared to the numerous assertions which Euclidean geometry makes about triangles, rectangles, circles, and other figures. In geometry as a consequence of the late 19th-century reorganization of other branches of mathematics, the many assertions are derived from a few fundamental ones, the axioms. Frege did precisely this for logic. His notation and axioms were complicated, and so we shall merely indicate verbally the approach to an axiomatic development of logic. (See also Chapter X.) One would certainly feel secure in adopting as an axiom the assertion "p implies p or q", for the meaning of p or q is that at least one, p or q, is true. But if we start with p being true, then surely one of the two, p and q, is true.

One also uses as an axiom that if some proposition (or combination of propositions) denoted by A is true and if A implies B, where B is another proposition (or combination), then we may assert B separately. This axiom, called a rule of inference, enables us to deduce and assert new propositions.

From axioms such as the above we can deduce, for example,

$$p \text{ is true or } p \text{ is false.}$$

This deduction constitutes the law of excluded middle.

One can also deduce the law of contradiction, which in verbal form states that it is not true that p and not p are both true, and so only one of the two possibilities can hold. The law of contradiction is used in what are called indirect proofs in mathematics. Therein if we suppose p is true and deduce that it is false, then we have p and not p. But both cannot hold. Hence p must be false. The indirect method often takes another form. We assume p and show it implies q. But q is known to be false. Hence, by a law of logic, p must be false. Many other commonly used laws of logic can be deduced from the axioms. This deductive organization of logic was begun by Frege in his *Concept-Writing* and continued in his *Fundamental Laws*.

Frege had a still more ambitious objective about which we shall say more in a later chapter (Chapter X). Briefly put for the moment, he sought in his work on logic to institute a new basis for number, algebra, and analysis, a basis still more rigorous than the critical movement was producing during the last few decades of the 19th century.

Another pivotal figure in the use of symbolic logic to improve the rigor of mathematics was Giuseppe Peano. Like Dedekind, he found in his teaching that the existing rigor was inadequate and he devoted his life to improving the logical foundations. He applied symbolic logic not only to the principles of logic but also to the expression of mathematical axioms and to the deduction of theorems by using symbolic logical principles to manipulate symbolic axioms. He was explicit and firm that we must renounce intuition, and this could be done only by working with symbols so that meaning could not play a role in mathematics proper. Symbols avoid the danger of appealing to the intuitive associations attached to common words.

Peano introduced his own symbolism for concepts, quantifiers, and connectives such as "and," "or," and "not." His own symbolic logic was rather rudimentary but his influence was great. The journal he edited, *Rivista di Matematica* (Mathematical Review, founded in 1891 and published until 1906) and the five-volume *Formulary of Mathematics* (1894–1908) are major contributions. It was in the *Formulary* that he gave the axioms for the whole numbers referred to earlier. Peano founded a school of logicians whereas Peirce's and Frege's work went largely unnoticed until Bertrand Russell discovered Frege's work in 1901. Russell had learned about Peano's work in 1900 and preferred Peano's symbolism to Frege's.

From Boole to Schröder, Peirce, and Frege, the innovations in logic consisted of the application of mathematical method: symbolism and deductive proof of logical principles from logical axioms. All of this work on formal logic or symbolic logic appealed to logicians and many mathematicians because the use of symbols avoids psychological, epistemological, and metaphysical meanings and connotations.

The system of logic which includes propositional functions, relationships such as "x loves y" or "A is between B and C," and quantifiers is now generally described as the first order predicate calculus or first order logic. Though according to some logicians it does not cover all the reasoning used in mathematics, for example, mathematical induction, it is the system most favored by modern logicians.*

The extension of the realm of logic to cover all types of reasoning used in mathematics, the greater precision given to statements by the distinction between propositions and propositional functions, and the use of quantifiers certainly added to the rigor mathematicians were seeking to institute in the 19th century. The axiomatization of logic was all the more in line with the movement of the times.

* To encompass all the reasoning used in mathematics some logicians require what is called second order logic and this calls for applying quantifiers to predicates. Thus, to express that $x = y$, we wish to assert that all predicates that apply to x apply to y and we must quantify the predicate by including "for all predicates" or in symbols $x = y \leftrightarrow (F)$ $(F(x) \leftrightarrow F(y))$.

In view of what we shall have to say later about the logical structure of mathematics, let us emphasize here that the rigorization of mathematics proper and logic was achieved by using the axiomatic approach first employed by Euclid. Several features of this method became clearer during the 19th-century axiomatization movement. Let us review them.

The first concerns the necessity for undefined terms. Since mathematics is independent of other subjects, a definition must be in terms of other mathematical concepts. Such a process would lead to an infinite regress of definitions. The resolution of this difficulty is that the basic concepts must be undefined. How then can they be used? How does one know what facts about them can be asserted? The answer is that the axioms make assertions about the undefined (and defined) concepts and so the axioms tell us what can be asserted. Thus, if point and line are undefined, the axiom that two points determine a unique line and the axiom that three points determine a plane furnish assertions that can be used to deduce further results about point, line, and plane. Though Aristotle in the *Organon,* Pascal in *Treatise on the Geometrical Spirit,* and Leibniz in *Monadology* had emphasized the need for undefined terms, mathematicians peculiarly overlooked this fact and consequently gave definitions that were meaningless. Joseph-Diaz Gergonne (1771–1859) in the early 19th century pointed out that the axioms tell us what we may assert about the undefined terms; they give what may be called an implicit definition. It was not until Moritz Pasch reaffirmed the need for undefined terms in 1882 that mathematicians took this matter seriously.

The fact that any deductive system must contain undefined terms that can be interpreted to be anything that satisfies the axioms, introduced into mathematics a new level of abstraction. This was recognized rather early by Hermann Grassman in his *Theory of Linear Extension* (1844), who pointed out that geometry must be distinguished from the study of physical space. Geometry is a purely mathematical structure which can apply to physical space but is not limited to that interpretation. The later workers in axiomatics, Pasch, Peano, and Hilbert, emphasized the abstraction. Though Pasch was clear that there are undefined terms and that only the axioms limit their meaning, he did have geometry in mind. Peano, who knew Pasch's work, was clearer in his article of 1889 that many other interpretations may be possible. Hilbert, in the *Foundations of Geometry* (1899), said that though the terms used were point, line, plane, etc., they could be beer mugs, chairs, or any other objects, provided merely that they obey the axioms involved. The possibility of multiple interpretations of a deductive system may indeed be a boon because it may permit more applications, but we shall find (Chapter XII) that it also has disturbing consequences.

Pasch had a fine understanding of modern axiomatics. He made the point, whose import was certainly not appreciated in the late 19th century, that the consistency of any set of axioms, that is, that they do not lead to contradictory theorems, must be established. The question of consistency had arisen with respect to the non-Euclidean geometries and for these the issue was satisfactorily settled. However, non-Euclidean geometry was strange. For the basic branches, such as the whole numbers or Euclidean geometry, any doubts about consistency seemed academic. Nevertheless, Pasch thought that the consistency of these systems of axioms should be established. In this matter he was seconded by Frege, who wrote in *Foundations of Arithmetic* (1884):

> It is common to proceed as if a mere postulation were equivalent to its own fulfillment. We postulate that it shall be possible in all cases to carry out the operation of subtraction, or of division, or of root extraction, and suppose that with that we have done enough. But why do we not postulate that through any three points it shall be possible to draw a straight line? Why do we not postulate that all the laws of addition and multiplication shall continue to hold for a three-dimensional complex number system just as they do for real numbers? Because this postulate contains a contradiction. Very well, then, what we have to do first is to prove that these other postulates of ours do not contain any contradiction. Until we have done that, all rigor, strive for it as we will, is so much moonshine.

Peano and his school also began in the 1890s to take somewhat seriously the question of consistency. Peano believed that ways of establishing it would readily be found.

The consistency of mathematics might well have been questioned in Greek times. Why did it come to the fore in the late 19th century? We have already noted that the creation of non-Euclidean geometry had forced the realization that mathematics is man-made and describes only approximately what happens in the real world. The description is remarkably successful, but it is not truth in the sense of representing the inherent structure of the universe and therefore not necessarily consistent. Indeed, the axiomatic movement of the late 19th century made mathematicians realize the gulf that separated mathematics from the real world. Every axiom system contains undefined terms whose properties are specified only by the axioms. The meaning of these terms is not fixed, even though intuitively we have numbers or points or lines in mind. To be sure, the axioms are supposed to fix properties so that these terms do indeed possess the properties that we intuitively associate with them. But can we be sure that we have done so and can we be sure that we have not allowed some property or implication to enter which we do not desire and which in fact may lead to a contradiction?

One more feature of the axiomatic method was also pointed out by Pasch. Preferably the axioms of any one branch of mathematics should be independent, that is, it should not be possible to derive any one of the axioms from the others of that branch, for if this can be done, the derivable axiom is a theorem. The method of establishing the independence of an axiom is to give an interpretation or model of the other axioms wherein these axioms are satisfied but the one in question is not. (The interpretation need not be consistent with the *negation* of the axiom in question.) Thus, to establish the independence of the parallel axiom of Euclidean geometry from the other axioms of that geometry, one can use the interpretation of hyperbolic non-Euclidean geometry in which all the Euclidean axioms other than the parallel axiom are satisfied but the Euclidean parallel axiom is not. An interpretation that satisfies the axiom in question and also satisfies a contradictory axiom would not be consistent. Hence, before one uses an interpretation or model to prove the independence of an axiom, one must know that the model is consistent. Thus, as we noted earlier, the independence of the Euclidean parallel axiom was established by providing a model of hyperbolic non-Euclidean geometry within Euclidean geometry.

Though much of our subsequent history will concern doubts, inadequacies, and deep problems raised by the axiomatization of mathematics, in the early part of the 20th century the axiomatic method was hailed as the ideal. No one praised it more than Hilbert, who was by that time the world's leading mathematician. In his article on "Axiomatic Thinking" (published in 1918), he declared:

> Everything that can be the object of mathematical thinking, as soon as the erection of a theory is ripe, falls into the axiomatic method and thereby directly into mathematics. By pressing to ever deeper layers of axioms . . . we can obtain deeper insights into the scientific thinking and learn the unity of our knowledge. By virtue especially of the axiomatic method mathematics appears called upon to play a leading role in all knowledge.

In 1922 he asserted again:

> The axiomatic method is indeed and remains the one suitable and indispensable aid to the spirit of every exact investigation no matter in what domain; it is logically unassailable and at the same time fruitful; it guarantees thereby complete freedom of investigation. To proceed axiomatically means in this sense nothing else than to think with knowledge of what one is about. While earlier without the axiomatic method one proceeded naïvely in that one believed in certain relationships as dogma, the axiomatic approach removes this naïveté and yet permits the advantages of belief.

One would think that mathematicians would welcome the establishment of their subject on a firm, rigorous basis. But mathematicians are very human. The precise formulation of basic concepts such as irrational numbers, continuity, integral, and derivative was not greeted enthusiastically by all mathematicians. Many did not understand the new technical language and regarded the precise definitions as fads, unnecessary for the comprehension of mathematics or even for rigorous proof. These men felt that intuition was good enough, despite the surprises of continuous functions without derivatives, and other logically correct but non-intuitive creations. Emile Picard (1856–1941) said in 1904 apropos of the rigor in partial differential equations, "True rigor is productive, being distinguished in this from another rigor which is purely formal and tiresome, casting a shadow over the problems it touches." Charles Hermite (1822–1901), in a letter to Thomas Jan Stieltjes of May 20, 1893, said, "I recoil with fear and loathing from that deplorable evil, continuous functions with no derivative." Poincaré (1854–1912), whose philosophy of mathematics will be examined in a later chapter, complained: "When earlier, new functions were introduced, the purpose was to apply them. Today, on the contrary, one constructs functions to contradict the conclusions of our predecessors and one will never be able to apply them for any other purpose."

Many men whose definitions and proofs were now seen to be faulty maintained that they meant exactly what the rigorization produced. Even the master Emile Borel defended himself in this manner. Others objected to what they called nit-picking. Godfrey H. Hardy in an article of 1934 said rigor was a matter of routine. Still others did not understand the rigor and so defensively deprecated it. Some spoke of anarchy in mathematics. New ideas, in the present case those contributing to the rigorization of mathematics, are not more open-mindedly received by mathematicians than by any other group of people.

The rigorization revealed another aspect of mathematical creation. Rigor filled a 19th-century need but the end result also teaches us something about the development of mathematics. The newly founded rigorous structures presumably guaranteed the soundness of mathematics but the guarantee was almost gratuitous. Not a theorem of arithmetic, algebra, or Euclidean geometry was changed as a consequence, and the theorems of analysis had only to be more carefully formulated. Thus, if one wishes to use the derivative of a continuous function, he must add the hypothesis that the function is differentiable. In fact, all that the new axiomatic structures and rigor did was to substantiate what mathematicians knew had to be the case. Indeed, the axioms had to yield the existing theorems rather than different ones because the theorems were on the whole correct. All of which means that mathe-

matics rests not on logic but on sound intuitions. Rigor, as Jacques
Hadamard pointed out, merely sanctions the conquests of the intuition;
or, as Hermann Weyl stated, logic is the hygiene which the mathematician practices to keep his ideas healthy and strong.

At any rate, by 1900 rigor had reasserted its role in mathematics and
had secured, if belatedly, the findings of many centuries. Mathematicians could claim that they had fulfilled their obligations to the standard set by the Greeks, and they could rest easier in the knowledge that
except for relatively minor corrections the vast subject matter they had
built on an empirical or intuitive basis had been sanctioned by logic. In
fact, mathematicians were exultant, even smug. They could look back
on several crises, irrational numbers, the calculus, non-Euclidean geometry, and quaternions and congratulate themselves on having overcome the problems which each of these creations had raised.

At the Second International Congress, held in Paris in 1900, Poincaré, the rival of Hilbert for leadership, gave a major address. Despite
his scepticism about the value of some of the refinements introduced in
the mathematical foundations, he boasted:

> Have we at last attained absolute rigor? At each stage of its evolution
> our forerunners believed they too had attained it. If they were de-
> ceived are we not like them also deceived? . . . But in analysis today,
> if we care to take pains to be rigorous, there are only syllogisms or ap-
> peals to the intuition of pure number, the only one [intuition] that
> could not possibly deceive us. One may say today that absolute rigor
> has been attained.

Poincaré repeated this boast in an essay incorporated in his book *The
Value of Science* (1905). When one observes the keenness mathematicians displayed in rigorizing the many branches of their discipline, one
can see reason for gloating. Mathematics now had the foundation that
all mathematicians except a few laggards were happy to accept, and so
they rejoiced.

In Voltaire's ironic *Candide*, the philosopher Dr. Pangloss, even when
he is about to be hanged, says, "This is the best of all possible worlds."
So the mathematicians, who did not know they were soon to be hoist by
their own petard, said they had reached the best possible state. Actually, storm clouds were gathering and had the mathematicians attend-
ing the Congress of 1900 looked out the windows they would have noticed them but they were too absorbed in drinking toasts to each other.

There was, however, one man at the very same Congress of 1900
who was fully aware that all the difficulties in the foundations of mathematics had not been disposed of. At this Congress, David Hilbert presented a list of twenty-three problems whose solution he regarded as
most important for the advancement of mathematics. The first of these

problems contains two parts. Georg Cantor had introduced transfinite numbers to represent the number of objects in infinite sets. Apropos of this innovation Hilbert proposed the problem of proving that the next larger transfinite number after that of the whole numbers is the number of all real numbers. We shall return to this problem in Chapter IX.

The second part of the first problem asked for a method of reordering the real numbers so that the reordered set would be what is called well ordered. Though we shall say more about this concept later, perhaps at the moment it may suffice to say that the well ordering of the real numbers requires that any subset selected from the set of reordered numbers must have a first member. In the usual ordering of real numbers, if one selects all numbers greater than 5, say, this subset has no first member.

Hilbert's second problem was of more obvious and broader significance. We have already noted that the question of consistency had been raised in connection with the non-Euclidean geometries and proofs were given which presumed that Euclidean geometry was consistent. Hilbert showed through the medium of analytic geometry that Euclidean geometry is consistent if the science of arithmetic is consistent. Hence, in his second problem he asked for a proof that the science of arithmetic is consistent.

Cantor, it is true, had already noted both parts of Hilbert's first problem. And Pasch, Peano, and Frege had called attention to the consistency problem. However, no one but Hilbert in 1900 regarded these problems as outstanding, much less as momentous. There is no doubt that most of the mathematicians who heard Hilbert at the Second International Congress of 1900 considered these problems trivial, unimportant, or sheer curiosities, and attached far more importance to the other problems Hilbert proposed. As for the consistency of arithmetic, no one doubted it. That many doubted the consistency of non-Euclidean geometry—it was strange and even contrary to intuition—was understandable. But the real number system had been in use for over five thousand years and endless theorems about real numbers had been proved. No contradiction had been found. The axioms for real numbers had produced the well-known theorems. How could the axiom system be inconsistent?

Any doubts about Hilbert's wisdom in proposing the above-described problems and indeed placing them at the head of his list of twenty-three were soon dispelled. The storm clouds that gathered outside the building were rolling over each other, and some mathematicians began to hear the thunder, but even Hilbert may not have foreseen the maelstrom that was to follow.

IX

Paradise Barred:
A New Crisis of Reason

In mathematics there are no true controversies
GAUSS

Logic is the art of going wrong with confidence.
ANONYMOUS

After many centuries of wandering through intellectual fog, by 1900 mathematicians had seemingly imparted to their subject the ideal structure that had been delineated by Euclid in his *Elements*. They had finally recognized the need for undefined terms; definitions were purged of vague or objectionable terms; the several branches were founded on rigorous axiomatic bases; and valid, rigorous, deductive proofs replaced intuitively or empirically based conclusions. Even the principles of logic had been extended to accommodate the types of reasoning that mathematicians had been using informally and often implicitly, though as far as one could see in 1900 their usage had been sound. And so, as we have already reported, mathematicians had cause to rejoice. While they were congratulating themselves developments were underway that were to disturb their equanimity even more than the creation of non-Euclidean geometry and quaternions had in the first half of the 19th century. As Frege put it, "just as the building was completed, the foundation collapsed."

It is true that Hilbert had called attention to several unsolved problems (Chapter VIII) concerning the foundations of mathematics. Of these the problem of establishing the consistency of the various axiomatized branches was fundamental. He recognized that the axiomatic method necessitates the use of undefined terms and axioms about these terms. Intuitively these terms and axioms have very specific meanings. The words point, line, and plane, for example, have physical counterparts and the axioms of Euclidean geometry are intended to assert physical facts about these concepts. Yet, as Hilbert emphasized, the abstract, purely logical framework of Euclidean geometry did not require that point, line, and plane be tied to the one intended interpretation. As for the axioms, one assumes as little as possible in the axioms with the aim of deducing as much as possible. Though one tries to formulate the axioms so that they assert what seems to be physically true, there is the danger that as formulated, the axioms may not be consistent; that is, they may lead to contradictions. Pasch, Peano, and Frege had already recognized this danger and Hilbert's speech at the Paris Congress of 1900 gave emphasis to the problem.

The defects that may occur in an abstract formulation of physical reality may perhaps be more readily seen by a rough analogy. A crime has been committed (and many would agree that mathematics is a crime). A detective investigating the crime has undefined terms—a criminal, a time of the crime, etc. Whatever facts he can obtain he writes down. These are his axioms. He then deduces facts in the hope that he will be able to make some assertions about the crime. He may very likely make contradictory deductions because some of his assumptions, though based as far as possible on what actually happened, may go beyond or only approximate it. There is no contradiction in the actual physical situation. There was a crime and a criminal. But deduction may lead to his being thought both five feet tall and six feet tall.

It is doubtful that proof of the consistency of the several axiomatic structures would have been regarded as a key problem were it not for new developments. By 1900 mathematicians did recognize that they could no longer rely upon the physical truth of mathematics to be sure of its consistency. Previously, when Euclidean geometry was accepted as the geometry of physical space, it was inconceivable that continuing deduction of theorems in Euclidean geometry would ever lead to contradictions. But by 1900 Euclidean geometry was recognized to be just a logical structure erected on a set of twenty or so man-made axioms, and it was indeed possible that contradictory theorems could turn up. In that event, much of the earlier work would be meaningless because, if two contradictory theorems did turn up, both might have been used to prove other contradictions and the resulting theorems would be useless. But Hilbert had disposed of that dire possibility by proving that Euclidean geometry is consistent if the logical structure of arithmetic, that is, the real number system, is consistent, and about that there was little concern or urgency (Chapter VIII).

But, to the consternation of all, shortly after 1900 contradictions actually were discovered in a theory that both underlies and extends our knowledge about number. By 1904, an outstanding mathematician, Alfred Pringsheim (1850–1914), could rightly say that the truth that mathematics seeks is neither more nor less than consistency. And when Hilbert again stressed the problem in a paper of 1918, he had far more reason to do so than in his talk of 1900.

The new theory that gave rise to contradictions and opened men's eyes to contradictions in older branches was the theory of infinite sets. The rigorization of analysis had to take into account the distinction between infinite series that converged—had a finite sum—and those that diverged. Among series, infinite series of trigonometric functions, called Fourier series in honor of Joseph Fourier, had come to play a vital role and some questions, raised during the rigorization, were open when Georg Cantor (1845–1918) undertook to answer them. He was led to consider the theory of sets of numbers and in particular to introduce numbers for infinite sets such as the set of all odd numbers, the set of all rational numbers (positive and negative whole numbers and fractions), and the set of all real numbers.

In the very fact that Cantor regarded infinite sets as totalities, as entities that can be considered by the human mind, he broke with long-standing judgments. From Aristotle onward, mathematicians had made a distinction between an actual infinity of objects and a potential infinity. Consider the age of the earth. If one assumes it was created at some definite time, its age is potentially infinite in that at any time it is finite but keeps increasing. The set of (positive) whole numbers can also be regarded as potentially infinite in that, if one stops at one million, he can always consider one more, two more, and so forth. If, however, the earth has always existed in the past, its age at any time is actually infinite. Likewise, the set of whole numbers viewed as an existing totality is actually infinite.

The question of whether one should consider infinite sets as actual or potential has a long history. Aristotle in his *Physics* concluded, "The alternative then remains that the infinite has a potential existence. . . . There will not be an actual infinite." He maintained that mathematics does not need the latter. The Greeks generally regarded infinity as an inadmissible concept. It is something boundless and indeterminate. Later discussions were sometimes befuddled because many mathematicians spoke of infinity as a number but never clarified the concept or established its properties. Thus, Euler in his *Algebra* (1770) said that 1/0 is infinite (though he did not define infinite and merely represented it by the symbol ∞), and then said that without question 2/0 is twice as large as 1/0. More confusion was created by the use of the symbol ∞ in situations such as the limit, as n approaches ∞, of $1/n$ is 0. Here the symbol ∞ means merely that n can take larger and larger values and can be taken so large (but finite) as to make the difference between 0 and $1/n$ as small as one pleases. The actual infinite is not involved.

However, most mathematicians—Galileo, Leibniz, Cauchy, Gauss, and others—were clear about the distinction between a potentially infinite set and an actually infinite one and rejected consideration of the latter. If they did speak, for example, of the set of all rational numbers, they refused to assign a number to the set. Descartes said, "The infinite is recognizable but not comprehensible." Gauss wrote to Schumacher in 1831, "In mathematics infinite magnitude may never be used as something final; infinity is only a *façon de parler,* meaning a limit to which certain ratios may approach as closely as desired when others are permitted to increase indefinitely."

Hence, when Cantor introduced actually infinite sets, he had to advance his creation against conceptions held by the greatest mathematicians of the past. He argued that the potentially infinite in fact depends upon a logically prior actually infinite. He also gave the argument that the irrational numbers, such as $\sqrt{2}$, when expressed as decimals involved actually infinite sets because any finite decimal could only be an approximation. He realized that he was breaking sharply from his predecessors and said in 1883, "I place myself in a certain opposition to widespread views on the mathematical infinite and to oft-defended opinions on the essence of number."

In 1873 he not only undertook to consider infinite sets as existing totalities, as entities, but he set about distinguishing them. He introduced definitions which determined when two infinite sets contain the same or different numbers of objects, and his basic idea was one-to-one correspondence. Just as we recognize that 5 books and 5 marbles can be represented by the same number 5 because we can pair off one and only one book with one and only one marble, so Cantor applied one-to-one correspondence to infinite sets. Now, one can set up the following one-to-one correspondence between the whole numbers and the even numbers:

$$
\begin{array}{cccccc}
1 & 2 & 3 & 4 & 5 & \cdots \\
2 & 4 & 6 & 8 & 10 & \cdots
\end{array}
$$

That is, each whole number corresponds to precisely one even number, its double, and each even number corresponds to precisely one whole number, its half. Hence Cantor concluded that the two sets contain the same number of objects. It was such a correspondence, the fact that the entire set of whole numbers can be put into one-to-one correspondence with a part of that set, that seemed so unreasonable to earlier thinkers and caused them to reject any efforts to deal with infinite sets. But Cantor was not deterred. Infinite sets, he foresaw, could obey new laws that do not apply to finite collections or sets, just as quaternions, for example, could obey new laws that did not hold for real numbers. In fact he defined an infinite set as one that can be put into one-to-one correspondence with a proper subset of itself.

Actually, Cantor was astonished by what his use of one-to-one correspondence led to. He showed that there is a one-to-one correspondence between the points of a line and the points of a plane (and even n-dimensional space), and he wrote to Richard Dedekind in 1877, "I see it, but I don't believe it." However, he did believe it, and he stuck to his one-to-one correspondence principle in assigning equality to infinite sets.

Cantor also defined what one means by one infinite set being larger than another. If set A can be put into one-to-one correspondence with a part or subset of set B but set B cannot be put into one-to-one correspondence with A or a subset of A, then set B is larger than set A. This definition merely extends to infinite sets what is immediately obvious for finite sets. If there are 5 marbles and 7 books, one can set up a one-to-one correspondence between the marbles and some of the books but all of the books cannot be so related to some or all of the marbles. By using his definitions of equality and inequality, Cantor was able to establish the surprising result that the set of whole numbers is equal to the set of rational numbers (all positive and negative whole numbers and fractions) but less than the set of all real numbers (rational and irrational).

Just as it is convenient to have the number symbols 5, 7, 10, etc. to denote the number of objects in a finite collection so Cantor decided to use symbols to denote the number of objects in infinite sets. The set of whole numbers and sets that can be put into one-to-one correspondence with it have the same number of objects and this number he denoted by $\aleph_0$ (aleph-null). Since the set of all real numbers proved to be larger than the set of whole numbers, he denoted this set by a new number, c.

Further, Cantor was able to show that for any given set there is always a still larger one. Thus, the set of all subsets of a given set is larger than the original set. We need not pursue the proof of this theorem but the reasonableness of the fact may be seen by considering a finite set. Thus, if one has 4 objects, one can form 4 different sets of 1 object, 6 different sets of 2 objects, 4 different sets of 3 objects, and 1 set of 4 objects. If we add the empty set, we find that the number of all subsets can be compactly stated as 2^4, which of course is larger than 4. In particular, by considering all the possible subsets of the set of whole numbers Cantor was able to show that $2^{\aleph_0} = c$, where c is the number of all real numbers.

When Cantor undertook to consider infinite sets in the 1870s and for a while thereafter, this theory could have been regarded as peripheral.

The theorems on trigonometric series he did prove were not fundamental. However, by 1900 his theory of sets was heavily employed in other areas of mathematics. Also, he and Richard Dedekind perceived its usefulness in founding a theory of whole numbers (finite and transfinite), in analyzing the concepts of curve and dimension, and even in serving as a foundation for all of mathematics. Other mathematicians, Borel and Henri Lebesgue (1875–1941), were already working on a generalization of the integral that depends upon Cantor's theory of infinite sets.

Hence, it was no minor matter when Cantor himself discovered a difficulty. He had shown that there were larger and larger transfinite sets and corresponding transfinite numbers. In 1895 Cantor thought to consider the set of all sets. Its number should be the largest that can exist. However, Cantor had shown that the set of all subsets of a given set must have a larger transfinite number than the set itself. Hence, there must be a larger transfinite number than the largest one. Cantor decided at that time that one must distinguish what he called consistent and inconsistent sets and wrote accordingly to Richard Dedekind in 1899. That is, one could not consider the set of all sets or its number.

When Bertrand Russell first ran across Cantor's conclusion about the set of all sets, he did not believe it. He wrote in an essay of 1901 that Cantor must have been "guilty of a very subtle fallacy, which I hope to explain in some future work." It is obvious, he added, that there has to be a greatest transfinite set because if everything is taken there is nothing left to add. Russell meditated on this matter and added to the problems of the times his own "paradox," which we shall deal with shortly. Sixteen years later when Russell reprinted his essay in *Mysticism and Logic,* he added a footnote apologizing for his mistake.

In addition to the transfinite numbers already described, which are called transfinite *cardinal* numbers, Cantor introduced transfinite *ordinal* numbers. The distinction is rather delicate. If one considers a collection, say, of pennies, what is usually most important is the number of them no matter how they are assembled. If however one ranks students in terms of the grades received on a test, there is a first, a second, a third, etc. If there are, say, ten students, their ranks compose the set from first to tenth and this is a set of ordinal numbers. Although some earlier civilizations did distinguish between cardinals and ordinals, they used the same symbol for the ordered set of ten objects as they did for the unordered collection. This practice was continued by later civilizations including our own. Thus, after a tenth person has been ordered, the number of people so ordered is also ten and both the ordered set of ten and the unordered set are denoted by 10. However, for infinite sets the distinction between cardinals and ordinals is of greater consequence and so different symbols are used. Thus for the infinite set of ordered whole numbers 1, 2, 3, $\cdots$, the ordinal number ω was used by Cantor. Accordingly the ordered set

$$1, 2, 3, \cdots, 1, 2, 3$$

was (and is) denoted by $\omega + 3$. Cantor introduced a hierarchy of transfinite ordinal numbers. This hierarchy extended to $\omega \cdot \omega$, ω^n, ω^ω, and beyond.

After creating the theory of transfinite ordinal numbers, Cantor realized in 1895 that there is a difficulty with these numbers, too, and he mentioned it to Hilbert in that year. The first man to publish the difficulty was Cesare Burali-Forti (1861–1931) in 1897. Cantor believed that the set of ordinal numbers can be ordered in some suitable fashion just as the familiar real numbers are ordered as to magnitude. Now, one theorem about transfinite ordinals states that the ordinal number of the set of all ordinals up to and including any one, α say, is larger than α. Thus the ordinal number of the ordinals 1, 2, 3, $\cdots$, ω is $\omega + 1$. Hence, the set of *all* ordinals should have a larger ordinal than the largest in the set. In fact, Burali-Forti noted, one could add one to the largest ordinal and obtain a larger ordinal. But this is a contradiction because the set includes all ordinals. Burali-Forti concluded that only a partial ordering of the ordinal numbers was possible.

With only these two difficulties to face, most mathematicians would no doubt have been content to live in the paradise which the late 19th-century rigorization of mathematics had created. The questions about whether there are a largest transfinite cardinal and ordinal could be dismissed. After all, there is no largest whole number and this fact does not disturb any one.

However, Cantor's theory of infinite sets provoked a host of protests. Despite the fact that, as already noted, the theory was being used in many areas of mathematics, some mathematicians still refused to accept actually infinite sets and their applications. Leopold Kronecker, who was also personally antipathetic to Cantor, called him a charlatan. Henri Poincaré thought the theory of infinite sets a grave malady and pathologic. "Later generations," he said in 1908, "will regard set theory as a disease from which one has recovered." Many other mathematicians tried to avoid using transfinite numbers even in the 1920s (Chapter X). Cantor defended his work. He said he was a Platonist and believed in ideas that exist in an objective world independent of man. Man had but to think of these ideas to recognize their reality. To meet criticisms from philosophers, Cantor invoked metaphysics and even God.

Fortunately, Cantor's theory was welcomed by others. Russell described Cantor as one of the great intellects of the 19th century. He said in 1910, "The solution of the difficulties which formerly surrounded the mathematical infinite is probably the greatest achievement of which our age has to boast." Hilbert affirmed, "No one shall drive us from the paradise which Cantor created for us." He also said in 1926 of Cantor's work, "This appears to me to be the most admirable flower of the mathematical intellect and one of the highest achievements of purely rational human activity."

The reason for the controversy created by set theory was rather wittily described by Felix Hausdorff in his *Foundations of Set Theory* (1914) when he characterized the subject as "a field in which nothing is self-evident, whose true statements are often paradoxical, and whose plausible ones are false."

However, most mathematicians were disturbed as a consequence of Cantor's work for a reason totally different from the acceptability of infinite sets of various sizes. The contradictions Cantor had discovered in attempting to assign a number to the set of all sets and to the set of all ordinals caused mathematicians to recognize that they had been using similar concepts not only in the newer creations but also in the supposedly well-established older mathematics. They preferred to call these contradictions paradoxes because a paradox can be resolved and the mathematicians wanted to believe these could be resolved. The technical word commonly used now is antinomy.

Let us note some of these paradoxes. A non-mathematical example is the statement, "All rules have exceptions." This statement is a rule and must then have an exception. Hence, there is a rule without exceptions. Such statements refer to themselves and deny themselves.

The most widely known of the non-mathematical paradoxes is called the liar paradox. It was discussed by Aristotle and many later logicians. The classic version concerns the sentence, "This sentence is false." Let us denote the statement in quotes by S. If S is true, then what it says is true, and so S is false. If S is false, then this is what it says and so S is true.

There are many variants of this paradox. A man may say, perhaps in reference to some assertion he has made, "I am lying." Is the statement, "I am lying," true or false? If the man is indeed lying, he is speaking the truth, and if he is speaking the truth, he is lying. Some variants involve self-reference less directly. Thus, the two sentences "The next sentence is false. The previous sentence is true" also involve a contradiction, for if the second sentence is true, the first one says it is false. But if the second sentence is false, as the first one asserts, then the second one is true.

Kurt Gödel (1906–1978), the foremost logician of this century, gave a somewhat different version of the above contradictory statements: On May 4, 1934, A makes the single statement, "Every statement that A makes on May 4, 1934 is false." This statement cannot be true because it says of itself that it is false. But it also cannot be false because, to be false, A would have had to make a true statement on May 4. But he made only the one statement.

The first of the really troublesome mathematical contradictions was noted by Bertrand Russell (1872–1970) and communicated to Gottlob Frege in 1902. Frege at that time was just publishing the second volume of his *Fundamental Laws* in which he was building up a new approach to the foundations of the number system. (We shall say more about this approach in the next chapter.) Frege used a theory of sets or classes which involved the very contradiction Russell noted in his letter to Frege and published in his *Principles of Mathematics* (1903). Russell had studied the paradox of Cantor's set of all sets and then generated his own version.

Russell's paradox deals with classes. A class of books is not a book and so does not belong to itself, but a class of ideas is an idea and does belong to itself. A catalogue of catalogues is a catalogue. Hence, some classes belong to (or are included in) themselves and some do not. Consider N, the class of all classes that do not belong to themselves. Where does N belong? If N belongs to N, it should not by the definition of N. If N does not belong to N, it should by the definition of N. When Russell first discovered this contradiction, he thought the difficulty lay somewhere in the logic rather than in mathematics itself. But this contradiction strikes at the very notion of classes of objects, a notion used throughout mathematics. Hilbert noted that this paradox had a catastrophic effect on the mathematical world.

Russell's antinomy was put in popular form by Russell himself in 1918 and this version is known as the barber paradox. A village barber advertised that he doesn't shave any people in the village who shave themselves, but he does shave all those who don't shave themselves. Of course, the barber was boasting that he had no competition, but one day it occurred to him to ask whether he should shave himself. If he does shave himself, then by the first half of his assertion—namely, that he doesn't shave those people who shave themselves—he should not shave himself, but if he doesn't, then, in accordance with his boast that he shaves all people who do not shave themselves, he should. The barber is in a logical predicament.

Another paradox representative of what occurs in mathematics was first stated by the mathematicians Kurt Grelling (1886–1941) and Leonard Nelson (1882–1927) in 1908, and concerns adjectives that describe themselves and those that do not. Thus, the adjectives short and English do describe or apply to themselves whereas long and French do not. Likewise polysyllabic is polysyllabic but monosyllabic is not monosyllabic. It seems fair to say that any one word either does or does not apply to itself. Let us call those adjectives that do apply to themselves autological and those that do not heterological. Now let us consider the word heterological itself. If heterological is heterological it applies to itself and so is autological. If heterological is autological, then it is not heterological. But if it is autological, then by the definition of autological it applies to itself and so "heterological" *is* heterological. Thus, either assumption about the word leads to a contradiction. In symbols the paradox states that x is heterological if x is not x.

In 1905, Jules Richard (1862–1956) presented another "paradox" using the very same procedure Cantor used to prove that the number of real numbers was larger than the number of whole numbers. The argument is somewhat complicated but the same contradiction was incorporated in a simplified version due to G. G. Berry of the Bodleian Library and sent to Russell, who published it in 1906. It is called the "word paradox." Each whole number is describable in many ways by words. Thus, five can be described by the single word "five" or by the phrase "the next integer after four." Consider now all possible descriptions made with 100 letters or fewer of the English alphabet. At most, 27^{100} descriptions are possible and so there are at most a finite number of whole numbers describable with all 27^{100} descriptions. There must then be some whole numbers not described by the 27^{100} descriptions. Consider "the smallest number not describable in 100 letters or fewer." This number has just been described in fewer than 100 letters.

Whereas many mathematicians of the early 1900s tended to disregard paradoxes such as the above because they involved set theory which was at the time new and peripheral, others, recognizing that they affected not only classical mathematics but reasoning generally, were disturbed. Some have tried to take the advice William James gave in his *Pragmatism,* "Whenever you meet a contradiction you must make a distinction." A few logicians, beginning with Frank Plumpton Ramsey (1903–1930), have tried to make a distinction between semantic and true contradictions, that is, logical ones. The "word paradox," the "heterological paradox," and the "liar paradox" they called semantic because these involve concepts such as truth and definability or ambiguous uses of a word. Presumably, strict definitions of these concepts, used accordingly, would resolve the paradoxes just mentioned. On the other hand Russell's paradox, Cantor's paradox of the set of all sets, and the Burali-Forti paradox are considered logical contradictions. Russell himself did not make this distinction. He believed that all the paradoxes arose from one fallacy which he called the vicious circle principle and which he described thus: "Whatever involves *all* of a collection must not be one of the collection." Put otherwise, if to define a collection of objects one must use the total collection itself, then the definition is meaningless. This explanation given by Russell in 1905 was accepted by Poincaré in 1906, who coined the term impredicative definition, that is, one wherein an object is defined (or described) in terms of a class of objects which contains the object being defined. Such definitions are illegitimate.

Consider the example offered by Russell himself in the *Principia Mathematica* (Chapter X). The law of excluded middle states that all propositions are true or false. But the law itself is a proposition. Hence, whereas its intent is to affirm a true law of logic, it is a proposition and so it too can be false. As Russell put it, this statement of the law is meaningless.

Some other examples may be helpful. Can an omnipotent being create an indestructible object? Of course, since he is omnipotent. But if he is omnipotent he can also destroy any object. In this example the word omnipotent ranges over an illegitimate totality. Such paradoxes, as the logician Alfred Tarski pointed out, though semantic, challenge language itself.

A number of other attempts have been made to resolve the paradoxes. The contradiction in "All rules have exceptions" is dismissed by some as meaningless. And, they add, there are grammatically correct English sentences that are logically meaningless or false, as is the sentence: "This sentence contains four words." Likewise, the original Russell version of the Russell paradox is swept aside on the ground that the class of all classes that are not members of themselves is meaningless or does not exist. The "barber paradox" is "resolved" by affirming that there is no such barber or by requiring that the barber exclude himself from the classes of people he does and does not shave, just as the statement that the teacher who teaches all those who attend the class does not include the teacher. Russell rejected this last explanation. As Russell put it in an article of 1908, "One might as well in talking to a man with a long nose, say, 'When I speak of noses I except such as are inordinately long,' which would not be a successful effort to avoid a painful topic."

It is true that the word "all" is ambiguous. According to some, several of the semantic paradoxes result from the use of the word "all." The Burali-Forti paradox deals with the class of all ordinals. Does this class include the ordinal of the entire class? Likewise, the heterological paradox defines a class of words. Does this class include the word "heterological" itself?

The Russell-Poincaré objection to impredicative definitions has been widely accepted. Unfortunately, such definitions have been used in classical mathematics. The instance that caused most concern is the notion of least upper bound. Consider the set of all numbers between 3 and 5. Upper bounds, that is, numbers larger than the largest number in the set are 5, 5½, 6, 7, 8, etc. Among these, there is a least upper bound, namely, 5. Hence, the least upper bound is defined in terms of a class of upper bounds that contains the very one being defined. Another example of an impredicative definition is that of the maximum value of a function in a given interval. The maximum value is the largest of the values which the function takes on in that interval. Both of these concepts are fundamental in mathematics and much of analysis depends on them. Further, many impredicative definitions are used in other mathematical contexts.

Though the impredicative definitions involved in the paradoxes do lead to contradictions, mathematicians were bewildered because, insofar as they could see, not all impredicative definitions seem to lead to contradictions. Statements such as "John is the tallest man on his team" and "This sentence is short" are surely innocuous even though impredicative. So is the statement "The largest number in the set 1, 2, 3, 4, 5 is 5." In fact, it is common to use impredicative statements. Thus, if one defines the class of all classes that contain more than five members, one has defined a class that contains itself. Likewise the set S of all sets definable in 25 words or less contains S. The abundance of such definitions in mathematics was indeed cause for alarm.

Unfortunately, there is no criterion to determine which impredicative definitions are and which are not innocuous. Hence, the danger existed that more impredicative definitions would be found to lead to contradictions. This problem was a pressing one from the very first discussions of it by Ernst Zermelo and Poincaré. Poincaré proposed a ban on all impredicative definitions. Hermann Weyl, a leading mathematician in the first half of this century, was concerned that some of the impredicative definitions might indeed be contradictory and he devoted considerable effort to rephrasing the least upper bound definition so as to avoid the impredicativeness. But he did not succeed. He remained uneasy and concluded that analysis is not well founded and that parts of it may have to be sacrificed. The injunction by Russell that "We cannot allow arbitrary conditions to determine sets and then indiscriminately allow the sets so formed to be members of other sets" certainly does not resolve the question of which impredicative definitions can be permitted.

Though the primary cause of the contradictions seemed to be evident, there remained the problem of how to build mathematics to eliminate them and, even more important, to make sure that new ones could not occur. We can see now why the problem of consistency became so urgent in the early 1900s. Mathematicians referred to the contradictions as the paradoxes of set theory. However, the work in set theory opened their eyes to possible contradictions even in classical mathematics.

In the effort to build solid foundations for mathematics, establishing consistency certainly became the most demanding problem. But other problems hardly less vital from the standpoint of assuring results already obtained were recognized in the early 1900s. The critical spirit had been sharpened during the late 19th century and mathematicians were reexamining everything that had been accepted earlier. They fixed upon a rather innocent looking assertion which had been used in many earlier proofs without attracting attention. This assertion is that, given any collection of sets, finite or infinite, one can select one object from each set and form a new set. Thus, from the people in all of the fifty states of the United States one can pick one person from each state and form a new set of people.

Recognition of the fact that the assertion really presupposes an axiom, called the axiom of choice, was impressed upon mathematicians by a paper of Ernst Zermelo (1871–1953) published in 1904. The history here is somewhat relevant. To be able to order his transfinite numbers as to size, Cantor needed the theorem that every set of real numbers can be well ordered. A set is well ordered if first of all it is ordered. Ordered means, as in the case of the whole numbers, that if a and b are any two members of the set, either a precedes b or b precedes a. Further, if a precedes b and b precedes c, then a precedes c. The set is well ordered if any subset, no matter how chosen, has a first element. Thus, the set of positive whole numbers in their usual order is well ordered. The set of real numbers arranged in the usual manner is ordered but not well ordered because the subset consisting of all numbers greater than zero does not have a first element. Cantor guessed that every set can be well ordered, a concept he introduced in 1883, and used but never proved. And, we may recall, Hilbert posed this problem of proving that the set of real numbers can be well ordered in his Congress talk of 1900. Zermelo did prove in 1904 that every set can be well ordered and in doing so he called attention to the fact that he used the axiom of choice.

As had happened many times in the past, mathematicians had used an axiom unconsciously and much later not only realized that they were using it but had to consider the ground for accepting such an axiom. Cantor had used the axiom of choice unwittingly in 1887 to prove that any infinite set contains a subset with cardinal number $\aleph_0$. It had also been used implicitly in many proofs of topology, measure theory, algebra, and functional analysis. It is used, for example, to prove that in a bounded infinite set one can select a sequence of numbers that converge to a limit point of the set. It is also used to construct the real numbers from Peano's axioms about the whole numbers, as fundamental a use as could be made. Still another use is in the proof that the power set of a finite set, that is, the set of all subsets of a finite set, is finite. In 1923 Hilbert described the axiom as a general principle that is necessary and indispensable for the first elements of mathematical inference.

Peano first called attention to the axiom of choice. In 1890 he wrote that one cannot apply an infinite number of times an arbitrary law that selects a member of a class from each of many classes. In the problem he dealt with (the integrability of differential equations), he gave a definite law of choice and so resolved the difficulty. The axiom was recognized as such by Beppo Levi in 1902 and was suggested to Zermelo by Erhardt Schmidt in 1904.

Zermelo's explicit use of the axiom of choice brought forth a storm of protest in the very next issue of the prestigious journal, *Mathematische Annalen* (1904). Papers by Emile Borel (1871–1956) and Felix Bernstein (1878–1956) criticized the use of the axiom. These criticisms were followed almost at once by letters to and from Emile Borel, René Baire (1874–1932), Henri Lebesgue (1875–1941), and Jacques Hadamard (1865–1963), all leading mathematicians, which were published in the *Bulletin de la Société Mathématique de France* of 1905.

The nub of the criticism was that, unless a definite law specified which element was chosen from each set, no real choice had been made, and so the new set was not really formed. The choice might be varied during the course of a proof, and so the proof would be invalid. As Borel put it, a lawless choice is an act of faith and the axiom is outside the pale of mathematics. Thus, to use an example Bertrand Russell gave in 1906, if I have one hundred pairs of shoes and say, I choose the left shoe from each pair, I have indicated a clear choice. But if I have one hundred pairs of stockings and now have to indicate which stocking I have chosen from each of the hundred pairs, I have no clear rule or law by which to do so. However, the defenders of the axiom of choice, though admitting there may be no law of choice, did not see the need for one. To them the choices are determined simply because one thinks of them as determined.

There were other objectors and grounds for objection. Poincaré admitted the axiom but not Zermelo's proof of the well-ordering because it used impredicative statements. Baire and Borel objected not only to the axiom but to the proof because it did not exhibit how the well-ordering could be accomplished; it proved only that it could be done.

Brouwer, whose philosophy we shall examine later (Chapter X), objected because he would not admit actually infinite sets. Russell's objection was that a set is defined by a property that all members of the set possess. Thus, one might define the set of all men who wear green hats by the property of wearing green hats. But the axiom of choice does not require that the selected elements have a definite property. It says merely that we can select a member from each of the given sets. Zermelo, himself, was satisfied to use the concept of set in an intuitive sense and so to him the choices of one element from each of the given sets clearly formed a new set.

Hadamard was Zermelo's only staunch defender. He urged that the axiom of choice be accepted on the same ground that he used to defend Cantor's work. For Hadamard, the assertion of the existence of objects did not require describing them. If the mere assertion of existence enables mathematics to make progress, then the assertion is acceptable.

To answer criticisms, Zermelo gave a second proof of well-ordering, which also used the axiom of choice and in fact proved that the two are equivalent. Zermelo defended the use of the axiom and said mathematics should continue to use it unless it led to contradictions. The axiom, he said, "has a purely objective character which is immediately clear." He agreed that it was not strictly self-evident because it dealt with choices from an infinite number of sets, but it was a scientific necessity because the axiom was being used to prove important theorems.

Many equivalent forms of the axiom of choice were devised. These are theorems if the axiom is adopted along with the other axioms of set theory. However, all attempts to replace the axiom by a less controversial one were unsuccessful and one acceptable to all mathematicians seems unlikely.

The key issue with respect to the axiom of choice was what mathematics means by existence. To some it covers any mental concept found useful that does not lead to contradictions, for example, an ordinary closed surface whose area is infinite. To others, existence means a specific, clear-cut identification or example of the concept, one which would enable anyone to point to or at least describe it. The mere possibility of a choice is not enough. These conflicting views were to be sharpened in the years to come and we shall say more about them in later chapters. What is relevant now is that the axiom became a serious bone of contention.

Despite this, many mathematicians continued to use it as mathematics expanded in the succeeding decades. A conflict continued to rage among mathematicians about whether it was legitimate, acceptable mathematics. It became the most discussed axiom next to Euclid's parallel axiom. As Lebesgue remarked, the opponents could do no better than insult each other because there was no agreement. He himself, despite his negative and distrustful attitude toward the axiom, employed it, as he put it, audaciously and cautiously. He maintained that future developments would help us to decide.

Still another problem began to plague mathematicians in the early 1900s. At the time, the problem did not seem momentous but as Cantor's theory of transfinite cardinal and ordinal numbers was utilized more and more, resolution became a major concern.

In his later work Cantor built the theory of transfinite cardinal numbers on the basis of the theory of ordinal numbers. For example, the cardinal number of the set of all possible sets of finite ordinals is $\aleph_0$. The cardinal number of all possible sets of ordinals which contain only a denumerable ($\aleph_0$) number of members is $\aleph_1$. Continuing in this fashion he obtained larger and larger cardinals which he denoted by $\aleph_0$, $\aleph_1$, $\aleph_2$, $\cdots$. Moreover, each aleph was the next larger one possible after the preceding one. But he had also shown very early in his work on transfinite numbers that the number of all real numbers is $2^{\aleph_0}$ or, more briefly denoted, c, and that $2^{\aleph_0}$ was greater than $\aleph_0$. The question he then raised was, where does c fit into the sequence of alephs? Since $\aleph_1$ was next after $\aleph_0$, c could be equal to or greater than $\aleph_1$. He conjectured that $c = \aleph_1$ and this conjecture, which he first stated in 1884 and published as such in 1884, is called the continuum hypothesis.* Another and somewhat simpler way of stating the hypothesis is that there is no transfinite number between $\aleph_0$ and c, or that any infinite subset of the real numbers has the cardinal number $\aleph_0$ or c.† In the first few decades of this century, the continuum hypothesis aroused much controversy which was not resolved. Beyond what new theorems could be proved with it, it had assumed great importance even for the understanding of infinite sets, one-to-one correspondence, and the choice of axioms that might be used to found set theory.

Thus, the mathematicians of the early part of this century faced several severe problems. The contradictions already discovered had to be resolved. More important the consistency of all of mathematics had to be proved to ensure that no new contradictions could arise. These problems were crucial. The axiom of choice was unacceptable to many mathematicians, and as a consequence many theorems of mathematics which depended on that axiom were in question. Could they be proved

---

* One can consider a set whose cardinal number is $\aleph_1$ and then consider the set of all subsets of that set. Its cardinal number is $2^{\aleph_1}$. Now $2^{\aleph_1} > \aleph_1$. One could then hypothesize that $2^{\aleph_1} = \aleph_2$, and $2^{\aleph_n} = \aleph_{n+1}$. This is the generalized continuum hypothesis.
† The last-mentioned version does not involve the axiom of choice.

by using a more acceptable axiom or could the axiom of choice be dispensed with entirely? The continuum hypothesis, whose importance became more evident with new developments, had to be proved or disproved.

Though the problems facing mathematicians in the early 1900s were serious ones, under other circumstances they might not have produced an upheaval. True, the contradictions had to be resolved but the ones actually known were in the theory of sets, a new branch which might be rigorized in time. As for the danger that new contradictions might be found in classical mathematics, perhaps because impredicative definitions had been used, by that time the consistency problem had been reduced to the question of the consistency of arithmetic, and no one could really doubt that. The real number system had been in use for over five thousand years and endless theorems about real numbers had been proved. No contradiction had been found. That an axiom, in the present case the axiom of choice, had been used implicitly and was to be used still more might not have disturbed many. The axiomatization movement of the late 19th century had revealed that many axioms had been used implicitly. The continuum hypothesis was at the time only a detail in Cantor's work, and some mathematicians derided Cantor's entire theory. Mathematicians had faced far more serious difficulties with equanimity. For example, in the 18th century, though fully aware of fundamental difficulties in the foundations of the calculus, they had proceeded nevertheless to build enormous branches of analysis on the basis of the calculus, and then analysis was rigorized on the basis of number.

The problems we have cited were like a match which lights a fuse, which in turn explodes a bomb. Some mathematicians still believed that mathematics proper is a body of truths. They hoped to establish this and Frege had already undertaken such a movement. Further, the objections to the axiom of choice were not based solely on what that axiom said. Mathematicians, Cantor in particular, had been introducing more and more constructs of the mind which, they maintained, had as much reality as, for example, the concept of a triangle. But others rejected such concepts as so tenuous that nothing solid could be built on them. The basic issue with respect to Cantor's work, the axiom of choice, and similar concepts was in what sense mathematical concepts exist. Must they correspond to or be idealizations of physically real objects? Aristotle had considered this problem and for him and for most Greeks real counterparts were a requirement. This is why Aristotle would not admit infinite sets as a totality, or a seven-sided regular polygon. The Platonists on the other hand, and Cantor was one, accepted ideas that they believed existed in some objective world independent of man. Man discovered these ideas or, as Plato put it, he recalled them.

Another facet of the existence question was the value of existence proofs. Gauss, for example, had proved that every nth degree polynomial equation with real or complex coefficients had at least one root. But the proof did not show how to calculate this root. Similarly, Cantor had proved that there were more real numbers than algebraic numbers (roots of polynomial equations). Hence, there must be transcendental irrational numbers. However, this existence proof did not enable one to name, much less calculate, even one transcendental number. Some early 20th-century mathematicians, Borel, Baire, and Lebesgue, regarded the mere proof of existence as worthless. Proof of existence should enable mathematicians to calculate the existing quantity to any desired degree of accuracy. Such proofs they called constructive.

Still another issue had been disturbing some mathematicians. The axiomatization of mathematics was a reaction to the intuitive acceptance of many obvious facts. It is true that this same movement removed contradictions and obscurities in, for example, the area of analysis. But it also insisted on explicit definitions, axioms, and proofs of facts that were obvious to the intuition, so obvious that no one had previously recognized the reliance upon intuition (Chapter VIII). The resulting deductive structures were indeed complicated and extensive. Thus, the development of the rational numbers and especially of the irrational numbers on the basis of axioms for the whole numbers was both detailed and complex. All of this struck some mathematicians, Leopold Kronecker (1823–1891) in particular, as highly artificial and unnecessary. Kronecker was the first of a distinguished group who felt that one could not build up more soundly by logical means what man's intuition assured him was sound.

Another bone of contention was the growing body of mathematical logic which made mathematicians aware that even the use of logical principles could no longer be informal and casual. The work of Peano and Frege required that mathematicians make sharp distinctions in their reasoning, such as the distinction between an object which belongs to a class and a class which is included in another. Such distinctions seemed pedantic and more of an obstacle than an aid.

More important by far, though not explicit in the late 1800s, was the fact that many mathematicians were beginning to be uneasy about the unrestricted applicability of the logical principles. What guaranteed their application to infinite sets? If the logical principles were a product of human experience, then surely there must be some question about whether they extended to mental constructs that had no basis in experience.

Long before 1900, mathematicians had begun to disagree on the basic issues we have just described. Thus, the new paradoxes merely aggravated disagreements that were already present. Years later, mathematicians were to look back longingly on the brief but happy period before contradictions were discovered, the period Paul du Bois-Reymond referred to as the time when "we still lived in Paradise."

X

Logicism versus Intuitionism

The discovery of the paradoxes of set theory and the realization that similar paradoxes might be present, though as yet undetected, in the existing classical mathematics, caused mathematicians to take seriously the problem of consistency. The issue of what is meant by existence in mathematics, raised in particular by the free use of the axiom of choice, also became a live one. The increasing use of infinite sets in rebuilding foundations and creating new branches brought to the fore the age-old disagreement on whether actually infinite sets are a legitimate concept. The axiomatic movement of the late 19th century did not treat these matters.

However, it was not just these issues and others treated in the preceding chapter that led mathematicians to reconsider the entire subject of the proper foundations. These issues were winds that caused smoldering fires to flare up into the white heat of controversy. Several new and radical approaches to mathematics had been voiced and somewhat elaborated before 1900. But they had not gained the limelight and most mathematicians did not take them seriously. In the first decade of this century, giants of mathematics came forth to do battle for new approaches to the foundations. They split up into opposing camps and declared war on their enemies.

The first of these schools of thought is known as the logistic school. Its thesis, briefly stated for the moment, is that all of mathematics is derivable from logic. In the early 1900s the laws of logic were accepted by almost all mathematicians as a body of truths. Hence the logicists contended that mathematics must also be. And since truth is consistent, so, they claimed, mathematics must be.

As in the case of all innovations, many people made contributions before this thesis gained definitive form and widespread attention. The thesis that mathematics is derivable from logic can be traced back to Leibniz. Leibniz distinguished truths of reason or necessary truths from truths of fact or contingent truths (Chapter VIII). In a letter to his friend Coste, Leibniz explained this distinction. A truth is *necessary* when the opposite implies contradiction, and when it is not necessary it is called *contingent.* That God exists, that all right angles are equal, etc. are necessary truths; but that I myself exist and that there are bodies in nature which possess an angle of exactly 90° are contingent truths. These could be true or false, for the whole universe might be otherwise. And God had chosen among an infinite number of possibles what he judged most fit. Because mathematical truths are necessary, they must be derivable from logic whose principles are also necessary and hold true in all possible worlds.

Leibniz did not carry out the program of deriving mathematics from logic, nor for almost two hundred years did others who expressed the same belief. For example, Richard Dedekind affirmed flatly that number is not derived from intuitions of space and time but is "an immediate emanation from the pure laws of thought." From number we gain precise concepts of space and time. He started to develop this thesis but did not pursue it.

Finally, Gottlob Frege, who contributed much to the development of mathematical logic (Chapter VIII) and was influenced by Dedekind, undertook to develop the logistic thesis. Frege believed that laws of mathematics are what is called analytic. They say no more than what is implicit in the principles of logic, which are a priori truths. Mathematical theorems and their proofs show us what is implicit. Not all of mathematics may be applicable to the physical world but certainly it consists of truths of reason. Having built up logic on explicit axioms in *Concept-Writing* (1879), Frege proceeded in *Foundations of Mathematics* (1884) and in his two-volume *Fundamental Laws of Mathematics* (1893, 1903) to derive the concepts of arithmetic and the definitions and laws of number from logical premises. From the laws of number, it is possible to deduce algebra, analysis, and even geometry because analytic geometry expresses the concepts and properties of geometry in algebraic terms. Unfortunately, Frege's symbolism was quite complex and strange to mathematicians. He therefore had little influence on his contemporaries. Rather ironic, too, is the oft-told story that just as the second volume of his *Fundamental Laws* was about to go to press in 1902 he received a letter from Bertrand Russell, informing him that his work involved a concept, the set of all sets, that can lead to a contradiction. At the close of Volume II, Frege remarked, "A scientist can hardly meet with anything more undesirable than to have the foundation give way just as the work is finished. A letter from Mr. Bertrand Russell put me in this position at the moment the work was nearly through the press." Frege had been unaware of the paradoxes that had already been noted during the period in which he was writing this book.

Independently, Bertrand Russell had conceived the same program and while developing it ran across Frege's work. Russell said in his *Autobiography* (1951) that he was influenced also by Peano whom he met at the Second International Congress in 1900:

> The Congress was a turning point in my intellectual life, because there I met Peano. I already knew him by name and had seen some of his work. . . . It became clear to me that his notation afforded an instrument of analysis such as I had been seeking for years, and that by studying him I was acquiring a new and powerful technique for the work I had long wanted to do.

He said further in his *Principles of Mathematics* (1st edition, 1903), "The fact that All Mathematics is Symbolic Logic is one of the greatest discoveries of our age. . . ."

In the early 1900s, Russell believed with Frege that if the fundamental laws of mathematics could be derived from logic, since logic was certainly a body of truths, then these laws would also be truths and the problem of consistency would be solved. He wrote in *My Philosophical Development* (1959) that he had sought to arrive "at a perfected mathematics which should leave no room for doubts."

Russell did know of course that Peano had derived the real numbers from axioms about the whole numbers, and he was also aware that Hilbert had given a set of axioms for the entire real number system. However, he remarked in his *Introduction to Mathematical Philosophy* (1919) apropos of a somewhat similar maneuver by Dedekind, "The method of postulating what we want has many advantages; they are the same as the advantages of theft over honest toil." Russell's real concern was that postulation of, say, ten or fifteen axioms about number does not ensure the consistency and truth of the axioms. As he put it, it gives unnecessary hostages to fortune. Whereas in the early 1900s Russell was sure that the principles of logic were truths and therefore consistent, Whitehead cautioned in 1907, "There can be no formal proofs of the consistency of the logical premises themselves."

For many years Russell held that the principles of logic and the objects of mathematical knowledge exist independently of any mind and are merely perceived by the mind. This knowledge is objective and unchanging. A clear statement of this position is in his 1912 book *The Problems of Philosophy*.

Russell's intent was to go even further than Frege insofar as truth is concerned. In his youth he believed that mathematics offered truths about the physical world. Among conflicting geometries, Euclidean and non-Euclidean, all of which fit the physical world (Chapter IV), he could not affirm which was the truth, but in his *Essay on the Foundations of Geometry* (1898) he did manage to find some mathematical laws, such as that physical space must be homogeneous (possess the same properties everywhere), which he then believed were physical truths. The three-dimensionality of space, by contrast, was an empirical fact. Nevertheless, there was an objective real world about which we can obtain exact knowledge. Thus Russell sought mathematical laws which must also be physically true. These laws should be derivable from logical principles.

In his *Principles* of 1903 Russell amplified his position about the physical truth of mathematics and there he said that "All propositions as to what actually exists, like the space we live in, belong to experimental or empirical science, not to mathematics; when they belong to applied mathematics, they arise from giving to one or more variables in a proposition of pure mathematics some constant value. . . ." Even in this version he still believed that some basic physical truths would be contained in the mathematics derived from logic. In reply to sceptics who asserted there is no absolute truth, Russell said, "Of such scepticism mathematics is a perpetual reproof; for its edifice of truths stands unshakable and inexpungable to all the weapons of doubting cynicism."

The ideas sketched by Russell in his *Principles* were developed in the detailed work by Alfred North Whitehead (1861–1947) and Russell, the *Principia Mathematica* (3 volumes; 1st edition, 1910–13). Since the *Principia* is the definitive version of the logistic school's position, let us concern ourselves with its contents.

The approach starts with the development of logic itself. Axioms of logic are carefully stated from which theorems are deduced to be used in subsequent reasoning. The development begins with undefined ideas, as any axiomatic theory must (Chapter VIII). Some of these undefined ideas are the notion of an elementary proposition, the assertion of the truth of an elementary proposition, the negation of a proposition, the conjunction and the disjunction of two propositions, and the notion of a propositional function.

Russell and Whitehead explained these notions, although, as they pointed out, this explanation was not part of the logical development. By proposition and propositional function, they meant what Peirce had already introduced. Thus, "John is a man" is a proposition, whereas "x is a man" is a propositional function. The negation of a proposition is intended to mean it is not true that the proposition holds, so that if p is the proposition that "John is a man," the negation of p, denoted by

$\sim p$, means "It is not true that John is a man," or "John is not a man." The conjunction of two propositions p and q, denoted by $p \cdot q$, means that both p and q must be true. The disjunction of two propositions p and q, denoted by $p \vee q$, means p or q. The meaning of "or" here is that intended in the sentence "Men or women may apply." That is, men may apply; women may apply; and both may apply. In the sentence "That person is a man or a woman" "or" has the more common meaning of either one or the other but not both. Mathematics uses "or" in the first sense, though sometimes the second sense is the only one possible. For example, "The triangle is isosceles or the quadrilateral is a parallelogram" illustrates the first sense. We also say that every number is positive or negative. Here additional facts about positive and negative numbers tell us that both cannot be true. Thus in the *Principia* the assertion p or q means that p and q are both true, or that p is false and q true, or p is true but q false.

A most important relationship between propositions is implication, that is, the truth of one proposition compelling the truth of another. In the *Principia*, implication, denoted by $\supset$, is defined. It means what Frege called material implication (Chapter VIII). That is, p implies q means that if p is true q must be true. However, if p is false then p implies q whether q is true or false. That is, a false proposition implies any proposition. This notion of implication is consistent at least with what can happen. Thus, if it is true that a is an even number, then $2a$ must be even. However, if it is false that a is an even number, then $2a$ might be even or (if a were a fraction) $2a$ might not be even. Either conclusion can be drawn from the falsity of the proposition that a is an even number.

Of course, to deduce theorems there must be axioms of logic. A few of these are:

A. Anything implied by a true elementary proposition is true.
B. If either p is true or p is true, then p is true.
C. If q is true, then p or q is true.
D. p or q implies q or p.
E. p or $(q$ or $r)$ implies q or $(p$ or $r)$.
F. The assertion of p and the assertion $p \supset q$ permit the assertion of q.

From such axioms the authors proceed to deduce theorems of logic. The usual syllogistic rules of Aristotle occur as theorems.

To see how even logic itself was formalized and made deductive, let us note a few theorems of the early part of *Principia Mathematica*. One theorem states that if the assumption of p implies that p is false, then p is false. This is the principle of *reductio ad absurdum*. Another theorem states that if q implies r, then if p implies q, p implies r. (This is one form of the Aristotelian syllogisms.) A basic theorem is the principle of excluded middle: for any proposition p, p is either true or false.

Having built up the logic of propositions, the authors proceed to propositional functions. These in effect represent classes or sets, for instead of naming the members of a class a propositional function describes them by a property. For example, the propositional function "x is red" denotes the set of all red objects. This method of defining a class enables one to define infinite classes as readily as finite classes of objects. It is called intensional as opposed to an extensional definition, which names the members of the class.

Russell and Whitehead wished of course to avoid the paradoxes that arise when a collection of objects is defined that contains itself as a member. The resolution by Russell and Whitehead of this difficulty was to require that "whatever involves all members of a collection must not itself be a member of the collection." To carry out this restriction in the *Principia* they introduce the theory of types.

The theory of types is somewhat involved. But the idea is straightforward. Individuals such as John or a particular book are of type 0. An assertion about a property of individuals is of type 1. A proposition about or concerning a property of individuals is of type 2. Every assertion must be of higher type than what it asserts of some lower type. Expressed in terms of sets, the theory of types states that individual objects are of type 0; a set of individuals is of type 1; a set of sets of individuals is of type 2; and so forth. Thus, if one says a belongs to b, b must be of higher type than a. Also, one cannot speak of a set belonging to itself. The theory of types is actually a little more complicated when applied to propositional functions. A propositional function cannot have as one of its arguments (values of the variables) anything defined in terms of the function itself. The function is then said to be of higher type than that of the variables. On the basis of this theory, the authors discuss the current paradoxes and show that the theory of types avoids them.

The virtue of the theory of types in avoiding contradictions is more evident from a non-mathematical example. Let us consider the contradiction posed by the statement "All rules have exceptions" (Chapter IX). This statement is about particular rules such as "All books have misprints." Whereas the statement about all rules is usually interpreted to apply to itself and so leads to the contradiction that there are rules without exceptions, in the theory of types the general rule is of higher type and what it says about particular rules cannot be applied to itself. Hence, the general rule need not have exceptions.

Similarly the heterological paradox—which defines a heterological word as a word that does not apply to itself—is a definition of all heterological words and so of higher type than any one heterological word. Therefore one cannot ask whether heterological is itself heterological. One can ask whether a particular word, short, for instance, is heterological.

The liar paradox also is resolved by the theory of types. As Russell puts it, the statement "I am lying" means "There is a proposition that I am affirming and it is false." Or "I assert a proposition p, and p is false." If p is of the nth order, then the assertion about p is of higher order. Hence if the *assertion* about p is true, then p itself is false, and if the *assertion* about p is false, then p itself is true. But there is no contradiction. The same use of the theory of types resolves the Richard paradox. All involve a statement of higher type about a statement of lower type.

Clearly the theory of types requires that statements be carefully distinguished by type. However, if one attempts to build mathematics in accordance with the theory of types, the development becomes exceedingly complex. For example, in the *Principia* two objects a and b are equal if every proposition and propositional function that applies to or holds for a also holds for b and conversely. But these various assertions are of different types. Accordingly the concept of equality is rather involved. Similarly, because irrational numbers are defined in terms of rational numbers and the rationals are defined in terms of the positive whole numbers, the irrationals are of higher type than the rationals and these are of higher type than the whole numbers. Thus the system of real numbers consists of numbers of different types. Hence one cannot assert a theorem about all the real numbers but must assert one about each type separately, because a theorem that applies to one type does not automatically apply to another.

The theory of types introduces a complication also in the concept of the least upper bound of a bounded set of real numbers (Chapter IX). The least upper bound is defined as the smallest of all the upper bounds. Thus the least upper bound is defined in terms of a set of real numbers. Hence it must be of higher type than the real numbers and so is not itself a real number.

To avoid such complexities, Russell and Whitehead introduced the rather subtle axiom of reducibility. The axiom of reducibility for propositions say that any proposition of higher type is equivalent to one of first order. The axiom of reducibility for propositional functions says any function of one or two variables is coextensive with a function of one type and of the same number of variables *whatever the type of the variables*. This axiom is also needed to support mathematical induction which is used in the *Principia*.

Having treated propositional functions the authors take up next the theory of relations. Relations are expressed by means of propositional functions of two or more variables. Thus "x loves y" expresses a relation. Following the theory of relations comes an explicit theory of classes or sets defined in terms of propositional functions. On this basis the authors are prepared to introduce the notion of natural (positive whole) number.

The definition of a natural number has of course considerable interest. It depends upon the previously introduced relation of one-to-one correspondence between classes. If two classes are in one-to-one correspondence, they are called similar. All similar classes possess a common property, their number. However, similar classes may have more than one common property. Russell and Whitehead get around this, as Frege did, by defining the *number of a class as the class of all classes which are similar to the given class.* Thus the number 3 is the class of all three-membered classes, and the denotation of all three-membered classes is $\{x,y,z\}$ with $x \neq y \neq z$. Since the definition of number presupposes the concept of one-to-one correspondence, it would seem as though the definition is circular. The authors point out, however, that a relation is one-to-one if when x and x' have the relation to y then x and x' are identical and, when x has the relation to y and y', then y and y' are identical. Hence the concept of one-to-one correspondence, despite the verbal expression, does not involve the number 1.

Given the natural numbers, it is possible to build up the real and complex number systems, functions, and all of analysis. Geometry can be introduced through the mathematics of number by using coordinates and equations of curves. However, to accomplish their objective Russell and Whitehead introduced two more axioms. To carry out the program of first defining the natural numbers (in terms of propositional functions) and then the successively more complicated rational and irrational numbers and to include transfinite numbers, Whitehead and Russell introduced the axiom that infinite classes exist (classes having been properly defined in logical terms) and the axiom of choice (Chapter IX), which is needed in the theory of types.

This is then the grand program of the logistic school. What it does with logic itself is a large story which we are skimming over here quite briefly. What it does for mathematics, and this we must emphasize, is to found mathematics on logic. Mathematics becomes no more than a natural extension of the laws and subject matter of logic.

The logistic approach to mathematics has received much criticism. The axiom of reducibility aroused opposition, for it seemed to many quite arbitrary. There is lack of evidence for it, though there is no proof of its falsity. Some have called it a happy accident, not a logical necessity. Frank Plumpton Ramsey, though sympathetic to the logistic theory, criticized the axiom with the words, "Such an axiom has no place in mathematics; and anything that cannot be proved without using it cannot be regarded as proved at all." Others have called it a sacrifice of the intellect. Hermann Weyl rejected the axiom unequivocally. Some critics declared that the axiom restores impredicative definitions. Perhaps the most serious question was whether it is an axiom of logic and hence whether the thesis that mathematics is founded on logic is really substantiated.

Poincaré said in 1909 that the axiom of reducibility is more questionable and less clear than the principle of mathematical induction, which is proved in effect by the axiom. It is a disguised form of mathematical induction. But mathematical induction is part of mathematics and is needed to establish mathematics. Hence we cannot prove consistency.

The justification that Russell and Whitehead gave for the axiom in the first edition of the *Principia* (1910) was that it was needed for certain results. They were apparently uneasy about using it. There the authors said in its defense:

> In the case of the axiom of reducibility, the intuitive evidence in its favor is very strong, since the reasonings which it permits and the results to which it leads are all such as appear valid. But although it seems very improbable that the axiom should turn out to be false, it is by no means improbable that it should be found to be deducible from some other more fundamental and more evident axiom.

Later Russell himself became more concerned about using the axiom of reducibility. In his *Introduction to Mathematical Philosophy* (1919), Russell said:

> Viewed from this strictly logical point of view, I do not see any reason to believe that the axiom of reducibility is logically necessary, which is what would be meant by saying that it is true in all possible worlds. The admission of this axiom into a system of logic is therefore a defect, even if the axiom is empirically true.

In the second edition of the *Principia* (1926), Russell rephrased the axiom of reducibility. But this created several difficulties such as not allowing higher infinities, omitting the theorem of least upper bound, and complicating the use of mathematical induction. Russell again said he hoped that the reducibility axiom could be derived from more evident axioms. But he again admitted the axiom is a logical defect. Russell and Whitehead agreed in the second edition of the *Principia* that "This axiom has a purely pragmatic justification; it leads to the desired results and no others. But clearly it is not the sort of axiom with which we can rest content." They realized that the fact that it leads to correct conclusions is not a cogent argument. Various attempts have been made to reduce mathematics to logic without an axiom of reducibility, but they have not been pursued in depth and some of them have been criticized on the grounds that they involve fallacious proofs.

Further criticism of the logistic foundation was directed against the axiom of infinity. The thrust was that the structure of the whole of arithmetic depends essentially on the truth of this axiom whereas there is not the slightest reason for believing in its truth and, worse yet, there is no way of reaching a decision regarding its truth. Further, there is the question of whether this axiom is an axiom of logic.

In all fairness to Russell and Whitehead, it should be pointed out that they hesitated to accept the axiom of infinity as an axiom of logic. They were disturbed by the fact that the content of the axiom had a "factual look." Not only its logicality but its truth also bothered them. One of the interpretations suggested for the term "individual" in *Principia Mathematica* is that of the ultimate particles or elements of which the universe is composed. The axiom of infinity, although it is couched in logical terms, thus seems to pose the problem of whether the universe is composed of a finite or infinite number of ultimate particles, a question which perhaps can be answered by physics, but certainly not by mathematics and logic. Nonetheless, if infinite sets were to be introduced, and if mathematical theorems in whose derivation the axiom of infinity was used were to be shown to be theorems of logic, then it seemed necessary to accept this axiom as an axiom of logic. In short, if mathematics was to be "reduced" to logic, then logic seemingly would have to include the axiom of infinity.

Russell and Whitehead also used the axiom of choice (Chapter IX) which they called the multiplicative axiom: Given a class of disjoint (mutually exclusive) classes, none of which is the null or empty class, then there exists a class composed of exactly one element from each class and of no other elements. As we know, this axiom has engendered more discussion and controversy than any other axiom, with the possible exception of Euclid's axiom of parallels. Russell and Whitehead were equally uneasy about the axiom of choice and could not persuade themselves to treat it as a logical truth on a par with the other logical axioms. Nonetheless, if parts of classical mathematics whose derivation needed the axiom of choice were to be "reduced" to logic, then it seemed that this axiom, too, had to be considered a part of logic.

The use of all three axioms, reducibility, infinity, and choice, challenged the entire logistic thesis that all of mathematics can be derived from logic. Where does one draw the line between logic and mathematics? Proponents of the logistic thesis maintained that the logic used in *Principia Mathematica* was "pure logic" or "purified logic." Others, having the three controversial axioms in mind, questioned the "purity" of the logic employed. Hence they denied that mathematics, or even any important branch of mathematics, had yet been reduced to logic. Some were willing to extend the meaning of the term logic so that it includes these axioms.

Russell, who took up the cudgels for the logistic thesis, for a while defended everything he and Whitehead had done in the first edition of the *Principia*. In his *Introduction to Mathematical Philosophy* (1919) he argued:

> The proof of this identity [of mathematics and logic] is, of course, a matter of detail; starting with premises which would be universally admitted to belong to logic and arriving by deduction at results which as obviously belong to mathematics, we find that there is no point at which a sharp line can be drawn with logic to the left and mathematics to the right. If there are still those who do not admit the identity of logic and mathematics, we may challenge them to indicate at what point, in the successive definitions and deductions of *Principia Mathematica*, they consider that logic ends and mathematics begins. It will then be obvious that any answer must be quite arbitrary.

In view of the controversies concerning Cantor's work and the axioms of choice and infinity, which reached high intensity during the early 1900s, Russell and Whitehead did not specify the two axioms as axioms of their entire system but used them (in the second edition, 1926) only in specific theorems where they explicitly call attention to the fact that these theorems use the axioms. However, they must be used to derive a large part of classical mathematics. In the second edition of his *Principles* (1937), Russell backtracked still more. He said that "The whole question of what are logical principles becomes to a very considerable extent arbitrary." The axioms of infinity and choice "can only be proved or disproved by empirical evidence." Nevertheless, he insisted that logic and mathematics are a unity.

However, the critics could not be stilled. In his *Philosophy of Mathematics and Natural Science* (1949), Hermann Weyl said the *Principia* based mathematics

> not on logic alone, but on a sort of logician's paradise, a universe endowed with an "ultimate furniture" of rather complex structure. . . . Would any realistically-minded man dare say he believes in this transcendental world? . . . This complex structure taxes the strength of our faith hardly less than the doctrines of the early Fathers of the Church or of the Scholastic philosophers of the Middle Ages.

Still another criticism has been directed against logicism. Though geometry was not developed in the three volumes of the *Principia*, it seemed clear, as previously noted, that by using analytic geometry, one could do so. Nevertheless, it is sometimes argued that the *Principia,* by reducing to logic a set of axioms for the natural numbers, thereby reduced arithmetic, algebra, and analysis to logic but did not reduce to logic the "non-arithmetical" parts of mathematics, such as geometry, topology, and abstract algebra. This is the view taken, for example, by the logician Carl Hempel who notes that, although it was possible in the case of arithmetic to give the customary meaning of the undefined or primitive concepts "in terms of purely logical concepts," an "analogous procedure is not applicable to those disciplines that are not outgrowths of arithmetic." On the other hand, the logician Willard Van Orman Quine, who holds that "mathematics reduces to logic," sees that for geometry "a method of reduction to logic is ready at hand" and that topology and abstract algebra "fit into the general structure of logic." Russell himself doubted that one could deduce all of geometry from logic alone.

A serious philosophical criticism of the entire logistic position is that, if the logistic view is correct, then all of mathematics is a purely formal, logico-deductive science whose theorems follow from the laws of thought. Just how such deductive elaboration of the laws of thought can represent wide varieties of natural phenomena, the uses of number, the geometry of space, acoustics, electromagnetics, and mechanics, seems unexplained. Weyl's criticism on this matter was that from nothing nothing follows.

Poincaré, whose views we shall say more about later, was equally critical of what he considered to be sterile manipulation of logical symbolism. He said in an essay of 1906, by which time Russell (and Hilbert) had given ample indication of their programs:

> This science [mathematics] does not have for its unique objective to eternally contemplate its own navel; it touches nature and some day it will make contact with it. On this day it will be necessary to discard the purely verbal definitions and not any more be the dupe of empty words.

In the same essay Poincaré also said:

> Logistic is to be remade and it is not clear how much of it can be saved. Needless to add that Cantorism and logistic alone are under consideration; real mathematics, that which is good for something, may continue to develop in accordance with its own principles without bothering about the storms which rage outside it, and go on step by step with its usual conquests which are final and which it never has to abandon.

Another serious criticism of the logistic program affirms that in the creation of mathematics, perceptual or imaginative intuition must supply new concepts whether or not derived from experience. Else, how could new knowledge arise? But in the *Principia* all concepts reduce to logical ones. Formalization apparently does not represent mathematics in any real sense. It is the husk, not the corn. Russell's own statement, made in another connection, that mathematics is the subject in which we never know what we are talking about nor whether what we are saying is true, can be turned against logicism.

The questions of how new ideas could enter mathematics and how mathematics can possibly apply to the physical world if its contents are derivable entirely from logic are not readily answered and were not answered by Russell or Whitehead. The argument that logicism does not explain why mathematics fits the physical world can be countered by the fact that mathematics is applied to basic physical principles. These are the premises as far as reality is concerned. Mathematical techniques draw out implication of physical principles, such as $pv = \text{const.}$ or $F = ma$. Yet the conclusions still apply to the physical world. There is a problem here: Why does the world conform to mathematical reasoning? We shall return to this topic (Chapter XV).

In the years after the second edition of *Principia Mathematica* was published, Russell continued to think about the logistic program. He himself admits in *My Philosophical Development* (1959) that this consisted of a piecemeal retreat from "Euclidianism," while striving to rescue as much certainty as he could. The criticism that the logistic philosophy received no doubt influenced Russell's later thinking. When Russell started his work in the early part of the century, he thought that the axioms of logic were truths. He abandoned this view in the 1937 edition of his *Principles of Mathematics*. He was no longer convinced that the principles of logic are a priori truths and hence neither is mathematics, since it is derived from logic.

If the axioms of logic are not truths, logicism leaves unanswered the overriding question of the consistency of mathematics. The dubious axiom of reducibility puts consistency in even greater jeopardy. Russell's reason for accepting the axiom of reducibility in the first and second editions of *Principia Mathematica*, "that many other propositions which are nearly indubitable can be deduced from it, and that no equally plausible way is known by which these propositions could be true if the axiom were false, and nothing which is probably false can be deduced from it" does not carry much weight. Material implication, accepted in the *Principia Mathematica* (and in many systems of logic), allows an implication to hold even when its first member is false. Hence if a false proposition p were introduced as an axiom, p implies q would hold in the system and q could still be true. Thus the contention that indubitable propositions can be deduced from the axiom is pointless since, in the logic of *Principia Mathematica,* any "indubitable" proposition could very well be deduced from the axiom even if it happened to be false.

The *Principia* has been criticized on various grounds which we have not explicitly considered above. The hierarchy of types has been found to be valid and useful but it is not certain that it fully accomplishes its purpose. The device of types was introduced in order to guard against antinomies and it has effectively barred the way to the known antinomies of set theory and logic. But there is no guarantee that new antinomies may not crop up against which even the hierarchy of types would be of no help.

Nevertheless, some outstanding logicians and mathematicians, Willard Van Orman Quine and Alonzo Church for instance, still advocate logicism even though they are critical of its present state. Many are working to eliminate defects. Others not necessarily defending all the theses of logicism insist that logic and therefore mathematics are analytic, that is, mere amplifications of what the axioms state. Thus the logistic program has enthusiastic supporters who seek to remove cause for objection and the cumbersomeness of some of its development. Others regard it as a pious hope; and still others, as we shall see in a moment, attack it as a totally false conception of mathematics. All in all, in view of the questionable axioms and the long, complicated development, the critics could with good reason say that logicism produces foregone conclusions from unwarranted assumptions.

The Russell-Whitehead work did make a contribution in another direction. The mathematization of logic had been initiated in the latter part of the 19th century (Chapter VIII). Russell and Whitehead carried out a thorough axiomatization of logic in entirely symbolic form and so advanced enormously the subject of mathematical logic.

Perhaps the final word on logicism was stated by Russell himself in his *Portraits from Memory* (1958):

> I wanted certainty in the kind of way in which people want religious faith. I thought that certainty is more likely to be found in mathematics than elsewhere. But I discovered that many mathematical demonstrations, which my teachers expected me to accept, were full of fallacies, and that, if certainty were indeed discoverable in mathematics, it would be in a new field of mathematics, with more solid foundations than those that had hitherto been thought secure. But as the work proceeded, I was continually reminded of the fable about the elephant and the tortoise. Having constructed an elephant upon which the mathematical world could rest, I found the elephant tottering, and proceeded to construct a tortoise to keep the elephant from falling. But the tortoise was no more secure than the elephant, and after some

twenty years of very arduous toil, I came to the conclusion that there was nothing more that I could do in the way of making mathematical knowledge indubitable.

In *My Philosophical Development* (1959), Russell confessed, "The splendid certainty which I had always hoped to find in mathematics was lost in a bewildering maze. . . . It is truly a complicated conceptual labyrinth." The tragedy is not just Russell's.

While logicism was in the making, a radically different and diametrically opposite approach to mathematics was undertaken by a group of mathematicians called intuitionists. It is a most interesting paradox of the history of mathematics that while the logicists were relying more and more on refined logic to secure a foundation for mathematics, others were turning away from and even abandoning logic. In one respect, both sought the same goal. Mathematics in the late 19th century had lost its claim to truth in the sense of expressing laws inherent in the design of the physical universe. The early logicists, notably Frege and Russell, believed that logic was a body of truths, and so mathematics proper if founded on logic would also be a body of truths, though ultimately they had to retreat from this position to logical principles that had only pragmatic sanction. The intuitionists also sought to establish the truth of mathematics proper by calling upon the sanction granted by human minds. Derivations from logical principles were less trustworthy than what can be intuited directly. The discovery of the paradoxes not only confirmed this distrust but accelerated the formulation of the definitive doctrines of intuitionism.

In a broad sense of the word, intuitionism can be traced back at least to Descartes and Pascal. Descartes in his *Rules for the Direction of the Mind* said:

> Let us now declare the means whereby our understanding can rise to knowledge without fear of error. There are two such means: intuition and deduction. By intuition I mean not the varying testimony of the senses, nor the misleading judgment of imagination naturally extravagant, but the conception of an attentive mind so distinct and so clear that no doubt remains to it with regard to that which it comprehends; or, what amounts to the same thing, the self-evidencing conception of a sound and attentive mind, a conception which springs from the light of reason alone, and is more certain, because more simple, than deduction itself, although as we have noted above the human mind cannot err in deduction either. Thus everyone can see by intuition that he exists, that he thinks, that a triangle is bounded by only three lines, a sphere by a single surface, and so on.
>
> . . .
>
> It may perhaps be asked why to intuition we add this other mode of knowing by deduction, that is to say, the process which, from some-

thing of which we have certain knowledge, draws consequences which necessarily follow therefrom. But we are obliged to admit this second step; for there are a great many things which, without being evident of themselves, nevertheless bear the marks of certainty if only they are deduced from true and incontestable principles by a continuous and uninterrupted movement of thought, with distinct intuition of each thing; just as we know that the last link of a long chain holds to the first, although we can not take in with one glance of the eye the intermediate links, provided that, after having run over them in succession, we can recall them all, each as being joined to its fellows, from the first up to the last. Thus we distinguish intuition from deduction, inasmuch as in the latter case there is conceived a certain progress or succession, while it is not so in the former. . . . whence it follows that primary propositions, derived immediately from principles, may be said to be known, according to the way we view them, now by intuition, now by deduction, although the principles themselves can be known only by intuition, the remote consequences only by deduction.

Pascal, too, placed great faith in intuition. In his mathematical work, Pascal was in fact largely intuitive; he anticipated great results, made superb guesses, and saw short cuts. Later in life he favored intuition as a source of all truths. Several of his declarations on this subject have become famous. "The heart has its own reasons, which reason does not know." "Reason is the slow and tortuous method by which those who do not know the truth discover it." "Humble thyself, impotent reason."

To a large extent intuitionism was anticipated by the philosopher Immanuel Kant (1724–1804). Though primarily a philosopher, Kant taught mathematics and physics at the University of Königsberg from 1755 to 1770. He granted that we receive sensations from a presumed external world. However, these sensations or perceptions do not provide significant knowledge. All perception involves an interaction between the perceiver and the object perceived. The mind organizes the perceptions and these organizations are intuitions of space and time. Space and time do not exist objectively but are the contributions of the mind. The mind applies its understanding of space and time to experiences which merely awaken the mind. Knowledge may begin with experience but does not really come from experience. It comes from the mind. Mathematics is the shining example of how far, independently of experience, we can progress in a priori or true knowledge. Moreover, it is what Kant called synthetic; that is, it offers new knowledge, whereas analytic propositions, such as "all bodies are extended," offer nothing new because, by the very nature of bodies, extension is a property. By contrast, the proposition that a straight line is the shortest distance between two points is synthetic.

Though Kant was wrong in asserting the a priori synthetic character of Euclidean geometry, it was the prevalent belief of all philosophers and mathematicians of his time. This error discredited his philosophy in the eyes of later philosophers and mathematicians. However, Kant's analysis of time as one form of intuition and his general thesis that the mind supplies the basic truths had a lasting influence.

Had mathematicians been more conversant with the views of men such as Descartes, Pascal, and Kant, they would not have been shocked by the intuitionist school of thought, which was regarded as radical at least at the outset. However, neither Descartes nor Pascal nor Kant had in mind an intuitionist approach to all of mathematics. As an approach to the foundations of mathematics it is a modern one.

The immediate forerunner of modern intuitionism is Leopold Kronecker (1823–1891). His epigram (delivered in an after-dinner speech), "God made the integers; all the rest is the work of man," is well known. The complicated logical derivation of the ordinary whole numbers such as Cantor and Dedekind presented through a general theory of sets seemed less reliable than direct acceptance of the integers. These were intuitively clear and needed no more secure foundation. Beyond the integers, all mathematical constructs must be built up in terms that have clear meaning for man. Kronecker advocated construction of the real number system on the basis of the integers and methods that would enable us to calculate the real numbers and not merely give general existence theorems. Thus he accepted irrational numbers that are roots of polynomial equations if these roots could be calculated.

Cantor proved that there are transcendental irrational numbers, that is, numbers that are not roots of algebraic equations, and in 1882 Ferdinand Lindemann proved that π is a transcendental irrational number. Apropos of this work Kronecker said to Lindemann, "Of what use is your beautiful investigation regarding π? Why study such problems since such irrationals do not exist?" Kronecker's objection was not to all irrationals but to proofs that did not in themselves permit the calculation of the number in question. Lindemann's proof was not constructive. Actually, π can be calculated to as many decimal places as desired by means of an infinite series expression but Kronecker would not accept the derivation of such a series.

Infinite sets and transfinite numbers Kronecker rejected because he accepted only a potential infinity. Cantor's work in this area was not mathematics but mysticism. Classical analysis was a game with words. He might well have added that, if God has a different mathematics, He should build it up for Himself. Kronecker stated his views but did not develop them. Possibly he did not take his own radical notions too seriously.

Borel, Baire, and Lebesgue, whose objections to the axiom of choice we have already encountered, were semi-intuitionists. They accepted the real number system as a foundation. The details of their views are rather of historical interest because these men, too, though they expressed themselves on specific issues, did not propound a systematic philosophy. Poincaré, like Kronecker, thought one does not have to define the whole numbers or construct their properties on an axiomatic foundation. Our intuition precedes such a structure. Poincaré also argued that mathematical induction does indeed permit generality of results and creation of new results; it is sound intuitively but one could not reduce this method to logic.

The nature of mathematical induction as Poincaré saw it warrants an examination because it remains a bone of contention today. In this method if one wants to prove, for example, that

$$(1) \qquad 1 + 2 + 3 + \cdots + n = \frac{n}{2}(n + 1)$$

for all positive integers n, one proves that it is true for $n = 1$ and then one proves that if it is true for any positive integer k, it is true for $k + 1$. Hence, Poincaré argued, this method involves an *infinite* number of arguments. It asserts that, since (1) is true for $n = 1$, it is true for $n = 2$. Since it is true for $n = 2$, it is true for $n = 3$; and so on for all the positive integers. No logical principles cover an infinite number of arguments, and the method cannot be deduced from such principles. Hence for Poincaré consistency could not be proved by the purported reduction of mathematics to logic.

As for infinite sets, Poincaré believed that "Actual infinity does not exist. What we call infinite is only the endless possibility of creating new objects no matter how many objects exist already."

Poincaré was totally antipathetic to the heavily symbolic logistic approach and in his *Science and Method* was even sarcastic. Referring to one such approach to the whole number advanced by Burali-Forti in an article of 1897 wherein one finds a maze of symbols that define the number 1, Poincaré remarked that this is a definition admirably suited to give an idea of the number 1 to people who never heard of it before. He said further, "I am very much afraid that this definition contains a *petitio principii* [begging the question] seeing that I notice the figure 1 in the first half and the word *un* in the second."

Poincaré then turns to the definition of zero given by Louis Couturat (1868–1914), an early proponent of logicism. Zero is "the number of elements in the null class. And what is the null class? It is that class containing no element." Couturat then reexpressed his definition in symbolism. Poincaré translated: "Zero is the number of objects that satisfy a condition that is never satisfied. But as never means *in no case* I do not see that any great progress has been made."

Poincaré next criticized Couturat's definition of the number 1. One,

Couturat said, is the number of a class in which any two elements are identical. "But I am afraid that if we asked Couturat what two is, he would be obliged to use the word one."

The initiators of intuitionism, Kronecker, Borel, Lebesgue, Poincaré, and Baire—a roster of stars—made a number of criticisms of the standard mathematical arguments and of the logistic approach and proposed new principles but their contributions were sporadic and fragmentary. Their ideas were all incorporated in a definitive version by Luitzen E. J. Brouwer (1881–1966), a Dutch professor of mathematics and the founder of the intuitionistic philosophy. In his doctoral dissertation *On the Foundations of Mathematics* (1907), Brouwer began to propound the intuitionist philosophy. From 1918 on he expanded and expounded his views in various journals.

His intuitionist position in mathematics stems from his philosophy. Mathematics is a human activity which originates and takes place in the mind. It has no existence outside of human minds; thus it is independent of the real world. The mind recognizes basic, clear intuitions. These are not sensuous or empirical but immediate certainties about some concepts of mathematics. They include the integers. The fundamental intuition is the recognition of distinct events in a time sequence. "Mathematics arises when the subject of twoness, which results from the passage of time, is abstracted from all special occurrences. The remaining empty form of the common content of all these twonesses becomes the original intuition of mathematics and repeated unlimitedly creates new mathematical subjects." By unlimited repetition Brouwer meant the formation of the successive natural numbers. The idea that the whole numbers derive from the intuition of time had been maintained by Kant, William R. Hamilton in his article "Algebra as a Science of Time," and the philosopher Arthur Schopenhauer.

Brouwer conceived of mathematical thinking as a process of mental construction which builds its own universe, independent of experience and restricted only insofar as it must be based upon the fundamental mathematical intuition. This fundamental intuitive concept must not be thought of as in the nature of an undefined idea, such as occurs in axiomatic theories, but rather as something in terms of which all undefined ideas which occur in the various mathematical systems are to be intuitively conceived, if they are indeed to serve in mathematical thinking. Moreover mathematics is synthetic. It composes truths rather than derives implications of logic.

Brouwer held that "in this constructive process, bound by the obligation to notice with reflection, refinement and cultivation of thinking which theses are acceptable to the intuition, self-evident to the mind, and which are not, the only possible foundation for mathematics is to be looked for." Intuition determines the soundness and acceptability of ideas, not experience or logic. It must of course be remembered that this statement does not deny the historical role that experience has played.

Beyond the natural numbers Brouwer insisted that addition, multiplication, and mathematical induction are intuitively clear. Further, having obtained the natural numbers 1, 2, 3, $\cdots$, the mind, utilizing the possibility of the unlimited repetition of the "empty form," the step from n to $n + 1$, creates infinite sets. However, such sets are only potentially infinite in that one can always add a larger number to any given finite set of numbers. Brouwer rejected Cantor's infinite sets whose elements are all present "at once," and accordingly rejected the theory of transfinite numbers, Zermelo's axiom of choice, and those portions of analysis which use actually infinite sets. In an address of 1912, he did accept the ordinal numbers up to ω and denumerable sets. He also allowed irrationals defined by sequences of rationals without any law of formation, "sequences of free choices." This definition was vague but did allow a non-denumerable set of real numbers. Geometry, on the other hand, involved space and so, unlike number, was not under the full control of our minds. Synthetic geometry belongs to the physical sciences.

In connection with the intuitionist notion of infinite sets, the intuitionist Weyl said in an article of 1946 that:

> The sequence of numbers which grows beyond any stage already reached . . . is a manifold of possibilities opening to infinity; it remains forever in the status of creation, but is not a closed realm of things existing in themselves. That we blindly converted one into the other is the true source of our difficulties, including the antinomies [the paradoxes]—a source of more fundamental nature than Russell's vicious circle principle indicated. Brouwer opened our eyes and made us see how far classical mathematics, nourished by a belief in the absolute that transcends all human possibilities of realization, goes beyond such statements as can claim real meaning and truth founded on evidence.

Brouwer then took up the relationship of mathematics to language. Mathematics is a wholly autonomous, self-sufficient activity. It is independent of language. Words or verbal connections are used only to communicate truths. Mathematical ideas are more deeply imbedded in the human mind than in language. The world of mathematical intuitions is opposed to the world of perceptions. To the latter world, not to mathematics, belongs language where it serves for the understanding of common dealings. Language evokes copies of ideas in man's mind by symbols and sounds. The distinction is analogous to the difference between climbing a mountain and describing it in words. But mathematical ideas are independent of the dress of language and in fact far richer. Thoughts can never be completely expressed even by mathematical language including symbolic language. Moreover, language deviates from the subject matter of true mathematics.

Even more drastic, especially in its opposition to logicism, is the intuitionist position on logic. Logic belongs to language. It offers a system of rules that permit the deduction of further verbal connections which are intended to communicate truths. However, these latter truths are not such before they are intuitively grasped, nor is it guaranteed that they can be so grasped. Logic is not a reliable instrument for uncovering truths and can deduce no truths that are not obtainable just as well in some other way. Logical principles are the regularity observed a posteriori in the language. They are a device for manipulating language, or they are the theory of representation of language. Logic is a structured verbal edifice but no more. The most important advances in mathematics are not obtained by perfecting the logical form but by modifying the basic theory itself. Logic rests on mathematics, not mathematics on logic. Logic is far less certain than our intuitive concepts and mathematics does not need the guarantees of logic. Historically, the principles of logic were abstracted from experience with finite collections of objects and then accorded an a priori validity and, in addition, applied to infinite sets.

Since Brouwer did not recognize any a priori obligatory logical principles, he also did not recognize the mathematical task of deducing conclusions from axioms. Hence he rejected the axiomatics of the late 19th century as well as logicism. Mathematics is not bound to respect the rules of logic. Knowing mathematics does not require knowing formal proofs, and for this reason the paradoxes are unimportant even if we were to accept the mathematical concepts and constructions they involve. Paradoxes are a defect of logic but not of true mathematics. Hence consistency is a hobgoblin. It has no point. Consistency is assured as a consequence of correct thought and the thoughts have meaning whose correctness we can judge intuitively.

However, in the realm of logic there are some clear, *intuitively* acceptable logical principles or procedures that can be used to assert new theorems from old ones. These principles are part of the fundamental mathematical intuition. Not all of the common logical principles are acceptable to the basic intuition and one must be critical of what has been accepted since the days of Aristotle. Because mathematicians have applied too freely these limited Aristotelian laws, they have produced antinomies. What is allowable or safe, the intuitionists ask, in dealing with mathematical constructions if temporarily one neglects the intuition and works with the verbal structure?

The intuitionists therefore proceed to analyze which logical principles are allowable in order *that the usual logic conform to and properly express the correct intuitions.* As a specific example of a logical principle that is applied too freely Brouwer cited the law of excluded middle. This principle, which asserts that every meaningful statement is true or false arose historically in the application of reasoning to finite sets and was abstracted therefrom. Then it was accepted as an independent a priori principle and was unjustifiably applied to infinite sets. Whereas for finite sets one can decide whether all elements possess a certain property by testing each one, this procedure is not possible for infinite sets. One may happen to know that an element of an infinite set does not possess the property or it may be that by the very construction of the set we know or can prove that every element has the property. In any case one cannot use the law of excluded middle to prove the property holds.

Thus if one proves that not all integers of an infinite set of whole numbers are even, the conclusion that there exists at least one integer which is odd was denied by Brouwer, because this argument applies the law of excluded middle to infinite sets. But this type of argument is used widely in mathematics to prove the existence of entities, as, for example, in proving that every polynomial equation has a root (Chapter IX). Hence many existence proofs are not accepted by the intuitionists. Such a proof, they claim, is far too vague about the entity whose existence is purportedly established. The law of excluded middle can be used in cases where only a finite number of elements is involved. Thus if one were dealing with a finite collection of integers and proved that not all are even, one could conclude that at least one is odd.

Weyl expanded on the intuitionist view of logic:

> According to his [Brouwer's] view and reading of history, classical logic was abstracted from the mathematics of finite sets and their subsets. . . . Forgetful of this limited origin, one afterwards mistook that logic for something above and prior to all mathematics, and finally applied it, without justification, to the mathematics of infinite sets. This is the Fall and original sin of set theory, for which it is justly punished by the antinomies. It is not that such contradictions showed up that is surprising, but they they showed up at such a late stage of the game.

Somewhat later Weyl added, "The principle of excluded middle may be valid for God who surveys the infinite sequence of natural numbers, as it were, with one glance, but not for human logic."

Brouwer in a paper of 1923 gave examples of theorems that are not established if we deny the application of the law of excluded middle to infinite collections.* In particular the Bolzano-Weierstrass theorem, which asserts that every bounded infinite set has a limit point, is not proved, nor is the existence of a maximum of a continuous function on a closed interval. Rejected also is the Heine-Borel theorem which asserts that, from any set of intervals that enclose or cover an interval of points, a finite number can be selected which cover the interval. Of course the consequences of these theorems are also unacceptable.

Beyond denying the unrestricted use of the law of excluded middle to establish the existence of mathematical entities, the intuitionists impose another requirement. They would not accept a set defined by an attribute characteristic of all its elements as, for example, the set defined by the attribute red. The concepts or objects the intuitionists will accept as legitimate for mathematical discussion—the objects that can truly be said to exist—must be constructible; that is, one must give a method of exhibiting the entity or entities in a finite number of steps, or a method of calculating them to any desired degree of accuracy.† Thus π is acceptable because we can calculate it to as many decimal places as we wish. If one proved merely the existence of integers x, y, z, and n satisfying $x^n + y^n = z^n$ for n greater than 2, but did not specify the integers, the intuitionist would not accept the proof. On the other hand, the definition of a prime number is constructive because a procedure can be applied to determine in a finite number of steps whether a number is prime.

Let us consider another example. Twin primes are two prime numbers of the form $l - 2$ and l, for example, 5 and 7, 11 and 13, and so on. It is an unsolved problem of mathematics as to whether an infinitude of such twin primes exists. Let us arbitrarily define the number l as the greatest prime such that $l - 2$ is also a prime, or $l = 1$ if such a number does not exist. The classicist accepts l as being perfectly well-defined whether or not we know if there exists a last pair of twin primes. For, by the law of the excluded middle, such a last pair either does or does not exist. In the first case, l is the greatest prime such that $l - 2$ is also a prime, and in the second case, $l = 1$. The fact that we cannot actually calculate l is irrelevant to non-intuitionists. But the intuitionist does not accept the above "definition" of l as meaningful until l can be computed, that is, until the problem of whether there exists an infinity of twin primes is solved. The insistence on a constructive pro-

---

* For our purposes the technical meaning of the theorems need not be pursued. They are mentioned merely to give specific examples.

† Poincaré was an exception. For him as for the formalists (Chapter XI), a concept was acceptable if it did not lead to contradiction.

cess applies especially to the determination of infinite sets. An infinitely large set constructed by means of the axiom of choice would not be acceptable. As some of the above examples show, some existence proofs are not constructive. Hence, beyond the fact that they may use the law of excluded middle, there is this additional ground for rejecting them.

Hermann Weyl said non-constructive existence proofs inform the world that a treasure exists without disclosing its location. Such proofs cannot replace construction without loss of significance and value. He also pointed out that adherence to the intuitionist philosophy means the abandonment of basic existence theorems of classical analysis. Cantor's hierarchy of transfinite numbers Weyl described as a fog on a fog. Analysis, he wrote in *Das Kontinuum* (1918), is a house built on sand. One can be certain only of what is established by intuitionist methods.

The denial of the law of excluded middle gives rise to a new possibility, undecidable propositions. The intuitionists maintain, with respect to infinite sets, that there is a third state of affairs, namely, there may be propositions which are neither provable nor disprovable. They give the following example of such a proposition: Let us define the kth position in the decimal expansion of π to be the position of the first zero which is followed by the integers 1 through 9 in sequence. Aristotelian logic says that k either does or does not exist, and mathematicians following Aristotle then proceed to argue on the basis of these two possibilities. Brouwer and the intuitionists generally reject all such arguments for we do not know whether we shall ever be able to prove that k does or does not exist. Thus according to the intuitionists there are sensible and substantial mathematical questions which may never be settled on the basis of any foundation of mathematics. These questions seem to us to be decidable but actually the basis of our belief is really nothing more than that they involve mathematical concepts and problems of the kind that have been decided in the past.

Under the intuitionist view, the classical and the logistic constructions of the system of real numbers, the calculus, the modern theory of real functions, the Lebesgue integral, and other subjects are not acceptable. Brouwer and his sympathetic contemporaries did not limit themselves to criticism but sought to build up mathematics on the basis of the constructions that they prescribed. They have succeeded in saving portions of the above-mentioned subjects but their constructions are very complicated, so much so that even the partisan Weyl complained about the unbearable awkwardness of the proofs. The intuitionists have also reconstructed elementary portions of algebra and geometry.

However, the reconstructions have proceeded at a slow pace. Hence in his 1927 paper on "The Foundations of Mathematics," Hilbert said, "Compared with the immense expanse of modern mathematics, what would the wretched remnants mean, the few isolated results, incomplete and unrelated, that the intuitionists have obtained." Of course, in 1927 the intuitionists had not made much progress in reconstructing classical mathematics by their standards. But the nettling from their philosophical opponents spurred them on. Since then, more and more intuitionists have taken a hand in the rebuilding of the foundations. Unfortunately, as in the case of logicism, intuitionists do not agree on what are acceptable bases. Some decided to eliminate all general set-theoretical notions and to limit themselves to concepts that can be effectively defined or constructed. Less extreme are the constructivists who do not challenge classical logic but rather use all of it. Some admit a class of mathematical objects and thereafter insist on constructive procedures. Thus there are many who admit at least a class of real numbers (which however does not extend to the entire continuum of real numbers); others admit just the integers and then entertain only concepts such as other numbers and functions that are computable. What is considered computable also varies from one group to another. Thus a number may be thought computable if it can be approximated more and more closely by some original set of admissible numbers, just as the usual irrationals can be approximated more and more closely by a finite decimal.

Unfortunately, the concept of constructivity is by no means sharp or unambiguous. Let us consider the following definition of a number N, namely,

$$N = 1 + \frac{(-1)^p}{10^p}.$$

For the moment let us suppose that $p = 3$. Then $N = 1 - .001$ or $.999$. If on the other hand $p = 2$, then $N = 1.01$. Now let us define p to be the first digit in the decimal expansion of π after which the sequence 123456789 occurs in the expansion. If there is no such p, we define N to be 1. If p exists and is an even number, then $N = 1.000\cdots$ until we get to the pth place where there is a 1 in the decimal part of N. If p is odd, then $N = .999\cdots$ wherein the 9's continue to the pth place. However, we do not know whether the p defined above exists. If it does not, then $N = 1$. If it does but does not occur, let us say, in the first thousand digits of the decimal expansion of π, then we cannot begin to write the value of N. Nevertheless, N is defined and even to any desired degree of approximation. Is N defined constructively?

Of course, proofs of existence that use the axiom of choice or the continuum hypothesis are not constructive and so are unacceptable not only to the intuitionists but even to many mathematicians who are not intuitionists.

Though the constructivists do differ among themselves, one can say that they have rebuilt a great deal of classical mathematics. Some of the rebuilt theorems do not assert quite as much as the classical theorems. To this the intuitionists reply that classical analysis, though useful, has less mathematical truth. In sum, their progress thus far has been limited and the prospects of extending their work to all of previously accepted mathematics are not good. Because the progress has been slow, even in 1960 the mathematicians of the Bourbaki school, about whom we shall say more later, remarked, "The memory of the intuitionist school will without doubt be destined to survive only as a historical curiosity." Critics of intuitionism could quote the versifier Samuel Hoffenstein:

> Little by little we subtract
> Faith and fallacy from fact,
> The illusory from the true
> And then starve upon the residue.

However, for the intuitionists if the price of a sound foundation is the sacrifice of parts of classical mathematics, even the "paradise" of Cantor's theory of transfinite numbers, the price is not too high.

Though the opponents of intuitionism may be too cavalier or too dogmatic in their dismissal of this philosophy of mathematics, there are criticisms of more sympathetic people that must be taken very seriously. One such criticism points out that the very theorems the intuitionists strive so fervently to reconstruct in a manner consistent with their principles were not suggested by and could hardly be considered guaranteed by human intuition. These theorems were discovered by all the methods mathematicians have ever used, reasoning of all sorts, guesses, generalization from special cases, and sudden insights whose origins are not explicable. Hence in practice the intuitionists are actually dependent upon the normal methods of creation and even on classical logic, much as all mathematicians are, even though the intuitionists seek to reconstruct the proofs to accord with their own principles. Intuitionists might reply that, though the normal methods of discovery should be used, the results must still be acceptable to human intuitions. Without denying the importance of other doctrines of intuitionism, it is nevertheless the case that many of the theorems accepted even by the intuitionists are so subtle and strange to the intuition that it is difficult to believe the mind can apprehend their truth directly.

The argument that normal modes of creation and of mathematical idealization and abstraction are essential was pushed further by Felix Klein and Moritz Pasch. Would intuition have discovered a continuous, nowhere differentiable function or a curve (the Peano curve) that covers a square? Such creations, even if suggested by intuition, must be refined by idealization and abstraction. Klein said the naïve intuition is not exact, while the refined intuition is not properly intuition at all but arises from the logical development based on axioms. To the demand for ultimate reliance upon logical deduction from axioms, Brouwer replied that a system of axioms must be proven consistent by using interpretations or models (Chapter VIII) which must themselves be known to be consistent. Do we always, he asked pointedly, have such interpretations and do we not rely upon an intuitive basis for accepting their consistency?

Weyl, too, challenged the assertion that the traditional modes of creation and proof are more powerful. In his *Mind and Nature* (1934), he said, "It is fond dreaming to expect that a deeper nature than that which lies open to intuition should be revealed to cognition."

Some of intuitionism's opponents agree that mathematics is a human creation, but believe that correctness or incorrectness can be objectively determined, whereas the intuitionists depend on self-evidency to fallible human minds. As Hilbert and Paul Bernays (1888–1978) argued in the first edition of their work on the foundations of mathematics, here we find the great weakness of the intuitionist philosophy. What concepts and reasoning may we rely upon if correctness were to mean self-evidency to the human mind? Where is truth, objectively valid for all human beings?

Another criticism of intuitionism is that it is not concerned with the application of mathematics to nature. Intuitionism does not relate mathematics to perception. Brouwer admitted that intuitionistic mathematics is useless for practical applications. In fact, Brouwer rejected human domination over nature. Whatever the criticisms, Weyl said in 1951, "I think everybody has to accept Brouwer's critique who wants to hold on to the belief that mathematical propositions tell the sheer truth, truth based on evidence."

The doctrines of the intuitionists have raised a related issue. They maintain, as we know, that sound and acceptable ideas can be and are conceived by the human mind. These ideas do not originate in verbal form. In fact, language is merely an imperfect device for communicating these ideas. The issue, which has been discussed at length, is whether thoughts can exist without words. On one side, there is the position which can be expressed as in the Gospel of St. John: "In the beginning there was the word." Though St. John did not have mathematics in mind, his statement is in accord with the Greek philosophical position and the views of some modern psychologists. On the other hand, Bishop Berkeley contended that words are an impediment to thought.

Euler in his letters to the Princess of Anhalt-Dessau, niece of Frederick the Great of Prussia (published 1768–72), discussed this question:

> Whatever aptitude a man may have to exercise the power of abstraction, and to furnish himself with general ideas, he can make no considerable progress without the aid of language, spoken or written. Both the one and the other contain a variety of words, which are only certain signs corresponding to our ideas, and whose signification is settled by custom, or the tacit consent of several men who live together.
>
> It would appear from this, that the only purpose of language to mankind is mutually to communicate their sentiments, and that a solitary man might do well without it; but a little reflection only is necessary to be convinced that men stand in need of language, as much to pursue and cultivate their own thoughts as to keep up a communication with others.

Jacques Hadamard in *The Psychology of Invention in the Mathematical Field* (1945) investigated the question of how mathematicians think, and his findings were that in the creative process practically all mathematicians avoid the use of precise language; they use vague images, visual or tactile. This mode of thinking was expressed by Einstein in a letter reproduced in Hadamard's book:

> The words or the language, as they are written or spoken, do not seem to play any role in my mechanism of thought. . . . The physical entities which seem to serve as elements in thought are certain signs and more or less clear images which can be voluntarily reproduced and combined. . . . The above-mentioned elements are, in my case, visual and some of muscular type. Conventional words or other signs have to be sought for laboriously only in a secondary state. . . .

Of course, visualization plays a major role in the creative act. That an infinite straight line divides the Euclidean plane into two parts derives from visualization. The issue then boils down to whether, as the intuitionists maintain, the mind's confidence in a fact, however obtained, can be so certain as to render unnecessary the need for expression in precise language and logical proof.

As a gesture toward improving communication with formal logicians, Arend Heyting (1898–   ), the leading exponent of intuitionism after Brouwer, published a paper in 1930 setting forth formal rules of intuitionist propositional logic. These include only a part of classical formal logic. For instance, Heyting allowed that the truth of p implies that it is false that p is false. However, if it is false that p is false, it does not follow that p is true because what p asserts may not be constructible. The law of excluded middle—p or not p must be true—is not used. But if proposition p implies proposition q, then the denial of q implies that p is false. This formalization is not considered to be fundamental by the intuitionists. It is an incomplete representation of the ideas. Heyting's formalization is, moreover, not the only one. The intuitionists differ among themselves about acceptable logical principles.

Despite the restrictions on mathematics intuitionists have imposed and the criticisms by others of the intuitionist philosophy, intuitionism has had a healthy influence. It has brought to the fore the question first seriously debated in connection with the axiom of choice. What does mathematical existence mean? Does it do any good, to paraphrase Weyl, to know that a number with special properties exists without any way of realizing or computing that number? The unrestricted, naïve extension of the use of the law of excluded middle certainly calls for reconsideration. Of the contributions of intuitionism perhaps the most valuable is its insistence on calculating numbers or functions whose existence was established merely by showing that non-existence leads to a contradiction. To know these numbers intimately is like living with a friend as opposed to knowing that one has a friend somewhere in the world.

The confrontation of logicists and intuitionists was merely the opening battle in the war to establish the proper foundations of mathematics. Other contestants entered the fray and we have yet to hear from them.

XI

The Formalist and Set-Theoretic Foundations

> For, compared with the immense expanse of modern
> mathematics, what would the wretched remnants mean,
> the few isolated results incomplete and unrelated, that the
> intuitionists have obtained.   DAVID HILBERT

The logicist and intuitionist philosophies, launched during the first de-
cade of this century and diametrically opposed in their views on the
proper foundations of mathematics, were just the first guns to be fired.
A third school of thought, called formalist, was fashioned and led by
David Hilbert, and a fourth, the set-theoretic school, was initiated by
Ernst Zermelo.

In his speech at the International Congress of 1900 (Chapter VIII),
Hilbert stressed the importance of proving the consistency of mathematics. He had also called for a method of well-ordering the real
numbers and we know, as a consequence of the work of Zermelo, that
well-ordering is equivalent to the axiom of choice. Finally, Hilbert had
also proposed that mathematicians undertake to prove the continuum
hypothesis, which states that there is no transfinite number between $\aleph_0$
and c. Even before the troublesome paradoxes became known and the
controversy about the axiom of choice arose, Hilbert foresaw that solving all these problems was vital.

Hilbert himself presented a sketch of his approach to the foundations, including a method of proving consistency, at the Third International Congress of 1904. However, he did not undertake more serious
work for some time. During the next fifteen years, the logicists and intuitionists spread their doctrines widely, and Hilbert was, to put it
mildly, dissatisfied with their solutions to the problems besetting the
foundations.

The logistic program he dismissed rather calmly. His fundamental
objection, as he put it in his 1904 talk and paper, was that in the long
and complicated development of logic, the whole numbers were al-

ready involved even though not named as such. Hence to build number on logic was circular reasoning. He also criticized defining sets by their properties: this necessitated distinguishing propositions and propositional functions by types and type theory required the questionable axiom of reducibility. He did agree with Russell and Whitehead that infinite sets should be included. But this required the axiom of infinity and Hilbert like others argued that this is not an axiom of logic.

The philosophy of the intuitionists, on the other hand, alarmed Hilbert because they rejected not only infinite sets but also vast portions of analysis such as those that depended on pure existence proofs, and he attacked their philosophy vehemently. In 1922, he said that intuitionism "seeks to break up and disfigure mathematics." In a paper of 1927, he protested that "Taking the principle of excluded middle from the mathematician would be the same, say, as proscribing the telescope to the astronomer or to the boxer the use of his fists. To deny existence theorems derived by using the principle of excluded middle is tantamount to relinquishing the science of mathematics altogether."

Weyl said in 1927, apropos of Hilbert's position on intuitionism, "that from this [intuitionist] point of view only a part, perhaps only a wretched part, of classical mathematics is tenable is a bitter but inevitable fact. Hilbert could not bear this mutilation."

Against both logicism and intuitionism, Hilbert held that neither proved consistency. In the 1927 paper, Hilbert exclaimed:

> To found it [mathematics] I do not need God, as does Kronecker, or the assumption of a special faculty of understanding attuned to the principle of mathematical induction, as does Poincaré [who had said that the consistency of a system using mathematical induction could not be proved], or the primal intuition of Brouwer, or finally, as do Russell and Whitehead, axioms of infinity, reducibility, or completeness which are indeed real and substantial propositions, but not capable of being established through a proof of consistency.

During the 1920s Hilbert formulated his own approach to the foundations and he worked on it for the rest of his life. Of the papers Hilbert published during the 1920s and early 1930s, the paper of 1925 is the major one for his ideas.* He said in this paper, "On the Infinite," "the goal of my theory is to establish once and for all the certitude of mathematical methods."

The first of his theses is that, since the development of logic really involves mathematical ideas and since extra-logical axioms such as the axiom of infinity must be introduced anyway if classical mathematics is

---

* A translation can be found in Paul Benacerraf and Hilary Putnam: *Philosophy of Mathematics*, Prentice Hall, 1964, 134–81.

to be preserved, the correct approach to mathematics must include concepts and axioms of both logic and mathematics. Moreover, logic must operate on something and that something consists of certain extra-logical concrete concepts, such as number, which are present in intuition before any logical development can be undertaken.

The logical axioms that Hilbert assumed are not essentially different from Russell's, though Hilbert assumed more because he was not as concerned with establishing an axiomatic basis for logic. However, because, according to Hilbert, one cannot deduce mathematics from logic alone—mathematics is not a consequence of logic but an autonomous discipline—each branch must have the appropriate axioms of both logic and mathematics. Moreover, the most reliable way to treat mathematics is to regard it not as factual knowledge but as a formal discipline, that is, abstract, symbolic, and without reference to meaning (though, informally, meaning and relation to reality do enter). Deductions are to be manipulations of symbols according to logical principles.

Hence to avoid ambiguities of language and unconscious use of intuitive knowledge, which are the cause of some of the paradoxes, to eliminate other paradoxes, and to achieve precision of proof and objectivity, Hilbert decided that all statements of logic and mathematics must be expressed in symbolic form. These symbols, though they may represent intuitively meaningful perceptions, are not to be interpreted in the formal mathematics that he proposed. Some symbols would even represent infinite sets since Hilbert wished to include them, but these have no intuitive meaning. These ideal elements, as he called them, are needed to build all of mathematics and so their introduction is justified, though Hilbert believed that in the real world only a finite number of objects exist. Matter is composed of a finite number of elements.

Hilbert's argument here can be understood by considering an analogy. The irrational number has no intuitive meaning as a number. Even though we can introduce lengths whose measures are irrational, the lengths themselves do not furnish any intuitive meaning for irrational *numbers*. Yet irrationals, as ideal elements, are necessary even for elementary mathematics and this is why mathematicians used them even without any logical basis until the 1870s. Hilbert made the same point with respect to complex numbers, that is, numbers involving $\sqrt{-1}$. These have no real immediate counterparts. Yet they make possible general theorems, such as that every polynomial equation of the nth degree has exactly n roots, as well as an entire theory of functions of a complex variable, which has proved immensely useful even in physical investigations. Whether or not the symbols represent intuitively meaningful objects, all signs and symbols of concepts and operations are freed of meaning. For the purpose of foundations the elements of mathematical thought are the symbols and the propositions, which are combinations or strings of symbols. Thus the formalists sought to buy certainty at a price, the price of dealing with meaningless symbols.

Luckily, the symbolism for logic had been developed during the late 19th and early 20th centuries (Chapter VIII) and so Hilbert had on hand what he needed. Thus, symbols such as ∼ for "not," · for "and," V for "or," → for "implies," and ∃ for "there exists" were available. These were undefined or primitive concepts. For mathematics proper the symbolism had, of course, long been available.

The axioms of logic that Hilbert chose were intended to yield in effect all the principles of Aristotelian logic. One could hardly question the acceptability of these axioms. For example, where X, Y, and Z are propositions, one axiom states that X implies XVY, which verbally means that, if X is true, then X or Y is true. Another, in verbal form, amounts to the statement that, if X implies Y, then Z or X implies Z or Y. A fundamental axiom is the rule of implication or rule of inference. The axiom states that, if formula A is true and if formula A implies formula B, then formula B is true. This principle of logic is called the *modus ponens* in Aristotelian logic. Hilbert also wished to use the law of excluded middle, and he introduced a technical device which expresses this law in symbolic form. This same device was used to express the axiom of choice, which is, of course, a mathematical axiom. It avoided the explicit use of the word "all" and thereby Hilbert expected to avoid any paradoxes.

In the branch of mathematics dealing with number, there are, in accordance with Hilbert's program, axioms of number. As an illustration, there is the axiom $a = b$ implies $a' = b'$ and this states that, if two integers a and b are equal, then their successors (intuitively, the next integers) are also equal. The axiom of mathematical induction is also included. Generally the axioms are at least relevant to experience with natural phenomena or to the world of mathematical knowledge already in existence.

If the formal system is to represent set theory, it must contain axioms (in symbolic form) stating which sets can be formed. Thus the axioms permit forming the set which is the sum of two sets and the set of all subsets of a given set.

With all logical and mathematical axioms expressed as formulas or collections of symbols, Hilbert was prepared to state what he meant by objective proof. It consists of the following process: the assertion of some formula; the assertion that this formula implies another; the assertion of the second formula. A sequence of such steps in which the final asserted formula is the implication of preceding axioms or conclusions constitutes the proof of a theorem. Also, substitution of one symbol or a group of symbols for another is a permissible operation. Thus, formulas are derived by applying the logical axioms to the manipulation of the symbols of previously established formulas or axioms.

A formula is true if and only if it can be obtained as the last of a sequence of formulas such that every formula of the sequence is either an axiom in the formal system or is itself derived by one of the rules of deduction. Everyone can check whether a given formula has been obtained by a proper sequence of derivations because the proof is essentially a mechanical manipulation of symbols. Thus, under the formalist view, proof and rigor are well defined and objective.

To the formalist, then, mathematics proper is a collection of formal systems, each building its own logic along with its mathematics, each having its own concepts, its own axioms, its own rules for deducing theorems, and its own theorems. The development of each of these deductive systems is the task of mathematics.

This was Hilbert's program for the construction of mathematics proper. However, are the conclusions derivable from the axioms free of contradictions? Since prior proofs of consistency of major branches of mathematics had been given on the assumption that arithmetic is consistent—indeed Hilbert himself had shown that the consistency of Euclidean geometry reduces to the consistency of arithmetic—the consistency of the latter became the crucial question. As Hilbert put it, "In geometry and physical theory the proof of consistency is accomplished by reducing it to the consistency of arithmetic. This method obviously fails in the proof for arithmetic itself." What Hilbert sought then is a proof of absolute as opposed to relative consistency. And this is the issue on which he concentrated. He said in this connection that we cannot risk unpleasant surprises in the future such as occurred in the early 1900s.

Now, consistency cannot be observed. One cannot foresee all the implications of the axioms. However, Hilbert, like almost all mathematicians concerned with foundations, used the concept of material implication (Chapter VIII) in accordance with which a false proposition implies any proposition. If there is a contradiction, then according to the law of contradiction one of the two propositions must be false and if there is a false proposition, it implies that $1 = 0$. Therefore, all one needed in order to establish consistency was to show that one could never arrive at the statement $1 = 0$. Then, said Hilbert in his article of 1925, "What we have experienced twice, first with the paradoxes of the infinitesimal calculus and then with the paradoxes of set theory, cannot happen a third time and will never happen again."

Hilbert and his students Wilhelm Ackermann (1896–1962), Paul Bernays (1888–1978), and John von Neumann (1903–1957) gradually evolved, during the years from 1920 to 1930, what is known as Hilbert's *Beweistheorie* [proof theory] or metamathematics, a method of establishing the consistency of any formal system. The basic idea of metamathematics may be understood through an analogy. If one wished to study the effectiveness or comprehensiveness of the Japanese language, to do so in Japanese would handicap the analysis because it might be subject to the limitations of Japanese. However, if English is an effective language, one might use English to study Japanese.

In metamathematics, Hilbert proposed to use a special logic that was to be free of all objections. The logical principles would be so obviously true that everyone would accept them. Actually, they were very close to the intuitionist principles. Controversial reasoning—such as proof of existence by contradiction, transfinite induction, actually infinite sets, impredicative definitions, and the axiom of choice—was not to be used. Existence proofs must be constructive. Since a formal system can be unending, metamathematics must entertain concepts and questions involving at least potentially infinite systems. However, there should be no reference either to an infinite number of structural properties of formulas or to an infinite number of manipulations of formulas. One can permit the consideration of formulas in which the symbols stand for actually infinite sets. But these are merely symbols within formulas. Mathematical induction over the natural numbers (positive whole numbers) is allowable because it proves a proposition up to any finite n but need not be understood to prove the proposition for the entire infinite set of natural numbers.

The concepts and methods of metamathematical proof he called finitary. Hilbert was somewhat vague about what finitary means. In his 1925 paper, he gave the following example. The statement "if p is a prime, there exists a prime larger than p" is non-finitary because it is an assertion about all the integers larger than p. However, the assertion "that if p is a prime, there exists a prime between p and $p! + 1$ (p factorial plus one)" is finitary because, for any prime p, all we have to check is whether there is a prime among the finite number of numbers between p and $p! + 1$.

In a book he published in 1934 with Paul Bernays, Hilbert described finitary thus:

> We shall always use the word "finitary" to indicate that the discussion, assertion, or definition in question is kept within the bounds of thorough-going producibility of objects and thorough-going practicability of processes, and may accordingly be carried out within the domain of concrete inspection.

Metamathematics would use the language of intuitive or informal mathematics with the help, where unquestionable, of some symbolism.

Speaking of his metamathematical program at the International Congress of Mathematicians (1928), Hilbert asserted confidently, "With this new foundation of mathematics, which one can properly call a proof theory, I believe I can banish from the world all the foundational problems." In particular he was sure he could settle consistency and the completeness problem. That is, all meaningful statements would be proven or disproven. There would no undecidable propositions.

One might anticipate that the formalist program would be criticized by its rivals. In the second edition of *The Principles of Mathematics* (1937), Russell objected that the axioms of arithmetic used by the formalists do not pin down the meaning of the symbols 0, 1, 2, $\cdots$. One could as well start with what we intuitively mean by 100, 101, 102, $\cdots$. Hence the statement "There were 12 Apostles" has no meaning in formalism. "The formalists are like a watchmaker who is so absorbed in making his watches look pretty that he has forgotten their purpose of telling the time, and has therefore omitted to insert any works." The logicist definition of number makes the connection with the actual world intelligible; the formalist theory does not.

Russell also struck at the formalist concept of existence. Hilbert had accepted infinite sets and other ideal elements, and had said that if the axioms of any one branch, which included the law of excluded middle and the law of contradiction, do not lead to a contradiction, the existence of the entities that satisfy the axioms is assured. Russell called this notion of existence metaphysical. Moreover, Russell said, there is no limit to the variety of non-contradictory systems of axioms that might be invented. We are, he continued, interested in systems that have application to empirical material.

Russell's criticisms remind one of the pot calling the kettle black. He had forgotten by 1937 what he had first written in 1901: "Mathematics may be defined as the subject in which we never know what we are talking about nor whether what we are saying is true."

The formalist program was unacceptable to the intuitionists. Beyond basic differences on infinity and the law of excluded middle, the intuitionists continued to stress that they relied on the meaning of the mathematics to determine its soundness, whereas the formalists (and the logicists) dealt with ideal or transcendental worlds which had no meaning. Brouwer in 1908 had already shown that logic and meaning were in blatant contradiction in some of the basic assertions of classical analysis, among them the Bolzano-Weierstrass theorem (a rather technical theorem that a bounded infinite set has at least one limit point). We must choose, Brouwer said, between our a priori concept of the positive integers and the free use of the principle of excluded middle when the latter is applied to any argument beyond what is finitely verifiable. The free use of Aristotelian logic led to formally valid but meaningless assertions. Classical mathematics abandons reality by abandoning meaning in many logical constructs.

Brouwer's critique caused many people to recognize the falsity of the previously unquestioned belief that the great theories of mathematics are a true expression of some underlying real content. They supposedly were idealizations of essentially real things or phenomena. However, in the 19th century especially, much of classical analysis had gone far beyond any intuitive meaning, apart from its logical unsatisfactoriness to the intuitionists. To accept Brouwer is to reject a great deal of classical mathematics on the ground that it lacks intuitive meaning.

The intuitionists say today that, even if Hilbert's proof of the consistency of formalized mathematics were to be made, the theory, the formalized mathematics, is pointless. Weyl complained that Hilbert "saved" classical mathematics "by a radical reinterpretation of its meaning," that is, by formalizing it and really robbing it of meaning, "thus transforming it in principle from a system of intuitive results into a game with formulas that proceeds according to fixed rules." "Hilbert's mathematics may be a pretty game with formulas, more amusing even than chess; but what bearing does it have on cognition, since its formulas admittedly have no material meaning by virtue of which they could express intuitive truths?" In defense of the formalist philosophy, one must point out that it is only for the purposes of proving consistency, completeness, and other properties that mathematics is reduced to meaningless formulas. As for mathematics as a whole, even the formalists reject the idea that it is simply a game; they regard it as an objective science.

Like Russell, the intuitionists objected to the formalist concept of existence. Hilbert had maintained that the existence of any entity was guaranteed by the consistency of the branch of mathematics in which it is introduced. This notion of existence is unacceptable to the intuitionists. Consistency does not ensure the truth of purely existential theorems. The argument had already been made two hundred years ago by Immanuel Kant in his *Critique of Pure Reason:* "For to substitute the logical possibility of the concept (namely, that the concept does not contradict itself) for the transcendental possibility of things (namely, that an object corresponds to the concept) can deceive and leave satisfied only the simple-minded."

A fiery dialogue between the formalists and intuitionists took place in the 1920s. In 1923, Brouwer blasted away at the formalists. Of course, he said, axiomatic, formalistic treatments may avoid contradictions, but nothing of mathematical value will be obtained in this way. "An incorrect theory, even if it cannot be rejected by any contradiction that would refute it, is nevertheless incorrect, just as a criminal act is nonetheless criminal whether or not any court could prevent it." Sarcastically he also remarked in an address of 1912 at the University of Amsterdam, "To the question, where shall mathematical rigor be found, the two parties give different answers. The intuitionist says, in the human intellect; the formalist says, on paper."

Hilbert in turn charged Brouwer and Weyl with trying to throw overboard everything that did not suit them and dictatorially promulgating an embargo. In his paper of 1925, he called intuitionism a treason against science. Yet in his metamathematics, as Weyl pointed out, he limited his principles to essentially intuitionistic ones.

Still another criticism of formalism is addressed to the principles of metamathematics. They are supposed to be acceptable to all. But the formalists did the choosing. Why should their intuition be the touchstone? Why not the intuitionist approach to all of mathematics? The ultimate test of whether a method is admissible in metamathematics must of course be whether it is convincing, but convincing to whom?

Though the formalists could not meet all the criticisms, they had, as of 1930, one weighty argument in their favor. By this time Russell and his fellow logicists had agreed that the axioms of logic were not truths, so that consistency was not assured, and the intuitionists could maintain only that the soundness of their intuitions guaranteed consistency. The formalists, on the other hand, had a well-thought-through procedure for establishing consistency, and successes with simple systems made them confident that they would succeed with the consistency of arithmetic and thereby all of mathematics. Let us therefore leave them in this relatively advantageous position to consider another rival approach to the foundations of mathematics.

The members of the set-theoretic school did not at the outset formulate a distinct philosophy, but they gradually gained adherents and an explicit program. Today certainly this school may be said to compete as much for favor among mathematicians as any of the three we have thus far described.

One can trace the beginnings of a set-theoretic school to the work of Dedekind and Cantor. Though both were primarily concerned with infinite sets, both also started to found even the ordinary whole numbers (the natural numbers) on the basis of set concepts. Of course, once the whole numbers were established, all of mathematics could then be derived (Chapter VIII).

When the contradictions in Cantor's set theory proper, those of the largest cardinal and the largest ordinal, and the contradictions such as Russell's and Richard's, which involve sets, became known, some mathematicians believed that these paradoxes were due to the rather informal introduction of sets. Cantor had boldly introduced radical ideas but his presentation was rather loose. He gave various verbal definitions of a set in 1884, 1887, and 1895. His notion of a set was essentially any collection of definite, distinguishable objects of our intuition or thought. Alternatively, a set was defined whenever, for each object x, we know whether or not x belongs to the set. These definitions are vague and Cantor's entire presentation of set theory is often described today as naïve. Hence the set-theorists thought that a carefully selected axiomatic foundation would remove the paradoxes of set theory, just as the axiomatization of geometry and of the number system had resolved logical problems in those areas.

Though set theory is incorporated in the logistic approach to mathematics, the set-theorists preferred to approach it directly through axioms. The axiomatization of set theory was first undertaken by Ernst Zermelo in a paper of 1908. He, too, believed that the paradoxes arose because Cantor had not restricted the concept of a set. Zermelo therefore hoped that clear and explicit axioms would clarify what is meant by a set and what properties sets should have. In particular, Zermelo sought to limit the size of possible sets. He had no philosophical basis but sought only to avoid the contradictions. His axiom system contained the undefined fundamental concepts of set and the relation of one set being included in another. These and the defined concepts were to satisfy the statements in the axioms. No properties of sets were to be used unless granted by the axioms. In the axioms, the existence of infinite sets and such operations as the union of sets and the formation of subsets were provided for. Zermelo also used the axiom of choice.

Zermelo's system of axioms was improved some years later (1922) by Abraham A. Fraenkel (1891–1965). Zermelo had failed to distinguish the property of a set and the set itself. These were used synonymously. The distinction was made by Fraenkel in 1922. The system of axioms used most commonly by set-theorists is known as the Zermelo-Fraenkel system. Both men presupposed the refined and sharper mathematical logic available in their time, but did not specify the logical principles. These they regarded as outside the pale of mathematics and to be confidently applied much as mathematicians applied logic before 1900.

Let us note some of the axioms of the Zermelo-Fraenkel set theory. We shall allow ourselves the liberty of stating them in verbal form.

1. Two sets are identical if they have the same members. (Intuitively this defines the notion of a set.)
2. The empty set exists.
3. If x and y are sets, then the unordered pair $\{x,y\}$ is a set.
4. The union of a set of sets is a set.

5. Infinite sets exist. (This axiom permits transfinite cardinals. It is crucial because it goes beyond experience.)
6. Any property that can be formalized in the language of the theory can be used to define a set.
7. One can form the power set of any set; that is, the collection of all subsets of any given set is a set. (This process can be repeated indefinitely; that is, consider the set of all subsets of any given set as a new set; the power set of this set is a new set.)
8. The axiom of choice.
9. x does not belong to x.

What is especially noteworthy about these axioms is that they do not permit the consideration of an all-inclusive set and so presumably avoid the paradoxes. Yet they are adequate to imply all the properties of set theory needed for classical analysis. The development of the natural numbers on the basis of set theory can readily be carried out. Cantor had in 1885 asserted that pure mathematics was reducible to set theory, and in fact this was done by Whitehead and Russell, even though their approach to sets was far more complicated. And from the mathematics of number all of mathematics follows, including geometry if it is founded on analytic geometry. Hence set theory serves as a foundation for all of mathematics.*

To repeat, the hope of avoiding contradictions rests in the case of the axiomatization of set theory on restricting the types of sets that are admitted while admitting enough to serve the foundations of analysis. The axioms of set theory avoid the paradoxes to the extent that as yet no one has been able to derive them within the theory. Zermelo declared that none can be derived. The later set-theorists were and are convinced beyond a shadow of a doubt that none can be derived because Zermelo and Fraenkel had carefully built up a hierarchy of sets that avoided the looseness in earlier work on sets and their properties. However, the consistency of axiomatized set theory has not been demonstrated. But the set-theorists have not been seriously concerned. Apropos of this open problem of consistency, Poincaré with one of his customary snide remarks said, "We have put a fence around the herd to

---

* Later Gödel (1940) and Bernays (1937) modified the Zermelo-Fraenkel system to distinguish sets and classes. Gödel and Bernays simplified a 1925 version due to von Neumann. Sets can belong to other sets. All sets are classes but not all classes are sets. Classes cannot belong to larger classes. This distinction between sets and classes means that no monstrously large collections are allowed to belong to other classes. Thus Cantor's inconsistent sets were eliminated. Any theorem in the Zermelo-Fraenkel system is a theorem in the Gödel-Bernays system and conversely. In fact, there are many variations on the systems of axioms for set theory. But there is no criterion for preferring one over the others.

protect it from the wolves but we do not know whether some wolves were already enclosed within the fence."

As in the cases of the other schools, the set-theorists were favored with criticism. The use of the axiom of choice was attacked by many. To other critics, the set-theorists' failure to be specific about their logical foundations was objectionable. Already in the first decade of this century, logic itself and its relationship to mathematics were under investigation, whereas the set-theorists were rather casual about the logical principles. Of course, their confidence about consistency was regarded by some as naïve, as naïve as Cantor himself had been until he discovered difficulties (Chapter IX). Still another criticism has been that the axioms of set theory are rather arbitrary and artificial. They were designed to avoid the paradoxes, but some are not natural or based on intuition. Why not then start with arithmetic itself since the logical principles are presupposed even by the set-theorists?

Nevertheless, the Zermelo-Fraenkel axioms of set theory are now used by some mathematicians as the desirable foundation on which to build all of mathematics. It is the most general and fundamental theory on which to build analysis and geometry. In fact, just as the other approaches to mathematics gained adherents as the respective leaders continued to develop and advance their philosophies, so did the set-theoretical approach. Some logicists, for example, Willard Van Orman Quine, would settle for set theory. Though many significant events have yet to be narrated, we should note in the present context that a group of prominent and highly respected mathematicians operating under the collective pseudonym of Nicholas Bourbaki undertook in 1936 to demonstrate in great detail what most mathematicians believed must be true, namely, that if one accepts the Zermelo-Fraenkel axioms of set theory, in particular Bernay's and Gödel's modification, and some principles of logic, one can build up all of mathematics on it. But to the Bourbakists, too, logic is subordinate to the axioms of mathematics proper. It does not control what mathematics is or what mathematicians do.

The Bourbakists expressed their position on logic in an article in the *Journal for Symbolic Logic* (1949): "In other words, logic, so far as we mathematicians are concerned, is no more and no less than the grammar of the language which we use, a language which had to exist before the grammar could be constructed." Future developments in mathematics may call for modifications of the logic. This had happened with the introduction of infinite sets and, as we shall see when we discuss non-standard analysis,* it would happen again. Thus, the Bourbaki school renounces Frege, Russell, Brouwer, and Hilbert. It uses the

---

* See Chapter XII.

axiom of choice and the law of excluded middle, though it derives them by using a technical device of Hilbert's. The Bourbaki group does not bother with the problem of consistency. On this issue, the Bourbakists say, "We simply note that these difficulties can be surmounted in a way which obviates all the objections and allows no doubt as to the correctness of the reasoning." Contradictions have arisen in the past and have been surmounted. This will be true of the future. "For twenty-five centuries mathematicians have been correcting their errors, and seeing their science enriched and not impoverished in consequence; and this gives them the right to contemplate the future with equanimity." The Bourbakists have put forth about thirty volumes in their development of the set-theoretic approach.

Thus by 1930, four separate, distinct, and more or less conflicting approaches to mathematics had been expounded, and the proponents of the several views were, it is no exaggeration to say, at war with each other. No longer could one say that a theorem of mathematics was correctly proven. By 1930 one had to add by whose standards it was deemed correct. The consistency of mathematics, the major problem which motivated the new approaches, was not settled at all if one excepts the intuitionist position that man's intuition guarantees consistency.

The science which in 1800, despite the failings in its logical development, was hailed as the perfect science, the science which establishes its conclusions by infallible, unquestionable reasoning, the science whose conclusions are not only infallible but truths about our universe and, as some would maintain, truths in any possible universe, had not only lost its claim to truth but was now besmirched by the conflict of foundational schools and assertions about correct principles of reasoning. The pride of human reason was on the rack.

The state of affairs in 1930 has been described by the mathematician Eric T. Bell:

> Experience has taught most mathematicians that much that looks solid and satisfactory to one mathematical generation stands a fair chance of dissolving into cobwebs under the steadier scrutiny of the next. . . . Knowledge in any sense of a reasonably common agreement on the fundamentals of mathematics seems to be non-existent. . . . The bald exhibition of the facts should suffice to establish the one point of human significance, namely, that equally competent experts have disagreed and do now disagree on the simplest aspects of any reasoning which makes the slightest claim, implicit or explicit, to universality, generality, or cogency.

What could the future bring? As we shall see, the future brought many more grievous problems.

XII

Disasters

For a charm of powerful trouble,
Like a hell-broth boil and bubble.
Double, double toil and trouble;
Fire burn and caldron bubble.

SHAKESPEARE, *Macbeth*

In retrospect, one could say that the state of the foundations of mathematics in 1930 was tolerable. The known paradoxes had been resolved, albeit in a manner peculiar to the several schools. True, there was no longer any unanimity about what was correct mathematics but a mathematician could adopt the approach that appealed to him. He could then proceed to create in accordance with the principles of that approach.

However, two problems continued to trouble the mathematical conscience. The overriding problem was to establish the consistency of mathematics, the very problem Hilbert had posed in his Paris speech of 1900. Though the known paradoxes had been resolved, the danger that new ones might be discovered was ever present. The second problem was what has been called completeness. Generally, completeness means that the axioms of any branch are adequate to establish the correctness or falsity of any meaningful assertion that involves the concepts of that branch.

On a very elementary level, the problem of completeness amounts to whether a reasonable conjecture of Euclidean geometry, for example, that the altitudes of a triangle meet in a point, can be proven (or disproven) on the basis of Euclid's axioms. On a more advanced level and in the realm of transfinite numbers, the continuum hypothesis (Chapter IX) serves as an example. Completeness would require that it be provable or disprovable on the basis of the axioms underlying the theory of transfinite numbers. Similarly, completeness would require that Goldbach's hypothesis—every even number is the sum of two primes—be provable or disprovable on the basis of the axioms of the theory of numbers. The problem of completeness actually included many other propositions whose proofs had defied mathematicians for decades and even centuries.

With respect to the problems of consistency and completeness, the several schools had adopted somewhat different attitudes. Russell had indeed abandoned his belief in the truth of the logical axioms used in the logistic approach and had confessed to the artificiality of his axiom of reducibility (Chapter X). His theory of types avoided the known paradoxes, and Russell was rather confident that it avoided all possible ones. Nevertheless, confidence is one thing and proof is another. The completeness problem he did not tackle.

Though the set-theorists were confident that no new contradictions could arise in their approach, proof of this belief was lacking. Completeness was a concern but not a primary one. The intuitionists, too, were indifferent to the problem of consistency. They were convinced that the intuitions accepted by the human mind were *eo ipso* consistent, and formal proof was unnecessary and even irrelevant to their philosophy. As for completeness, human intuition, they believed, was powerful enough to decide the truth or falsity of almost any meaningful proposition, though some might be undecidable.

However, the formalists, led by Hilbert, were not complacent. After some limited efforts in the early 1900s to resolve the problem of consistency, in 1920 Hilbert returned to this problem and to the problem of completeness.

In his metamathematics he had outlined the approach to proof of consistency. As for completeness, in his article of 1925, "On the Infinite," Hilbert reexpressed essentially what he had said in his Paris speech of 1900. In the latter he said, "Every definite mathematical problem must necessarily be susceptible of an exact settlement." In 1925 he amplified this assertion:

> As an example of the way in which fundamental questions can be treated I would like to choose the thesis that every mathematical problem can be solved. We are all convinced of that. After all, one of the things that attract us most when we apply ourselves to a mathematical problem is precisely that within us we always hear the call: here is the problem, search for the solution; you can find it by pure thought, for in mathematics there is no *ignorabimus* [we shall not know].

At the International Congress of 1928 in Bologna, Hilbert, in an address, criticized older proofs of completeness because they used principles of logic that are not allowed in his metamathematics, but he was very confident that his own system was complete: "Our reason does not carry any secret art but proceeds by quite definite and stateable rules which are the guarantee of the absolute objectivity of its judgment." Every mathematician, he said, shares the conviction that each definite mathematical problem must be capable of being solved. In his 1930 article, "Natural Knowledge and Logic," he said, "The real reason for Comte's failure to find an unsolvable problem is, in my opinion, that an unsolvable problem does not exist."

In "The Foundations of Mathematics," a paper read in 1927 and published in 1930, Hilbert elaborated on one he wrote in 1905. Referring to his metamathematical method (proof theory) of establishing consistency and completeness, he affirmed:

> With this new way of providing a foundation for mathematics, which we may appropriately call a proof theory, I pursue a significant goal, for I should like to eliminate once and for all the questions regarding the foundations of mathematics, in the form in which they are now posed, by turning every mathematical proposition into a formula that can be concretely exhibited and strictly derived, thus recasting mathematical derivations and inferences in such a way that they are unshakable and yet provide an adequate picture of the whole science. I believe I can attain this goal completely with my proof theory, even if a great deal of work must still be done before it is fully developed.

Clearly Hilbert was confident that his proof theory would settle the questions of consistency and completeness.

By 1930, several results on completeness had been obtained. Hilbert himself had constructed a somewhat artificial system which covered only a portion of arithmetic and had established its consistency and completeness. Other such limited results were soon obtained by other men. Thus relatively trivial axiomatic systems such as the propositional calculus were proven to be consistent and even complete. Some of these proofs were made by students of Hilbert. In 1930, Kurt Gödel (1906–1978), later a professor at the Institute for Advanced Study, proved the completeness of the first order predicate calculus which covers propositions and propositional functions.* These results delighted the formalists. Hilbert was confident that his metamathematics, his proof theory, would succeed in establishing consistency and completeness for all of mathematics.

But in the very next year Gödel published another paper that opened up a Pandora's box. This paper, "On Formally Undecidable Propositions of *Principia Mathematica* and Related Systems" (1931), con-

---

*It is also consistent and its axioms are independent. This was shown by Hilbert and other men.

tained two startling results. To the mathematical world, the more devastating assertion was that the consistency of any mathematical system that is extensive enough to embrace even the arithmetic of whole numbers cannot be established by the logical principles adopted by the several foundational schools, the logicists, the formalists, and the set-theorists. This applies especially to the formalist school because Hilbert had deliberately limited his metamathematical logical principles to those acceptable even to the intuitionists and so fewer logical tools were available. This result prompted Hermann Weyl to say that God exists because mathematics is undoubtedly consistent and the devil exists because we cannot prove the consistency.

The above result of Gödel's is a corollary of his other equally startling result, which is called Gödel's incompleteness theorem. It states that, if any formal theory T adequate to embrace the theory of whole numbers is consistent, then T is incomplete.* This means that there is a meaningful statement of number theory, let us call it S, such that neither S nor not S is provable in the theory. Now either S or not S is true; there is, then, a true statement of number theory which is not provable and so is undecidable. Though Gödel was not too clear on the class of axiom systems involved, his theorem does apply to the Russell-Whitehead system, the Zermelo-Fraenkel system, Hilbert's axiomatization of number theory, and in fact to all widely accepted axiom systems. Apparently the price of consistency is incompleteness. To add insult to injury, some of the undecidable statements can be shown to be true by arguments, that is, rules of reasoning, that transcend the logic used in the formal systems just mentioned.

As one might expect, Gödel did not obtain his amazing results readily. His overall scheme was to associate numbers with each symbol and sequence of symbols of, for example, the logistic and formalistic approaches to mathematics. Then to any proposition or set of propositions constituting a proof, he also attached a Gödel number.

Specifically his arithmetization consisted in assigning natural numbers to mathematical concepts. 1 is assigned 1. To the equals sign, he assigned 2; to Hilbert's negation symbol, he assigned 3; to the plus sign he assigned 5, and similarly for the other symbols. Thus for the collection of symbols $1 = 1$, he had the number symbols 1, 2, 1. However, Gödel now assigned to the formula $1 = 1$, not the symbols 1, 2, 1, but a single number which nevertheless still showed the component numbers. He took the first three prime numbers, 2, 3, 5,

---

* This result applies also to the second order predicate calculus (Chapter VIII). Incompleteness does not invalidate those theorems that can be proved.

and formed $2^1 \cdot 3^2 \cdot 5^1 = 90$. So to $1 = 1$ he assigned the number 90. Note that 90 can always be decomposed uniquely to $2^1 \cdot 3^2 \cdot 5^1$, so that we can recover the symbols 1, 2, 1.

To each formula of the systems he considered, Gödel assigned a number. And to an entire sequence of formulas which constitute a proof, he likewise assigned a number. The exponents of such a number are the numbers of formulas. They are not themselves prime but are attached to primes. Thus $2^{900} \cdot 3^{90}$ can be the number of a proof. This proof contains the formula 900 and the formula 90. Hence from the number of a proof, we can reconstruct the formulas of the proof.

Then Gödel showed that the concepts of metamathematics about the formulas of the formal systems he considered are also representable by numbers. Thereby each assertion in metamathematics has a Gödel number assigned to it. This number is the number of a metamathematical statement. It is also a number of some arithmetical statement. Thus metamathematics is also "mapped" into arithmetic.

In these arithmetical terms Gödel showed how to construct an *arithmetical assertion G* that says, in the verbal metamathematical language, that the statement with Gödel number m, say, is not provable. But G, as a sequence of symbols, has the Gödel number m. Thus, G says of itself that it is not provable. But if the entire *arithmetical assertion G* is provable, it asserts that it is not provable, and if G is not provable it affirms just that and so is not provable. However, since the arithmetical assertion is either provable or not provable, the formal system to which the arithmetical assertion belongs, if *consistent,* is incomplete. Nevertheless, the arithmetical statement G is true because it is a statement about integers that can be established by more intuitive reasoning than the formal systems permit.

The essence of Gödel's scheme may also be seen from the following example. If one considers the statement, "This sentence is not true," we have a contradiction. For if the entire sentence is true, then, as it asserts, it is false; and if the entire sentence is false, then it is true. Gödel substituted unprovable for false so that the sentence reads, "This sentence is unprovable." Now if the statement is not provable, then what it says is true. If, on the other hand, the sentence is provable, it is not true or, by standard logic, if true, it is not provable. Hence the sentence is true if and only if it is *not* provable. Thus the result is not a contradiction but a true statement which is unprovable or undecidable.

After exhibiting his undecidable statement Gödel constructed an arithmetical statement A that represents the metamathematical statement "Arithmetic is consistent," and he proved that A implies G. Hence if A were provable, G would be provable. But since G is undecidable, A is not provable. It is undecidable. This result establishes the impossibility of proving consistency by any method or set of logical principles that can be translated into the system of arithmetic.

It would seem that incompleteness could be avoided by adding to the logical principles or by adding a mathematical axiom to the formal system. But Gödel's method shows that, if the additional statement is also expressible in arithmetical terms by his scheme of assigning numbers to the symbols and formulas, then an undecidable statement can still be formulated. Put otherwise, undecidable statements can be avoided and consistency proved only by means of principles of reasoning that cannot be "mapped" into arithmetic. To use a somewhat loose analogy, if the principles of reasoning and mathematical axioms were in Japanese and the Gödel arithmetization were in English, then as long as the Japanese could be translated into English, the Gödel results would obtain.

Thus Gödel's incompleteness theorem asserts that no system of mathematical and logical axioms that can be arithmetized in some manner such as Gödel used is adequate to encompass all the truths of even that one system, to say nothing about all of mathematics, because any such axiom system is incomplete. There exist meaningful statements that belong to these systems but cannot be proved within the systems. They can nevertheless be shown to be true by non-formal arguments. This result, that there are limitations on what can be achieved by axiomatization, contrasts sharply with the late 19th-century view that mathematics is coextensive with the collection of axiomatized branches. Gödel's result dealt a death blow to comprehensive axiomatization. This inadequacy of the axiomatic method is not in itself a contradiction, but it is surprising because mathematicians, the formalists in particular, had expected that any true statement could certainly be established within the framework of some axiomatic system. Thus, while Brouwer made clear that what is intuitively certain falls short of what is proved in classical mathematics, Gödel showed that what is intuitively certain extends beyond mathematical proof. As Paul Bernays has said, it is less wise today to recommend axiomatics than to warn against an overevaluation of it. Of course, the above arguments do not exclude the possibility that new methods of proof may go beyond what the logical principles accepted by the foundational schools permit.

Both of Gödel's results were shattering. The inability to prove consistency dealt a death blow most directly to Hilbert's formalist philosophy because he had planned such a proof in his metamathematics and was confident it would succeed. However, the disaster extended far beyond Hilbert's program. Gödel's result on consistency says that we cannot prove consistency in any approach to mathematics by safe logical principles. No one of the approaches that had been put forth was excepted. The one distinguishing feature of mathematics that it might have claimed in this century, the absolute certainty or validity of its results, could no longer be claimed. Worse, since consistency cannot be proved, mathematicians risked talking nonsense because any day a contradiction could be found. If this should happen and the contradiction were not resolvable, then all of mathematics would be pointless. For, of two contradictory propositions, one must be false, and the logical concept of implication adopted by all the mathematical logicians, called material implication (Chapter VIII), allows a false proposition to imply any proposition. Hence mathematicians were working under a threat of doom. The incompleteness theorem was another blow. Here, too, Hilbert was directly involved, but the theorem applies to all formal approaches to mathematics.

Though mathematicians generally had not expressed themselves so confidently as Hilbert had, they certainly expected to solve any clearcut problem. Indeed, the efforts to prove, for example, Fermat's last "theorem," which asserts that there are no integers that satisfy $x^n + y^n = z^n$ when n is greater than 2, had even by 1930 produced hundreds of long and deep papers. Perhaps all of these efforts were in vain because the assertion may very well be undecidable.

Gödel's incompleteness theorem is to an extent a denial of the law of excluded middle. We believe a proposition is true or false, and in modern foundations this means provable or disprovable by the laws of logic and any axioms of the particular subject to which the proposition belongs. But Gödel showed that some are neither provable or disprovable. This is an argument for the intuitionists who argued against the laws on other grounds.

There is a possibility of proving consistency if one can show, unlike Gödel's approach, that a system contains an undecidable proposition, for, by the argument noted earlier, concerning material implication, if there were a contradiction, every proposition could be proved. But thus far this has not been done.

Hilbert was not convinced that he had failed. He was an optimist. He had unbounded confidence in the power of man's reason and understanding. This optimism gave him courage and strength, but it barred him from granting that there could be undecidable mathematical problems. For Hilbert, mathematics was a domain in which the researcher could find no bounds other than his own personal power.

Gödel's 1931 results were published between the writing of the first volume (actually published in 1934) and the second volume (1939) of a basic work on foundations by Hilbert and Paul Bernays. Hence in the preface to the second volume the authors agreed that one must enlarge the methods of reasoning in metamathematics. They included transfinite induction.* These new principles, Hilbert thought, might still be intuitively sound and universally acceptable. He persisted in this direction but produced no new results.

Developments after the crucial year of 1931 have only complicated the picture and have further frustrated any attempt to define mathematics and what correct results are. One such development, though relatively minor, should at least be mentioned. Gerhard Gentzen (1909–1945), a member of Hilbert's school, did loosen the methods of proof allowed in Hilbert's metamathematics, for example, by using transfinite induction, and managed in 1936 to establish the consistency of number theory and restricted portions of analysis.

Gentzen's consistency proof is defended and accepted by some Hilbertians. These formalists say that Gentzen's work does not exceed the limits of readily acceptable logic. Thus to defend formalism one has to go from finitary Brouwerian logic to the transfinite Gentzenian logic. Opponents of Gentzen's method argue that it is amazing how sophisticated "acceptable" logic can be and that, if we are in doubt about the consistency of arithmetic, then our doubts will not be relieved by using a metamathematical principle which is as doubtful. The issue of transfinite induction had been controversial even before Gentzen's use of it, and some mathematicians had made efforts to eliminate it from proofs wherever possible. It is not an intuitively convincing principle. As Weyl put it, such principles lower the standard of valid reasoning to the point where what is trustworthy becomes vague.

The incompleteness theorem of Gödel has given rise to subsidiary problems that are worthy of mention. Since in any branch of mathematics of any appreciable complexity, there are assertions that can be neither proved nor disproved, the question does arise whether one can determine if any particular assertion can be proved or disproved. In the literature this question is known as the decision problem. It calls for effective procedures, perhaps such as computing machines employ, which enable one to determine in a finite number of steps the provability (truth or falsity) of a statement or a class of statements.

Just to concretize the notion of a decision procedure, let us consider some trivial examples. To decide whether one whole number divides another exactly, we can use the process of division and if there is no remainder the answer is yes. The same process applies to answer the question of whether one polynominal divides exactly into another. Sim-

---

* The usual mathematical induction proves that a theorem is true for all finite positive integers. Transfinite induction uses the same method but extends it to the well ordered sets of transfinite cardinal numbers.

ilarly, to decide whether the equation $ax + by = c$, wherein a, b, and c are integers, has integral solutions for x and y, there is a clear-cut method.

It so happens that in his talk at the International Congress of 1900 in Paris, Hilbert had raised a very interesting problem, his tenth, concerning Diophantine equations, and had asked whether one could give an effective procedure to determine whether such equations are solvable in whole numbers. Thus, whereas the class of equations $ax + by = c$ consists of Diophantine equations because each involves two unknowns with but one equation and the solutions must be integers, Hilbert's Diophantine problem was more general. In any case the decision problem is a vastly more complicated one than Hilbert's problem presents, but those working on the decision problem like to refer to Hilbert's problem because the very fact that one has results on a problem posed by Hilbert grants status to the result.

The notion of what an effective procedure is, was defined by Alonzo Church (1903–   ), a professor at Princeton University, to be that of a recursive function or, one might say, a computable function. Let us consider a simple example of recursiveness. If one defines $f(1)$ to be 1 and $f(n + 1)$ to be $f(n) + 3$, then $f(2) = f(1) + 3$, or $1 + 3$, or 4. Further, $f(3)$ is $f(2) + 3$, or $4 + 3$, or 7. And so we can compute the successive values of $f(n)$. The function $f(x)$ is said to be recursive. Church's notion of recursiveness was more general but amounts to computability.

Church in 1936, using his newly developed notion of recursive function, showed that in general no decision procedure was possible. Thus, given a specific assertion, one cannot always find an algorithm to determine whether it is provable or disprovable. In any particular case one might find a proof, but no advance test of whether such a proof can be found is available. Thus, mathematicians might waste years in trying to prove what is not provable. In the case of Hilbert's tenth problem, Yuri Matyasevich proved in 1970 that there is no algorithm to determine whether or not integers satisfy the relevant Diophantine equations. The problem may not be undecidable; but no effective procedure, which means to most mathematicians today a recursive procedure (not necessarily the one described above), can tell us in advance whether it is solvable.

The distinction between undecidable propositions and problems for which no decision procedure is available is somewhat subtle but sharp. The undecidable propositions are undecidable in a particular system of axioms, and they exist in any axiomatic structure of significance. Thus Euclid's parallel postulate is not decidable on the basis of the other Euclidean axioms. Another example is the assertion that the real numbers are the smallest set satisfying the usual axiomatic properties of the real numbers.

Problems that have not been solved may be decidable but this cannot always be known in advance. The problem of trisecting an angle with straightedge and compass might, at least for centuries, have been regarded erroneously as undecidable. But trisection turned out to be impossible. Church's theorem states that it is impossible to decide *in advance* whether a proposition is provable or disprovable. It may be either. It may also be neither and so undecidable, but this is not apparent as in the case of known undecidable propositions. Goldbach's hypothesis that every even number is the sum of two primes is at present unproved. It may turn out to be undecidable on the basis of the axioms of number, but it is not now evidently undecidable whereas Gödel's examples are. Hence some day it may be proved or disproved.

The shocks to mathematicians that Gödel's work on incompleteness and the inability to prove consistency caused had not been absorbed when, about ten years later, new shocks threatened the course of mathematics. Again it was Gödel who launched another series of investigations that resulted in throwing into still greater confusion the question of what sound mathematics is and what directions it can take. We may recall that one of the approaches to mathematics initiated early in the century was to build mathematics on set theory (Chapter XI), and for this purpose the Zermelo-Fraenkel system of axioms was developed.

In *The Consistency of the Axiom of Choice and of the Generalized Continuum Hypothesis with the Axioms of Set Theory* (1940), Gödel proved that, if the Zermelo-Fraenkel system of axioms without the axiom of choice is consistent, then the system obtained by adjoining this axiom is consistent; that is, the axiom cannot be disproved. Likewise, Cantor's conjecture, the continuum hypothesis—that there is no cardinal number between $\aleph_0$ and $2^{\aleph_0}$ (which is the same as c, the cardinal number of all real numbers), or that there is no non-denumerable set of real numbers that has a cardinal number less than $2^{\aleph_0}$—(and even the generalized continuum hypothesis)* is consistent with the Zermelo-Fraenkel system even if the axiom of choice is included. In other words these assertions cannot be disproved. To prove his results Gödel constructed models in which these assertions hold.

The consistency of these two assertions, the axiom of choice and the continuum hypothesis, was somewhat reassuring. One could use them,

---

* The generalized continuum hypothesis says that the set of all subsets of a set that has the cardinal number $\aleph_n$, that is, $2^{\aleph_n}$, is $\aleph_{n+1}$. Cantor had proved that $2^{\aleph_n} > \aleph_n$.

then, at least as confidently as one could use the other Zermelo-Fraenkel axioms.

But the complacency of mathematicians, if they had become complacent, was shattered by the next development. Gödel's results did not exclude the possibility that either or both of the axiom of choice and the continuum hypothesis could be proved on the basis of the other Zermelo-Fraenkel axioms. The thought that at least the axiom of choice cannot be proved on this basis was voiced as far back as 1922. In that year and in subsequent years, several men, including Fraenkel himself, gave proofs that the axiom of choice is independent, but each found it necessary to add some subsidiary axiom to the Zermelo-Fraenkel system in order to make the proof. Later proofs by several men were subject to about the same objection. Gödel conjectured in 1947 that the continuum hypothesis is also independent of the Zermelo-Fraenkel axioms and the axiom of choice.

Then in 1963, Paul Cohen (1934–), a professor of mathematics at Stanford University, proved that both the axiom of choice and the continuum hypothesis are independent of the other Zermelo-Fraenkel axioms if the latter are consistent; that is, the two assertions cannot be proved on the basis of the other Zermelo-Fraenkel axioms. Moreover, even if the axiom of choice is retained in the Zermelo-Fraenkel system, the continuum hypothesis (and certainly then the generalized continuum hypothesis) could not be proved. (However, the Zermelo-Fraenkel axioms without the axiom of choice but with the generalized continuum hypothesis do imply the axiom of choice.) The two independence results mean that in the Zermelo-Fraenkel system the axiom of choice and the continuum hypothesis are undecidable. In particular, for the continuum hypothesis Cohen's result means that there can be a transfinite number between $\aleph_0$ and $2^{\aleph_0}$ or c, even though no set of objects is known which might have such a transfinite number.

In principle Cohen's method, called the forcing method, was no different from other independence proofs. We may recall that to show the Euclidean parallel axiom is independent of the other axioms of Euclidean geometry, one finds an interpretation or model which satisfies the other axioms but does not satisfy the one in question.* The model must be known to be consistent, else it may also satisfy the axiom in question. Cohen's proof improved on earlier ones of Fraenkel, Gödel, and others, in that he used only the Zermelo-Frankel axioms with no subsidiary

---

* The commutative axiom of multiplication in group theory is independent of the other axioms of a group. There are models of groups which do satisfy it, as, for example, the ordinary positive and negative integers, and there are models that do not, for example, quaternions.

conditions. Whereas there had been earlier proofs, though less satisfactory, of the independence of the axiom of choice, the independence of the continuum hypothesis was an open question before Cohen's work.

Thus, if one builds mathematics on set theory (or even on the logistic or formalist bases), one can take several positions. One can avoid using the axiom of choice and the continuum hypothesis. This decision would restrict the theorems that can be proved. *Principia Mathematica* does not include the axiom of choice among its logical principles but does use it in the proofs of some theorems where it is explicitly stated. It is, in fact, basic in modern mathematics. Alternatively, one can adopt either one or both of the axioms. And one can deny either one or both. To deny the axiom of choice one can assume that even for a countable collection of sets there is no explicit choice. To deny the continuum hypothesis one can assume $2^{\aleph_0} = \aleph_2$ or $2^{\aleph_0} = \aleph_3$. This was, in fact, done by Cohen and he gave a model.

There are then many mathematics. There are numerous directions in which set theory (apart from other foundations of mathematics) can go. Moreover, one can use the axiom of choice for only a finite number of sets, or a denumerable number of sets, or of course for any number of sets. Mathematicians have taken each of these positions on the axiom.

With Cohen's independence proofs, mathematics reached a plight as disturbing as at the creation of non-Euclidean geometry. As we know (Chapter VIII), the fact that the Euclidean parallel axiom is independent of the other Euclidean axioms made possible the construction of several non-Euclidean geometries. Cohen's results raise the issue: which choice among the many possible versions of the two axioms should mathematicians make? Even if one considers only the set-theoretic approach, the variety of choices is bewildering.

The decision as to which of the many choices to adopt cannot be made lightly because there are both positive and negative consequences in each case. To refrain from using either axiom, that is, to avoid any affirmation or denial would, as noted, limit severely what can be proved and force exclusion of much that has been regarded as basic in the existing mathematics. Even to prove that any infinite set S has a denumerable or countable infinite subset requires the axiom of choice. The theorems that require the axiom of choice are fundamental in modern analysis, topology, abstract algebra, the theory of transfinite numbers, and other areas. Thus, not to accept the axiom hobbles mathematics.

On the other hand by adopting the axiom of choice one can prove theorems which to say the least defy intuition. One of these is known as the Banach-Tarski paradox. One may describe this in the following way: Given any two solid spheres, one the size of a baseball and the other the size of the earth, both the ball and the earth can be divided into a finite number of non-overlapping little solid pieces, so that each part of one is congruent to one and only one part of the other. Alternatively, one can state the paradox thus: One can divide the entire earth up into little pieces and merely by rearranging them make up a sphere the size of a ball. A special case of this paradox discovered in 1914 is that a sphere's surface may be decomposed into two parts which can be reassembled to give two complete spherical surfaces, each of the same radius as the original sphere. These paradoxes, unlike the ones encountered in set theory of the early 1900s, are not contradictions. They are logical consequences of the axioms of set theory and the axiom of choice.

Stranger consequences result from denial of the general axiom of choice. One technical result, perhaps more meaningful to professionals, is that every linear set is measurable. Or, since the axiom of choice implies the existence of non-measurable sets, one can deny the axiom of choice by assuming that every linear set is measurable. Other strange consequences hold for transfinite cardinal numbers. As for the continuum hypothesis, here one ventures into the unknown and, whether one affirms or denies it, significant consequences are not known as yet. But if one assumes that $2^{\aleph_0} = \aleph_2$, then every set of real numbers is measurable. One can also derive many other new conclusions. But they are not crucial.

Just as the work on the parallel axiom led to the parting of the ways for geometry, so Cohen's work on these two axioms about sets leads to a manifold parting of the ways for all of mathematics based especially on set theory, though it also affects other foundational approaches. It opens up several directions that mathematics can take but provides no obvious reason for preferring one over another. Actually, since Cohen's work of 1963, many more propositions undecidable in the Zermelo-Fraenkel set theory have been discovered so that the variety of choices one can make using the basic Zermelo-Fraenkel axioms and one or more of the undecidable propositions is bewildering. The independence proofs of the axiom of choice and the continuum hypothesis amount to showing a builder that by changing his plans slightly he can build a castle instead of an office building.

Current workers in set theory hope that they can modify the axioms of set theory in some sound fashion so as to determine whether the axiom of choice or the continuum hypothesis or both can be deduced from a set of axioms widely accepted by mathematicians. In Gödel's opinion, these possibilities should surely be realizable. The efforts have been numerous but thus far unsuccessful. Perhaps some day there will be a consensus on what axioms to use.

Gödel's, Church's, and Cohen's works were not the only ones to bewilder mathematicians. Their troubles were multiplied during the passing years. The research begun in 1915 by Leopold Löwenheim (1878–c. 1940), and simplified and completed by Thoralf Skolem (1887–1963) in a series of papers from 1920 to 1933, disclosed new flaws in the structure of mathematics. The substance of what is now known as the Löwenheim-Skolem theory is this. Suppose one sets up axioms, logical and mathematical, for a branch of mathematics or for set theory as a foundation for all of mathematics. The most pertinent example is the set of axioms for the whole numbers. One intends that these axioms should completely characterize the positive whole numbers and only the whole numbers. But, surprisingly, one discovers that one can find interpretations—models—that are drastically different and yet satisfy the axioms. Thus, whereas the set of whole numbers is countable, or, in Cantor's notation, there are only $\aleph_0$ of them, there are interpretations that contain as many elements as the set of all real numbers, and even sets larger in the transfinite sense. The converse phenomenon also occurs. That is, suppose one adopts a system of axioms for a theory of sets and one intends that these axioms should permit and indeed characterize non-denumerable collections of sets. One can, nevertheless, find a countable (denumerable) collection of sets that satisfies the system of axioms and other transfinite interpretations quite different from the one intended. In fact, every consistent system of axioms has a countable model.

What does this mean? Suppose one were to set down a list of the features which, he believes, characterize Americans and only Americans. But, surprisingly, someone discovers a species of animal which possesses all the characteristics on the list but also totally different features from Americans. In other words, axiom systems that are designed to characterize a unique class of mathematical objects do not do so. Whereas Gödel's incompleteness theorem tells us that a set of axioms is not adequate to prove all the theorems belonging to the branch of mathematics that the axioms are intended to cover, the Löwenheim-Skolem theorem tells us that a set of axioms permits many more essentially different interpretations than the one intended. The axioms do not limit the interpretations or models. Hence mathematical reality cannot be unambiguously incorporated in axiomatic systems.*

---

* Older texts did "prove" that the basic systems were categorical; that is, all the interpretations of any basic axiom system are isomorphic—they are essentially the same but differ in terminology. But the "proofs" were loose in that logical principles were used that are not allowed in Hilbert's metamathematics and the axiomatic bases were not as carefully formulated then as now. No set of axioms is categorical, despite "proofs" by Hilbert and others.

One reason that unintended interpretations are possible is that each axiomatic system contains undefined terms. Formerly, it was thought that the axioms "defined" these terms implicitly. But the axioms do not suffice. Hence the concept of undefined terms must be altered in some as yet unforeseeable way.

The Löwenheim-Skolem theorem is as startling as Gödel's incompleteness theorem. It is another blow to the axiomatic method which from 1900 even to recent times seemed to be the only sound approach, and is still the one employed by the logicists, formalists, and set-theorists.

The Löwenheim-Skolem theorem is not totally surprising. Gödel's incompleteness theorem does say that every axiomatic system is incomplete. There are undecidable propositions. Suppose p is one such proposition. Then neither p nor not p follows from the axioms. It is independent. Hence one could adopt a larger system of axioms, the original set and p, or the original set and not p. These two axiom systems would not be categorical because the interpretations could not be isomorphic. That is, incompleteness implies non-categoricalness. But the Löwenheim-Skolem theorem denies categoricalness in a far stronger or more radical way. It establishes the existence of interpretations or models of a given axiom system that, without adding any new axiom, are radically different. Of course, incompleteness must be present, else radically different interpretations would not be possible. Some meaningful statement of one interpretation must be undecidable, else it would hold in both interpretations.

After contemplating his own result, Skolem in a paper of 1923 deprecated the axiomatic method as a foundation for set theory. Even von Neumann agreed in 1925 that the axioms of his own and other axiom systems for set theory bore "the stamp of unreality. . . . No categorical axiomatization of the theory of sets exists. . . . Since there exists no axiom system for mathematics, geometry, etc., which does not assume the theory of sets, so will there certainly be no categorical axiomatic infinite systems." This circumstance, he continued, "appears to me to be an argument for intuitionism."

Mathematicians have tried to calm themselves by recalling the history of non-Euclidean geometry. When after many centuries of struggle with the parallel axiom, Lobatchevsky and Bolyai produced their non-Euclidean geometry and Riemann pointed at another such geometry, mathematicians at first were inclined to dismiss these new geometries on several grounds, one being that they must be inconsistent. However, interpretations made later showed that they were consistent. For example, Riemann's double elliptic geometry, though intended to apply to figures in the ordinary flat plane, was interpreted to refer to figures on the surface of a sphere, a quite different interpretation from the one originally intended (Chapter VIII). However, the discovery of this model or interpretation was welcomed. It proved consistency. Moreover, it did not introduce a discrepancy between the number of objects, points, lines, planes, triangles, etc. intended by Riemann and those in the interpretation. The two interpretations, in the language of mathematics, were isomorphic. However, the interpretations of axiom systems covered by the Löwenheim-Skolem theorem are not isomorphic; they are drastically different.

Poincaré, referring to the abstractness of mathematics, once said that mathematics is the art of giving the same name to different things. Thus, the notion of a group represents properties of the whole numbers, matrices under addition, and geometric transformations. The Löwenheim-Skolem theorem supports Poincaré's statement but controverts its meaning. Whereas the group axioms were not intended to specify that all interpretations were of the same extent and character— the group axioms are not categorical nor is Euclidean geometry if we omit the parallel axiom—the axiom systems to which the Löwenheim-Skolem theorem applies were intended to specify *one particular interpretation*, and the discovery that they apply to radically different ones confounds their purpose.

Whom the gods would destroy they first make mad. Perhaps because the gods were not sure that the work of Gödel, Cohen, and Löwenheim and Skolem would do the trick, they set in motion one more development that would be more likely to send mathematicians over the brink. In his approach to the calculus, Leibniz introduced quantities called infinitesimals (Chapter VI). An infinitesimal for Leibniz was a quantity not zero but smaller than .1, .01, .001, $\cdots$, and any other positive number. Moreover, he maintained, one could operate with such infinitesimals as one operates with ordinary numbers. They were ideal elements or fictions, but they were useful. In fact, the quotient of two such infinitesimals was for Leibniz the derivative, the fundamental concept of the calculus. Leibniz also used infinitely large numbers as ordinary numbers.

Throughout the 18th century, mathematicians struggled with the concept of infinitesimals, used them in accordance with arbitrary and even illogical rules, and ended up by rejecting them as nonsense. Cauchy's work not only banished them but disposed of any need for them. However, the issue of whether infinitesimals can be legitimized kept cropping up. When Gösta Mittag-Leffler (1846–1927) asked Cantor if there might be other kinds of numbers between the rationals and the reals, Cantor answered emphatically with a no. In 1887 he published a proof of the logical impossibility of infinitesimals that depended essentially on what is called the Archimedean axiom, which asserts that, given any real number a, there is a whole number n such that na is larger than any other given real number b. Peano, too, published a proof that demonstrated the non-existence of infinitesimals. Bertrand Russell in his *Principles of Mathematics* (1903) agreed with this conclusion.

However, the convictions of even truly great men should not be accepted too readily. In Aristotle's days and long thereafter, the notion that the earth is spherical was dismissed by many thinkers as nonsense because people living on the opposite side of the sphere would have their heads hanging in the air. Yet sphericity proved to be the proper concept. Likewise, despite the proofs that Leibniz's infinitesimals must be banished, a number of men persisted in trying to build up a logical theory of such entities.

Paul du Bois-Reymond, Otto Stolz, and Felix Klein did think that a consistent theory based on infinitesimals was possible. In fact, Klein identified the very axiom of real numbers, the Archimedean axiom, that would have to be abandoned to obtain such a theory. Skolem himself in 1934 began the introduction of new numbers—hyperintegers—which were different from the ordinary real numbers, and he established some of their properties. The culmination of a series of papers by several mathematicians was the creation of a new theory which legitimizes infinitesimals. The most important contributor was Abraham Robinson (1918–1974).

The new system, called non-standard analysis, introduces hyperreal numbers, which include the old real numbers and infinitesimals. An infinitesimal is defined practically as Leibniz did; that is, a positive infinitesimal is a number less than any ordinary positive real number but greater than zero and, similarly, a negative infinitesimal is greater than any negative real number but less than zero. These infinitesimals are fixed numbers, not variables in Leibniz's sense nor variables which approach zero, which is the sense in which Cauchy sometimes used the term. Moreover, non-standard analysis introduces new infinite numbers, which are the reciprocals of the infinitesimals but not the transfinite numbers of Cantor. Every finite hyperreal number r is of the form $x + \alpha$ where x is an ordinary real number and α is an infinitesimal.

With the notion of infinitesimal, one can speak of two hyperreal numbers being infinitely close. This means merely that their difference is an infinitesimal. Every hyperreal number is also infinitely close to an (ordinary) real number, the difference being an infinitesimal. We can operate with the hyperreals just as we operate with the ordinary real numbers.*

With this new hyperreal number system, we can introduce functions whose values may be ordinary or hyperreal numbers. In terms of these numbers, we can redefine the continuity of a function. Thus $f(x)$ is continuous at $x = a$ if $f(x) - f(a)$ is an infinitesimal when $x - a$ is. We can use the hyperreals to define also the derivative and other calculus concepts and prove all the results of analysis. The major point is that with the system of hyperreals we can give precision to an approach to the calculus that had previously been banished as unclear and even nonsense.†

Will the use of this new number system increase the power of mathematics? At the present time, no new results of any consequence have been produced through this means. What is significant is that another new road has been opened up which some mathematicians will travel eagerly; in fact, books on non-standard analysis are already appearing. Others will condemn the new analysis on one ground or another. Physicists are relieved because they continued to use infinitesimals as a convenience even though they knew that infinitesimals had been banished by Cauchy.

The developments in the foundations of mathematics since 1900 are bewildering, and the present state of mathematics is anomalous and deplorable. The light of truth no longer illuminates the road to follow. In place of the unique, universally admired and universally accepted body of mathematics whose proofs, though sometimes requiring emendation, were regarded as the acme of sound reasoning, we now have conflicting approaches to mathematics. Beyond the logicist, intuitionist, and formalist bases, the approach through set theory alone gives many options. Some divergent and even conflicting positions are possible even within the other schools. Thus the constructivist movement within the intuitionist philosophy has many splinter groups. Within formalism there are choices to be made about what principles of metamathematics

---

* The proofs of Cantor and Peano are correct if one uses the usual axiomatic properties of real numbers. The property that must be altered to allow hyperreal numbers is the axiom of Archimedes described above. R^*, the system of hyperreals, is non-Archimedean in the usual sense. But it is Archimedean if we allow infinite multiples of a number a^* of the system of hyperreal numbers.

† For example, in non-standard analysis the quotient of infinitesimals dy/dx exists in the system R^*, and dy/dx for $y = x^2$ is $2x + dx$ where dx is an infinitesimal. That is, dy/dx is a hyperreal number. But the derivative is the standard part of this hyperreal number, namely, $2x$. Similarly, the definite integral is the standard part of a sum of an infinite number of infinitesimals, the number of summands being itself a non-standard natural number.

may be employed. Non-standard analysis, though not a doctrine of any one school, permits an alternative approach to analysis which may also lead to diverging and even conflicting views. At the very least what was considered to be illogical and to be banished is now accepted by some schools as logically sound.

The efforts to eliminate possible contradictions and establish the consistency of the mathematical structures have thus far failed. There is no longer any agreement on whether to accept the axiomatic approach—or, if so, with which axioms—or the non-axiomatic intuitionist approach. The prevalent concept of mathematics as a collection of structures each based on its own set of axioms is inadequate to embrace all that mathematics should embrace, and on the other hand embraces more than it should. Disagreement now extends even to the methods of reasoning. The law of excluded middle is no longer an unquestionable principle of logic, and existence proofs which do not permit calculation of the quantities whose existence is being established, whether or not the proofs use the law of excluded middle, are bones of contention. The claim therefore to impeccable reasoning must be abandoned. Clearly, different bodies of mathematics will result from the multiplicity of choices. The recent research on foundations has broken through frontiers only to encounter a wilderness.

The only mathematicians who could retain some composure and smugness from 1931 on, during which time the results we have described were breaking the hearts of the logicists, formalists, and set-theorists, were the intuitionists. All the play with logical symbols and principles which taxed the minds of intellectual giants was to them nonsense. The consistency of mathematics was clear because the intuitive meaning guaranteed it. As for the axioms of choice and the continuum hypothesis, they were unacceptable, and Brouwer had said as much in 1907. Incompleteness and the existence of undecidable propositions not only did not disturb them, they could justifiably say I told you so. However, even the intuitionists were unwilling to expunge all of those portions of mathematics erected before 1900 that did not satisfy their standards. They had affirmed that it was unacceptable to establish the existence of mathematical entities by use of the law of excluded middle, and that only constructions that permitted one to calculate as accurately as one wishes the quantity whose existence is being affirmed are satisfactory. Hence they are still struggling with constructive existence proofs.

In short, no school has the right to claim that it represents mathematics. And unfortunately, as Arend Heyting remarked in 1960, since 1930 the spirit of friendly cooperation has been replaced by a spirit of implacable contention.

In 1901, Bertrand Russell said, "One of the chiefest triumphs of modern mathematics consists in having discovered what mathematics really is." These words strike us as naïve today. Beyond the differences in what is accepted as mathematics today by the several schools, one may expect more in the future. The existing schools have been concerned with justifying the existing mathematics. But if one looks at the mathematics of the Greeks, of the 17th century, and of the 19th century one sees dramatic and drastic changes. The several modern schools seek to justify the mathematics of 1900. Can they possibly serve for the mathematics of the year 2000? The intuitionists do think of mathematics as growing and developing. But would their "intuitions" ever generate or give forth what had not been historically developed? Certainly this was not true even in 1930. Hence it seems that revisions of the foundations will always be needed.

The developments in this century bearing on the foundations of mathematics are best summarized in a story. On the banks of the Rhine, a beautiful castle had been standing for centuries. In the cellar of the castle, an intricate network of webbing had been constructed by industrious spiders who lived there. One day a strong wind sprang up and destroyed the web. Frantically the spiders worked to repair the damage. They thought it was their webbing that was holding up the castle.

XIII

The Isolation of Mathematics

> I have resolved to quit only abstract geometry, that is to
> say, the consideration of questions that serve only to exer-
> cise the mind, and this, in order to study another kind of
> geometry, which has for its object the explanation of the
> phenomena of nature. RENÉ DESCARTES

The history of mathematics is crowned with glorious achievements but
is also a record of calamities. The loss of truth is certainly a tragedy of
the first magnitude, for truths are man's dearest possessions and a loss
of even one is cause for grief. The realization that the splendid show-
case of human reasoning exhibits a by no means perfect structure but
one marred by shortcomings and vulnerable to the discovery of disas-
trous contradictions at any time is another blow to the stature of mathematics. But these are not the only grounds for distress. Grave misgivings and cause for dissension among mathematicians stem from the
direction which research of the past one hundred years has taken. Most
mathematicians have withdrawn from the world to concentrate on
problems generated within mathematics. They have abandoned
science. This change in direction is often described as the turn to pure
as opposed to applied mathematics. But these terms pure and applied,
though we shall use them, do not quite describe what has been happen-
ing.

What had mathematics been? To past generations, mathematics was
first and foremost man's finest creation for the investigation of nature.
The major concepts, broad methods, and almost all the major theorems
of mathematics were derived in the course of this work. Science had
been the life blood and sustenance of mathematics. Mathematicians
were willing partners with physicists, astronomers, chemists, and engineers in the scientific enterprise. In fact, during the 17th and 18th centuries and most of the 19th, the distinction between mathematics and
theoretical science was rarely noted. And many of the leading mathe-

maticians did far greater work in astronomy, mechanics, hydrodynamics, electricity, magnetism, and elasticity than they did in mathematics proper. Mathematics was simultaneously the queen and the handmaiden of the sciences.

We have related (Chapters I–IV) the long succession of efforts from Greek times onward to wrest from nature its mathematical secrets. This devotion to the study of nature did not limit all applied mathematics to the solution of physical problems. The great mathematicians often transcended the immediate problems of science. Because they were great and fully understood the traditional role of mathematics, they could discern directions of investigation which would prove significant in the scientific enterprise or shed light on the concepts already instrumental in the investigation of nature. Thus Poincaré (1854–1912), who devoted many years to astronomy and whose most celebrated work is the three-volume *Celestial Mechanics,* saw the need to pursue new themes in differential equations that might ultimately advance astronomy.

Some mathematical investigations round out or complete a subject which has been shown to be useful. If the same type of differential equation occurs in several applications, certainly the general type should be investigated either to discover an improved or general method of solution or to learn as much as possible about the entire class of solutions. It is characteristic of mathematics that its very abstractness enables it to represent quite diverse physical happenings. Thus, water waves, sound waves, and radio waves are all represented by one partial differential equation, known, in fact, as the wave equation. The additional mathematical knowledge gained by further investigation of the wave equation itself, which first arose in the investigation of sound waves, might prove useful in questions that arise in the investigation, say, of radio waves. The rich fabric of creations inspired by problems of the real world can be strengthened and illuminated by the recognition of identical mathematical structures in dissimilar situations and their common abstract basis.

The establishment of existence theorems of differential equations, first undertaken by Cauchy, was intended to guarantee that the mathematical formulations of physical problems do have a solution, so that one could confidently seek that solution. Hence, though this work is totally mathematical, it does have ulterior physical significance. Cantor's work on infinite sets, which did lead to many investigations in pure mathematics, was motivated at the outset by his desire to answer open questions about the enormously useful infinite series called Fourier series.

The development of mathematics has suggested and even demanded the pursuit of problems independent of science. We saw (Chapter VIII) that 19th-century mathematicians had recognized the vagueness of many concepts and the looseness of their arguments. The movement to instill rigor, an extensive one, was certainly not in itself the pursuit of scientific problems, nor were the subsequent attempts of the several schools to rebuild the foundations. All of this work, though devoted to mathematics proper, was certainly a response to urgent needs of the entire mathematical structure.

In short, there are many purely mathematical investigations that round out or shore up old areas, or even explore new areas that give promise of being essential to the pursuit of applications. All of these directions of work can be regarded as applied mathematics in the broad sense.

Was there then before one hundred years ago no mathematics created for its own sake, with no relevance at all to applications? There was. The outstanding example is the theory of numbers. Though for the Pythagoreans the study of whole numbers was the study of the constitution of material objects (Chapter I), the theory of numbers soon became of interest for its own sake—a major subject with Fermat. Projective geometry, initiated by Renaissance artists to achieve realism in painting and taken up by Girard Desargues and Pascal to provide superior methodology in Euclidean geometry, became in the 19th century a purely aesthetic interest, though even then pursued also because it has important connections with non-Euclidean geometry. Many other themes were studied solely because mathematicians found them interesting or challenging.

However, pure mathematics totally unrelated to science was not the main concern. It was a hobby, a diversion from the far more vital and intriguing problems posed by the sciences. Though Fermat was the founder of the theory of numbers, he devoted most of his efforts to the creation of analytic geometry, to problems of the calculus, and to optics (Chapter VI). He tried to interest Pascal and Huygens in the theory of numbers but failed. Very few men of the 17th century took any interest in that subject.

Euler did devote some time to the theory of numbers. But Euler was not only the supreme mathematician of the 18th century; he was the supreme mathematical physicist. His work ranged from deep mathematical methodologies for solving physical problems, such as the solution of differential equations, to astronomy, the motion of fluids, the design of ships and sails, artillery, map-making, the theory of musical instruments, and optics.

Lagrange also spent some time on the theory of numbers but he, too, devoted most of his life to analysis, the mathematics vital to applications

(Chapter III), and his masterpiece was *Mécanique analytique* (Analytical Mechanics), the application of mathematics to mechanics. In fact, in 1777 he complained that "The arithmetical researches are those which have cost me most trouble and are perhaps the least valuable." Gauss, too, did masterful work in the theory of numbers. His *Disquisitiones arithmeticae* (Arithmetical Dissertations, 1801) is a classic. If one were to note just this work of Gauss, one might be convinced that Gauss was a pure mathematician. But his major efforts were in applied mathematics (Chapter IV). Felix Klein in his history of 19th-century mathematics refers to the *Disquisitiones* as a work of Gauss's youth.

Though Gauss did return to the theory of numbers in his later years, he clearly did not regard this subject as most important. The problem of proving Fermat's last "theorem," that no integers satisfy $x^n + y^n = z^n$ when n is greater than 2, was often thrown at him. But Gauss in a letter to Wilhelm Olbers of March 21, 1816 wrote that the Fermat hypothesis was an isolated theorem and so had little interest. He added that there are many such conjectures that can neither be proved nor disproved and he was so busy with other matters that he did not have time for the type of work he had done in his *Disquisitiones*. He did hope that perhaps the Fermat hypothesis might be proven on the basis of other work he had done, but it would be one of the least interesting corollaries.

Gauss's statement, "Mathematics is the queen of the sciences, and arithmetic is the queen of mathematics. She often condescends to render service to astronomy and other natural sciences, but under all circumstances the first place is her due," is often quoted to show his preference for pure mathematics. But Gauss's lifework does not bear out this statement; he may have made it in an off moment. His motto was "Thou, nature, art my goddess: to thy laws my services are bound." It is ironic that his extraordinary scrupulousness about the accord of mathematics with nature proved, through his work on non-Euclidean geometry, to have the profound and dramatic consequence of discrediting the truth of mathematical laws. Apropos of all the mathematics created prior to 1900, one can make the broad generalization that there was some pure mathematics but no pure mathematicians.

Several developments altered radically the mathematicians' attitude toward their own work. The first was the realization that mathematics is not a body of truths about nature (Chapter IV). Gauss had made this clear in geometry, and the creation of quaternions and matrices forced the realization, which Helmholtz drove home, that even the mathematics of ordinary number is not a priori applicable. Though the applicability of mathematics remained flawless, the search for truths no longer justified mathematical efforts.

Moreover, such vital developments as non-Euclidean geometry and quaternions, though motivated by physical considerations, were seemingly at variance with nature, and yet the resulting creations had proved applicable. The realization that even man-made creations, as well as what seemed inherent in the design of nature, proved to be extraordinarily applicable soon became an argument for a totally new approach to mathematics. Why should this not happen with future free creations of the mind? Hence, many mathematicians concluded, it was not necessary to undertake problems of the real world. Man-made mathematics, concocted solely from ideas springing up in the human mind, would surely prove useful. In fact, pure thought, unhindered by adherence to physical happenings, might do far better. Human imagination, freed of any restrictions, might create even more powerful theories that would also find application to the understanding and mastery of nature.

Several other forces have influenced mathematicians to break from the real world. The vast expansion of mathematics and science has made it far more difficult to be at home in both fields. Moreover, the open problems of science, worked on by giants of the past, are far more difficult. Why not stick to pure mathematics and make research easier?

Another factor induced many mathematicians to tackle problems of pure mathematics. Problems of science are rarely solved *in toto*. Better and better approximations are achieved but not a final solution. Basic problems—for example the three-body problem, that is, the motion of three bodies such as the sun, earth, and its moon, each attracting the other two with the force of gravitation—are still unsolved. As Francis Bacon remarked, the subtlety of nature is far greater than human wit. On the other hand, pure mathematics permits clear-cut, circumscribed problems to which complete solutions can be attained. There is a fascination to clear-cut problems as opposed to problems of never-ending complexity and depth. Even the few that have thus far resisted conquest, such as Goldbach's hypothesis, possess a simplicity of statement that is enticing.

Another inducement to take up problems of pure mathematics is the pressure on mathematicians from institutions such as universities to publish research. Since applied problems require vast knowledge in science as well as in mathematics and since the open ones are more difficult, it is far easier to invent one's own problems and solve what one can. Not only do professors select problems of pure mathematics that are readily solved but they assign such problems to their doctoral students so that they can produce theses quickly. Also, the professors can more readily help them overcome any difficulties they encounter.

A few examples of the direction modern pure mathematics has taken may make clear the distinction between pure and applied themes. One area is abstraction. After Hamilton introduced quaternions, for which he had physical applications in mind, other mathematicians, recognizing that there can be many algebras, undertook to pursue all possible algebras whether or not they had potential applicability. This direction of work flourishes today in the field of abstract algebra.

Another direction of pure mathematics is generalization. The conic sections—ellipse, parabola, and hyperbola—are represented algebraically by equations of the second degree. A few curves represented by equations of the third degree are useful for applications. Generalization jumps at once to curves whose equations are of the nth degree and studies their properties in detail even though none of these curves is ever likely to appear in natural phenomena.

Generalization and abstraction undertaken solely because research papers characterized by them can be written are usually worthless for application. In fact, most of these papers are devoted to a reformulation in more general or more abstract terms or in new terminology of what had previously existed in more concrete and specific language. And this reformulation provides no gain in power or insight to one who would apply the mathematics. The proliferation of terminology, largely artificial and with no relation to physical ideas but purporting to suggest new ideas, is certainly not a contribution but rather a hindrance to the use of mathematics. It is new language but not new mathematics.

A third direction that pure mathematical research has taken is specialization. Whereas Euclid considered and answered the question of whether there is an infinity of prime numbers, now it is "natural" to ask whether there is a prime number in any seven consecutive integers. The Pythagoreans had introduced the notion of amicable numbers. Two numbers are amicable if the sum of the divisors of one number equals the other. Thus 284 and 220 are amicable. Leonard Dickson, one of the best men in the theory of numbers, introduced the problem of amicable triple numbers. "We shall say that three numbers form an amicable triple if the sum of the proper divisors of each one equals the sum of the remaining two numbers," and he raised the question of finding such triples. Still another example concerns powerful numbers. A powerful number is a positive integer which, if divisible by a prime p, is also divisible by p^2. Are there positive integers (other than 1 and 4) which are representable in an infinite number of ways as the difference of two relatively prime powerful numbers?

These examples of specialization, chosen because they are easy to state and comprehend, do not do justice to the complexity and depth of such problems. However, specialization has become so widespread and the problems so narrow that what was once incorrectly said of the theory of relativity, namely, that only a dozen men in the world understood it, does apply to most specialties.

The spread of specialization has become so great that the school of major mathematicians operating under the pseudonym of Nicolas Bourbaki, a school certainly not devoted to applied mathematics, felt obliged to be critical:

> Many mathematicians take up quarters in a corner of the domain of mathematics, which they do not intend to leave; not only do they ignore almost completely what does not concern their special field, but they are unable to understand the language and the terminology used by colleagues who are working in a corner remote from their own. Even among those who have the widest training, there are none who do not feel lost in certain regions of the immense world of mathematics; those who, like Poincaré or Hilbert, put the seal of their genius on almost every domain, constitute a very great exception even among the men of greatest accomplishment.

The price of specialization is sterility. It may well call for virtuosity but rarely does it offer significance.

Abstraction, generalization, and specialization are three types of activity undertaken by pure mathematicians. A fourth is axiomatics. There is no question that the axiomatic movement of the late 19th century was helpful in shoring up the foundations of mathematics, even though it did not prove to be the last word in settling foundational problems. But many mathematicians then undertook trivial modifications of the newly created axiom systems. Some could show that by rewording an axiom, it could be stated more simply. Others showed that by complicating the wording three axioms might be encompassed in two. Still others chose new undefined terms and by recasting the axioms accordingly arrived at the same body of theorems.

Not all axiomatization, as we have noted, is worthless. But the minor modifications that can be made are on the whole insignificant. Whereas the solution of real problems demands the best of human capabilities because the problems must be faced, axiomatics permits all sorts of liberties. It is fundamentally man's organization of deep results, but whether he chooses one set of axioms rather than another or fifteen instead of twenty is almost immaterial. In fact, the numerous variations to which even prominent mathematicians devoted time have been denounced as "postulate piddling."

So much time and effort were devoted to axiomatics in the first few decades of this century that Hermann Weyl in 1935, though fully conscious of the value of axiomatization, complained that the fruits of axiomatization were exhausted and pleaded for a return to problems of substance. Axiomatics, he pointed out, merely gives precision and organization to substantial mathematics; it is a cataloguing or classifying function.

One cannot characterize all abstractions, generalizations, specialized problems, and axiomatics as pure mathematics. We have pointed out the value of some of this work as well as of foundational studies. One must look into the motivation for the work. What is characteristic of pure mathematics is its irrelevance to immediate or potential application. The spirit of pure mathematics is that a problem is a problem is a problem. Some pure mathematicians argue that there is potential usefulness in any mathematical development and no one can foresee its actual future application. Nevertheless, a mathematical theme is like a piece of oil-bearing land. Dark puddles on the surface may suggest that a particular spot be explored for oil and, if this is discovered, the value of the land is established. The proven worth of the land warrants further drilling in the expectation that more oil will be found if the drilling sites are not too far removed from the location of the original strike. Of course, one might choose a very distant site because the drilling is easier there and still claim that he will strike oil. But human effort and ingenuity are limited and should therefore be devoted to good risks. If potential application is the goal, then as the great physical chemist Josiah Willard Gibbs remarked, the pure mathematician can do what he pleases but the applied mathematician must be at least partially sane.

Criticism of pure mathematics, mathematics for its own sake, can be found as far back as Francis Bacon's *The Advancement of Learning* (1620). There he objected to pure, mystic, and self-sufficient mathematics that was "entirely severed from matter and from axioms of natural philosophy" and served only "to satisfy [an] appetite for expatiation and meditation [which were] incidental to the human mind." He understood the value of applied mathematics:

> For many parts of nature can neither be invented with sufficient subtlety, nor demonstrated with sufficient perspicuity, nor accommodated to use with sufficient dexterity without the aid and intervention of mathematics, of which sort are perspective, music, astronomy, cosmography, architecture, machinery and some others. . . . For as physics advances farther and farther every day and develops new axioms, it will require fresh assistance from mathematics in many things, and so the parts of mixed [applied] mathematics will be more numerous.

In Bacon's time the concern of mathematicians with physical studies needed no prompting. But today the break from science is factual. In the last one hundred years, a schism has developed between those who would cleave to the ancient and honorable motivations for mathematical activity, the motivations which have thus far supplied the substance and fruitful themes, and those who sailing with the wind investigate what strikes their fancy. Today mathematicians and physical scientists go their separate ways. The newer mathematical creations have little application. Moreover, mathematicians and scientists no longer understand each other, and it is little comfort that, because of the intense specialization, mathematicians do not even understand other mathematicians.

The divergence from "reality," the study of mathematics for its own sake, provoked controversy almost from the outset. In his classic work *The Analytical Theory of Heat* (1822), Fourier waxed enthusiastic about the application of mathematics to physical problems:

> The profound study of nature is the most fertile source of mathematical discoveries. This study offers not only the advantages of a well determined goal but the advantage of excluding vague questions and useless calculations. It is a means of building analysis itself and of discovering the ideas which matter most and which science must always preserve. The fundamental ideas are those which represent the natural happenings. . . .
>
> Its chief attribute is clarity; it has no symbols to express confused ideas. It brings together the most diverse phenomena and discovers hidden analogies which unite them. If matter evades us, such as the air and light, because of its extreme thinness, if objects are located far from us in the immensity of space, if man wishes to understand the performance of the heavens for the successive periods which separate a large number of centuries, if the forces of gravity and of heat be at work in the interior of a solid globe at depths which will be forever inaccessible, mathematical analysis can still grasp the laws of these phenomena. It renders them present and measurable and seems to be a faculty of the human reason destined to make up for the brevity of life and for the imperfection of the senses; and what is more remarkable still, it follows the same method in the study of all phenomena; it interprets them by the same language as if to attest the unity and simplicity of the plan of the universe and to make still more manifest the immutable order which rules all natural matter.

Though Carl Gustav Jacob Jacobi (1804–1851) had done first-class work in mechanics and in astronomy, he took issue with what he regarded as at best a one-sided declaration. He wrote to Adrien-Marie Legendre on July 2, 1830: "It is true that Fourier has the opinion that the principal object of mathematics is public utility and the explanation of natural phenomena; but a scientist like him ought to know that the unique object of science is the honor of the human spirit and on this basis a question of [the theory of] numbers is worth as much as a question about the planetary system."

Of course, the mathematical physicists did not side with Jacobi's views. Lord Kelvin (William Thomson, 1824–1907) and Peter Guthrie Tait (1831–1901) in 1867 claimed that the best mathematics is suggested by application. This yields "astonishing theorems of pure mathematics, such as rarely fall to the lot of those mathematicians who confine themselves to pure analysis or geometry instead of allowing themselves to be led into the rich and beautiful fields of mathematical truth which lie in the way of physical research."

Many mathematicians also deplored the new trend to pure studies. In 1888 Kronecker wrote to Helmholtz, who had contributed to mathematics, physics, and medicine, "The wealth of your practical experience with sane and interesting problems will give to mathematicians a new direction and a new impetus. . . . One-sided and introspective mathematical speculations lead into sterile fields."

In 1895 Felix Klein, by that time a leader in the mathematical world, also felt the need to protest against the trend to abstract, pure mathematics:

> We cannot help feeling that in the rapid development of modern thought our science is in danger of becoming more and more isolated. The intimate mutual relation between mathematics and theoretical natural science which, to the lasting benefit of both sides, existed ever since the rise of modern analysis, threatens to be disrupted.

In his *Mathematical Theory of the Top* (1897), Klein returned to this theme:

> It is the great need of the present in mathematical science that the pure science and those departments of physical sciences in which it finds its most important applications should again be brought into the intimate association which proved so fruitful in the works of Lagrange and Gauss.

Poincaré in his *Science and Method,* despite his ironic remarks about some of the purely logical creations of the late 19th century (Chapter VIII), did grant that axiomatics, unusual geometries, and peculiar functions "show us what the human mind can create when it frees itself more and more from the tyranny of the external world." Nevertheless, he insisted "it is toward the other side, the side of nature that we must direct the bulk of our army." In *The Value of Science,* he wrote:

> It would be necessary to have completely forgotten the history of science not to remember that the desire to understand nature has had on the development of mathematics the most important and happiest

influence. . . . The pure mathematician who should forget the existence of the exterior world would be like a painter who knows how to harmoniously combine colors and forms, but who lacked models. His creative power would soon be exhausted.

Somewhat later, in 1908, Felix Klein spoke out again. Fearing that the freedom to create arbitrary structures was being abused, he emphasized that arbitrary structures are "the death of all science." The axioms of geometry are "not arbitrary but sensible statements which are, in general, induced by space perception and are determined as to their precise content by expediency." To justify the non-Euclidean axioms Klein pointed out that visualization requires the Euclidean parallel axiom only within certain limits of accuracy. On another occasion he pointed out that "whoever has the privilege of freedom should also bear responsibility." By responsibility Klein meant service to the investigation of nature.

Toward the end of his life Klein, who was by then the head of the mathematics department at what was the world center of mathematics, the University of Göttingen, felt compelled to make one more protest. In his *Development of Mathematics in the Nineteenth Century* (1925), he recalled Fourier's interest in solving practical problems with the best mathematics available and contrasted this with the purely mathematical refinement of tools and abstraction of concrete ideas. He then said:

> The mathematics of our day seems to be like a great weapons factory in peace time. The show window is filled with parade pieces whose ingenious, skillful, eye-appealing execution attracts the connoisseur. The proper motivation for and purpose of these objects, to battle and conquer the enemy, has receded to the background of consciousness to the extent of having been forgotten.

Richard Courant, who succeeded Klein in the leadership of mathematics at Göttingen and subsequently headed what has since been named the Courant Institute of Mathematical Sciences of New York University, also deplored the emphasis on pure mathematics. In the Preface to the first edition of Courant's and Hilbert's *Methods of Mathematical Physics,* written in 1924, Courant opened with the observation:

> From olden times onward mathematics won powerful stimuli from the close relations which existed between the problems and methods of analysis and the intuitive ideas of physics. The last few decades have witnessed a loosening of this connection in that mathematical research has severed itself a great deal from the intuitive starting points and especially in analysis has mostly concerned itself with the refinement of its methods and the sharpening of its concepts. Thus it has come about that many leaders of analysis have lost the knowledge of the

connection between their discipline and physics as well as other domains, while on the other side the understanding by physicists of the
problems and methods of the mathematicians and even of their entire
spheres of interest and language has been lost. Without doubt there
lies in this tendency a threat to all of science. The current of scientific
development is in danger of further and further ramification, trickling
away and drying up. To escape this fate we must direct a great deal of
our power to uniting what is separated in that we lay clear under a
common viewpoint the inner connections of the manifold facts. Only
in this way will the student master an effective control of the material
and the researcher be prepared for an organic development.

Again in 1939 Courant wrote:

A serious threat to the very life of science is implied in the assertion
that mathematics is nothing but a system of conclusions drawn from
the definitions and postulates that must be consistent but otherwise
may be created by the free will of mathematicians. If this description
were accurate, mathematics could not attract any intelligent person. It
would be a game with definitions, rules, and syllogisms without motivation or goal. The notion that the intellect can create meaningful
postulational systems at its whim is a deceptive half-truth. Only under
the discipline of responsibility to the organic whole, only guided by intrinsic necessity, can the free mind achieve results of scientific value.

George David Birkhoff (1884–1944), the leading United States mathematician of his day, made the same point in the *American Scientist* of
1943:

It is to be hoped that in the future more and more theoretical physicists will command a deep knowledge of mathematical principles; and
also that mathematicians will no longer limit themselves so exclusively
to the aesthetic development of mathematical abstractions.

John L. Synge, admittedly a mathematical physicist, described the situation in 1944 at some length in a Shavian preface to a technical article:

Most mathematicians work with ideas which, by common consent,
belong definitely to mathematics. They form a closed guild. The initi-
ate foreswears the things of the world, and generally keeps his oath.
Only a few mathematicians roam abroad and seek mathematical sustenance in problems arising directly out of other fields of science. In
1744 or 1844 this second class included almost the whole body of
mathematicians. In 1944 it is such a small fraction of the whole that it
becomes necessary to remind the majority of the existence of the minority, and to explain its point of view.
    The minority does not wish to be labelled "physicist" or "engineer,"
for it is following a mathematical tradition extending through more

than twenty centuries and including the names Euclid, Archimedes, Newton, Lagrange, Hamilton, Gauss, Poincaré. The minority does not wish to belittle in any way the work of the majority, but it does fear that a mathematics which feeds solely on itself will in time exhaust its interest.

Apart from its effect on the future of mathematics proper, the isolation of mathematicians has robbed the rest of science of a support on which it counted in all previous epochs. . . . Out of the study of nature there have originated (and in all probability will continue to originate) problems far more difficult than those constructed by mathematicians within the circle of their own ideas. Scientists have relied on the mathematician to throw his energies against these problems. They know that the mathematician is not merely a skilled user of certain tools already made—they can use those tools with no inconsiderable skill themselves. Rather they look to the qualities peculiar to the mathematician—his logical penetration and capacity to see the general in the particular and the particular in the general. . . .

In all this the mathematician was the directing and disciplining force. He gave science its methods of calculation—logarithms, calculus, differential equations, and so on—but he gave it much more than this. He gave it a blue-print. He insisted that thought be logical. As each new science came up, he gave it—or tried to give it—the firm logical structure that Euclid gave to Egyptian land surveying. A subject came to his hands a rough stone, trailing irrelevant weeds; it left his hands a polished gem.

At present science is humming as it never hummed before. There are no obvious signs of decay. Only the most observant have noticed that the watchman has gone off duty. He has not gone to sleep. He is working as hard as ever, but now he is working solely for himself. . . .

In brief, the party is over—it was exciting while it lasted. . . . Nature will throw out mighty problems, but they will never reach the mathematician. He may sit in his ivory tower waiting for the enemy with an arsenal of guns, but the enemy will never come to him. Nature does not offer her problems ready formulated. They must be dug up with pick and shovel, and he who will not soil his hands will never see them.

Change and death in the world of ideas are as inevitable as change and death in human affairs. It is certainly not the part of a truth-loving mathematician to pretend that they are not occurring when they are. It is impossible to stimulate artificially the deep sources of intellectual motivation. Something catches the imagination or it does not, and, if it does not, there is no fire. If mathematicians have really lost their old universal touch—if, in fact, they see the hand of God more truly in the refinement of precise logic than in the motion of the stars—then any attempt to lure them back to their old haunts would not only be useless—it would be denial of the right of the individual to intellectual freedom. But each young mathematician who formulates his own philosophy—and all do—should make his decision in full possession of the facts. He should realize that if he follows the pattern of modern mathematics he is heir to a great tradition, but only part heir. The rest of the legacy will have gone to other hands, and he will never get it back. . . .

Our science started with mathematics and will surely end not long after mathematics is withdrawn from it (if it is withdrawn). A century hence there will be bigger and better laboratories for the mass-production of facts. Whether these facts remain mere facts or become science will depend on the extent to which they are brought into contact with the spirit of mathematics.

John von Neumann was sufficiently alarmed to issue a warning. In the often quoted but little heeded essay "The Mathematician" (1947), he stated:

> As a mathematical discipline travels far from its empirical source, or still more, if it is a second and third generation only indirectly inspired from "reality" it is beset with very grave dangers. It becomes more and more pure aestheticizing, more and more purely *l'art pour l'art*. This need not be bad if the field is surrounded by correlated subjects which still have closer empirical connections, or if the discipline is under the influence of men with exceptionally well-developed taste. But there is a grave danger that the subject will develop along the line of least resistance, that the stream so far from its source will separate into a multitude of insignificant branches, and that the discipline will become a disorganized mass of details and complexities. In other words, at a great distance from its empirical source, or after much abstract in-breeding, a mathematical subject is in danger of degeneration. At the inception the style is usually classical; when it shows signs of becoming baroque, the danger signal is up. . . . In any event, whenever this stage is reached, the only remedy seems to me to be the rejuvenating return to the source: the reinjection of more or less directly empirical ideas. I am convinced that this was a necessary condition to conserve the freshness and vitality of the subject and that this will remain equally true in the future.

But the trend to pure creations has not been stemmed. Mathematicians have continued to break away from science and to pursue their own course. Perhaps to placate their own consciences, they have tended to look down on the applied mathematicians as lowly draftsmen. They complain that the sweet music of pure mathematics is being drowned out by the trumpets of technology. However, they have also felt the need to counter the criticisms such as those reproduced above. The outspoken ones, either ignorant of or deliberately distorting the history, now argue that many of the major creations of the past were motivated solely by intellectual interest and yet proved to be immensely important for applications later. Let us look at some of the examples these pure mathematicians adduce from history. How pure was the mathematics they describe as pure?

The commonest example offered is the Greek work on the parabola, ellipse, and hyperbola. The argument of the pure mathematicians is that these curves were investigated by the Greeks, notably Apollonius, simply to satisfy a mathematical interest. Yet eighteen hundred years later the ellipse was found by Kepler to be just the curve he needed to describe the motions of the planets around the sun. Though we do not know the very early history of the conic sections, one theory, proposed by the authoritative historian Otto Neugebauer, is that they arose in work on the construction of sundials. Ancient sundials using conics are known. Moreover, the fact that the conic sections can be used to focus light was known long before Apollonius devoted his classic work to that subject (Chapter I). Hence the physical uses of the conic sections for optics, a subject to which the Greeks devoted considerable time and effort, certainly motivated some of the work on the conic sections.

The conic sections were also investigated long before Apollonius's time to solve the problem of constructing the side of a cube whose volume was double that of a given cube, a problem of importance to Greek geometry for which construction was the method of proving existence.

That Apollonius also proved hundreds of theorems about the conics that have no application immediate or potential is undoubtedly true. In this work he did not differ from moderns who have found a substantial theme and proceed to elaborate on it either for the reasons cited above of learning more about a vital subject or for reasons such as intellectual challenge.

The second most often cited example of pure mathematics later found to be applicable is non-Euclidean geometry. Presumably mathematicians merely speculating about what would result from changing the parallel axiom of Euclid created non-Euclidean geometry. This assertion ignores two thousand years of history. Euclid's axioms were taken to be self-evident truths about physical space (Chapter I). However, the parallel axiom, rather deliberately and peculiarly worded by Euclid to avoid the outright assumption of parallel lines, was not as self-evident as the other axioms. Hence the many efforts to find a more acceptable version led ultimately to the realization that the parallel axiom was not necessarily true and that a differing axiom on parallels—and consequently a non-Euclidean geometry—could be as good a representation of physical space. The main point is that the efforts to guarantee the truth of Euclid's parallel axiom were not the "amusement of speculative brains" but an attempt to secure the *truth* of geometry which underlies thousands of theorems of mathematics proper and of applied mathematics.

The pure mathematicians often cite the work of Riemann, who generalized on the non-Euclidean geometry known in his time and introduced a large variety of non-Euclidean geometries, now known as Riemannian geometries. Here, too, the pure mathematicians contend that Riemann created his geometries merely to see what could be done. Their account is false. The efforts of mathematicians to put beyond question the physical soundness of Euclidean geometry culminated, as we have just noted, in the creation of a non-Euclidean geometry which proved as useful in representing the properties of physical space as Euclidean geometry was. This unexpected fact gave rise to the question, since these two geometries differ, what are we really sure is true about physical space? This question was Riemann's explicit point of departure and in answering it in his paper of 1854 (Chapter IV) he created more general geometries. In view of our limited physical knowledge, these could be as useful in representing physical space as Euclidean geometry. In fact, Riemann foresaw that space and matter would have to be considered together. Is it to be wondered then that Einstein found a Riemannian geometry useful? Riemann's foresight concerning the relevance of this geometry does not detract from the ingenious use which Einstein made of it; its suitability was the consequence of work on the most fundamental physical problem which mathematicians have ever tackled, the nature of physical space.

Perhaps one more example should be considered. A subject very actively studied in modern mathematics is group theory, which, the pure mathematicians contend, was tackled for its intrinsic interest. Group theory was created primarily by Evariste Galois (1811–1832), though there was preceding work by Lagrange and Paolo Ruffini. The problem Galois tackled was in essence the simplest and most practical problem of all mathematics, namely, the solvability of simple polynomial equations such as the second degree equation

$$3x^2 + 5x + 7 = 0,$$

the third degree equation

$$4x^3 + 6x^2 - 5x + 9 = 0,$$

and higher degree equations. Such equations arise in thousands of physical problems. Mathematicians had succeeded by Galois's time in showing how to solve the general equations from the first to the fourth degree, and Niels Henrik Abel (1802–1829) had shown that it is impossible to solve *algebraically* general equations of the fifth and higher degree such as

$$ax^5 + bx^4 + cx^3 + dx^2 + ex + f = 0$$

where a, b, c, d, e and f can be any real (or complex) numbers. Galois thereupon set out to show why such general equations of degree five or higher are not solvable algebraically and which special ones may be solvable. In his work he created the theory of groups. Is it any wonder that a subject derived from so basic a problem as the solution of polynomial equations should be applicable to many other mathematical and physical problems? Certainly group theory was not inspired independently of real problems.

Moreover, the motivation for group theory does not rest solely on Galois's work. The pure mathematicians seem to overlook the work of Auguste Bravais (1811–1863) on the structure of crystals such as quartz, diamonds, and rocks. These substances are composed of different atoms arranged in a pattern that repeats itself throughout the crystal. Moreover, crystals such as salt and common minerals have very special types of atomic arrangements. For example, in the simplest case, the atoms though contiguous may be thought of as located at the vertices of a cube. Bravais studied, from 1848 on, the possible transformations that permit one to rotate a crystal about some axis or to translate or reflect the crystal about some axis so that the crystal is transformed into itself. These transformations form various types of groups. Camille Jordan (1838–1922) noted this work of Bravais, extended it himself in a paper of 1868, and included it among other motivations for the study of finite groups in his most influential *Traité des substitutions* (Treatise on Substitutions, 1870).

Bravais's work also suggested to Jordan the study of infinite groups—groups of translations and rotations. The subject of infinite groups was brought into prominence by Felix Klein's talk and paper of 1872 in which he distinguished the different geometries known in his time by the groups of movements possible in those geometries and by what remains invariant under these movements. Thus Euclidean geometry is characterized by the fact that it studies properties of figures that are invariant under rotations, translations, and similarity transformations. The problem in 1872 of distinguishing the various known geometries and the problem of which geometry fits physical space—then uppermost in the minds of mathematicians—can hardly be dismissed as pure mathematics. Many more works involving infinite continuous and discontinuous groups to characterize methods of solving differential equations had yet to enter mathematics before the modern notion of an abstract group was formulated in the 1890s.*

An investigation of other subjects claimed to be the product of pure mathematics, such as matrices, tensor analysis, and topology, would reveal the same story. For example, all of modern abstract algebra owes its origins to Hamilton's creation of quaternions (Chapter IV). The motivations were either directly or indirectly physical and the men involved were vitally concerned with the uses of mathematics. In other words, the subjects purportedly created as pure mathematics and found to be applicable later were, as a matter of historical fact, created in the study of real physical problems or those bearing directly on physical problems. What does often happen is that good mathematics, originally motivated by physical problems, finds new applications that were not foreseen. Thereby mathematics pays its debt to science. Such uses are to be expected. Are we surprised to find that the hammer, which may have been invented originally to crush rocks, may also be used to drive nails into wood? The unexpected scientific uses of mathematical theories arise because the theories are physically grounded to start with and are by no means due to the prophetic insight of all-wise mathematicians who wrestle solely with their souls. The continuing successful use of these creations is by no means fortuitous.

Godfrey H. Hardy (1877–1947), one of Britain's leading mathematicians, is reported to have made the toast: "Here's to pure mathematics! May it never have any use." Leonard Eugene Dickson (1874–1954), an authority at the University of Chicago, used to say: "Thank God that number theory is unsullied by any applications."

In an article on mathematics in wartime (1940), Hardy said:

> I had better say at once that by "mathematics" I mean *real* mathematics, the mathematics of Fermat and Euler and Gauss and Abel, and not the stuff that passes for mathematics in an engineering laboratory. I am not thinking only of "pure" mathematics (though naturally that is my first concern); I count Maxwell and Einstein and Eddington and Dirac among "real" mathematicians.

One would believe from this quotation, which Hardy repeated in *A Mathematician's Apology* (1967 edition), that Hardy was willing to accept some at least of applied mathematics. But then he went on:

> I am including the whole body of mathematical knowledge which has permanent aesthetic value; as for example, the best Greek mathemat-

---

* Arthur Cayley did suggest a definition of an abstract group in papers of 1849 and 1854. But the significance of an abstract concept was not appreciated until the many more concrete uses described above had been made.

ics has, the mathematics which is eternal because the best of it may,
like the best literature, continue to cause emotional satisfaction to
thousands of people after thousands of years.

Hardy and Dickson can rest in peace because they have the assurance
of history. Their pure mathematics, like all mathematics created for its
own sake, will almost certainly not have any use. However, the possibility is not out of the question. A child who dabs paint at random on a
canvas may rival Michelangelo (though more likely modern art) and, as
Arthur Stanley Eddington pointed out, a monkey who types letters at
random may produce a play of Shakespearean quality. As a matter of
fact, with thousands of pure mathematicians at work, an occasional
result bearing on useful mathematics can appear. A man who looks for
gold coins in the street may find nickels and dimes. But intellectual efforts not tempered by reality are almost sure to prove sterile. As
George David Birkhoff remarked, "It will probably be the new mathematical discoveries suggested through physics that will always be the
most important, for from the beginning Nature has led the way and established the pattern which mathematics, the language of Nature, must
follow." However, nature never blazons her secrets with a loud voice
but always whispers them. And the mathematician must listen atten-
tively and then attempt to amplify and proclaim them.

Despite the evidence of history, some mathematicians still assert the
future applicability of pure mathematics and in fact declare that independence from science will improve the prospects. This thesis was
reaffirmed not too long ago (in 1961) by Marshall Stone, a professor at
Harvard, Yale, and Chicago. Though he begins his article entitled
"The Revolution in Mathematics" with a tribute to the importance of
mathematics in science, Stone continues:

> While several important changes have taken place since 1900 in our
> conception of mathematics or in our points of view concerning it, the
> one which truly involves a revolution in ideas is the discovery that
> mathematics is entirely independent of the physical world. . . . Math-
> ematics is now seen to have no necessary connection with the physical
> world beyond the vague and mystifying one implicit in the statement
> that thinking takes place in the brain. The discovery that this is so may
> be said without exaggeration to mark one of the most significant intel-
> lectual advances in the history of mathematics. . . .
>   When we stop to compare the mathematics of today with mathemat-
> ics as it was at the close of the nineteenth century we may well be
> amazed to note how rapidly our mathematics has grown in quantity
> and complexity, but we should also not fail to observe how closely this
> development has been involved with an emphasis on abstraction and
> an increasing concern with the perception and analysis of broad math-

ematical patterns. Indeed, upon closer examination we see that this new orientation, made possible only by the divorce of mathematics from its applications, has been the true source of its tremendous vitality and growth during the present century. . . .

A modern mathematician would prefer the positive characterization of his subject as the study of general abstract systems, each one of which is an edifice built of specified abstract elements and structured by the presence of arbitrary but unambiguously specified relations among them. . . . He would maintain that neither these systems nor the means provided by logic for studying their structural properties have any directly immediate or necessary connection with the physical world. . . . For it is only to the extent that mathematics is freed from the bonds which have attached it in the past to particular aspects of reality that it can become the extremely flexible and powerful instrument we need to break into areas beyond our ken. The examples which buttress this argument are already numerous. . . .

Then Stone mentions genetics, game theory, and the mathematical theory of communication. Actually these are a poor defense of his thesis. They have come about by applying sound classical mathematics.

In an article which appeared in the *SIAM Review* [Society for Industrial and Applied Mathematics] of 1962, Courant rebutted Stone's thesis: *

The article [by Stone] asserts that we live in an era of great mathematical successes, which outdistance everything achieved from antiquity until now. The triumph of "modern mathematics" is credited to one fundamental principle: abstraction and conscious detachment of mathematics from physical and other substance. Thus, the mathematical mind, freed from ballast, may soar to heights from which reality on the ground can be perfectly observed and mastered.

I do not want to distort or belittle the statements and the pedagogical conclusions of the distinguished author. But as a sweeping claim, as an attempt to lay down a line for research and before all for education, the article seems a danger signal, and certainly in need of supplementation. The danger of enthusiastic abstractionism is compounded by the fact that this fashion does not at all advocate nonsense, but merely promotes a half truth. One-sided half truths must not be allowed to sweep aside the vital aspects of the balanced whole truth.

Certainly mathematical thought operates by abstraction; mathematical ideas are in need of abstract progressive refinement, axiomatization, crystallization. It is true indeed that important simplification becomes possible when a higher plateau of structural insight is reached. Certainly it is true—and has been clearly emphasized for a long time—

* Reprinted with permission from *SIAM Review,* October 1962, pp. 297–320. Copyright by SIAM, 1962.

that basic difficulties in mathematics disappear if one gives up the metaphysical prejudice of mathematical concepts as descriptions of a somehow substantive reality.

Yet, the life blood of our science rises through its roots; these roots reach down in endless ramification deep into what might be called reality, whether this "reality" is mechanics, physics, biological form, economic behavior, geodesy, or, for that matter, other mathematical substance already in the realm of the familiar. Abstraction and generalization are not more vital for mathematics than individuality of phenomena and, before all, not more than inductive intuition. Only the interplay between these forces and their synthesis can keep mathematics alive and prevent its drying out into a dead skeleton. We must fight against attempts to push the development one-sidedly towards the one pole of the life-spending antinomy.

We must not accept the old blasphemous nonsense that the ultimate justification of mathematical science is "the glory of the human mind." Mathematics must not be allowed to split and to diverge towards a "pure" and an "applied" variety. It must remain, and be strengthened as, a unified vital strand in the broad stream of science and must be prevented from becoming a little side brook that might disappear in the sand.

The divergent tendencies are immanent in mathematics and yet prove an ever-present danger. Fanatics of isolationist abstractionism are dangerous indeed. But so are conservative reactionaries who do not discriminate between hollow pretense and dedicated aspirations.

Courant did not deny the value of abstraction. However, he said in an article of 1964 that mathematics must take its motivation from concrete specific substance and aim at some layer of reality. If a flight into abstraction is necessary, it must be something more than an escape. Reentry to the ground is indispensable, even if the same pilot cannot handle all phases of the trajectory.

Mathematics has often been compared with a tree whose roots are firmly and deeply grounded in rich natural soil. The central trunk is number and geometrical figure, and the outgrowths from this trunk of the multiplicity of branches represent the many developments. Some are stout and nourish many vital offshoots. Other branches have given rise to minor offshoots which contribute little to the size and strength of the entire structure. Still others are dead. But what is most relevant is that the tree is rooted in solid earth. Each branch draws on that reality through the roots and trunk. Recent efforts to remove the soil entirely and still sustain the tree, roots, trunk, and superstructure, can not succeed. The many branches will thrive only as the roots go deeper into fertile soil. Those who would graft on new branches having no nourishment in reality produce still-born stalks that will never acquire life.

These would-be branches may, with enough effort, be made to look like real ones and may be made to intertwine with the live branches and seem to grow out of the trunk but they are nonetheless dead and can be lopped off with not the slightest harm to the live growth.

There are other arguments which seem to undermine Stone's claim that the freedom to pursue pure mathematics will strengthen and contribute new approaches to applied mathematics. The pursuit of pure mathematics, no matter what the level of sophistication or how eminent the practitioner, must diminish a man's power to apply mathematical reasoning to practical situations. If he spends time and energy on abstract mathematics, he will inevitably be influenced by its atmosphere and by the attitudes of mind necessary to its successful practice. He will also have less time to learn about the needs in applications and to fashion the tools for handling them. The applied mathematicians can usefully take note of what the abstract mathematicians are up to and what they achieve; but too much attention is a harmful diversion of resources. Disregard for applications leads to isolation and possibly to the atrophy of mathematics as a whole.

Judged by historical evidence, Stone is certainly wrong. As von Neumann pointed out in his essay "The Mathematician" (1947):

> It is undeniable that some of the best inspirations in mathematics—in those parts of it which are as pure mathematics as one can imagine—have come from the natural sciences. . . . The most vitally characteristic fact about mathematics is, in my opinion, its quite peculiar relationship to the natural sciences, or, more generally, to any science which interprets experience on a higher than purely descriptive level.

Laurent Schwartz, a leading French mathematician, did not hesitate to say that today's most active fields, abstract algebra and algebraic topology, have no application. Some papers dress up concrete applications in the language and concepts of these fields but make no advance in the solution of applied problems.

However, the proponents of pure, abstract mathematics have not yielded. Professor Jean Dieudonné, a leading analyst, in an article of 1964 rejected the contention that a mathematics which feeds on itself will die for lack of nourishment:

> I would like to stress how little recent history has been willing to conform to the pious platitudes of the prophets of doom, who regularly warn us of the dire consequences mathematics is bound to incur by cutting itself off from the applications to other sciences. I do not intend to say that close contact with other fields, such as theoretical physics, is not beneficial to all parties concerned; but it is perfectly clear that of all the striking progress I have been talking about, not a single

one, with the possible exception of distribution theory, had anything
to do with physical applications; and even in the theory of partial differential equations, the emphasis is now much more on "internal" and
structural problems than on questions having a direct physical significance. Even if mathematics were to be forcibly separated from all
other channels of human endeavor, there would remain food for centuries of thought in big problems we still have to solve within our
science.

Though Dieudonné could see unending problems of pure mathematics, to be fair to the man one must add that he did not swallow the
argument that any creation of pure mathematics will ultimately be useful. He cited many investigations of pure mathematics, particularly in
the theory of numbers, of which he said, "It is inconceivable that such
results might be applied to any physical problem." Moreover, though
he defended pure mathematics generally, he also argued that the
boasting by mathematicians of the value of their pure mathematics to
science is "a mild kind of swindle." Pure mathematicians, he pointed
out, will take great trouble to prove the uniqueness of a solution to a
problem but won't try to find the solution. The physicist knows there is
a unique solution—the earth does not move on two different paths—
but he wants the actual path.

More realistic about the value of the kind of mathematics that should
be pursued is a man of equal stature who has worked in pure mathematics. At the International Congress of Mathematicians in 1958, he
was frank. Said Lars Gårding:

> I cannot go here into many important parts of the subject like dif-
> ference equations, the theory of systems and applications to quantum
> mechanics and differential geometry. My subject has been the general
> theory of partial differential operators. It has grown out of classical
> physics, but has no really important applications to it. Still physics
> remains its main source of interesting problems. I have the feeling
> that general talks of the kind I have given are perhaps less useful than
> periodical surveys of unsolved problems in physics which seem to
> require new mathematical techniques. Such surveys will of course not
> be news to specialists, but they will provide many mathematicians with
> worthwhile problems. Planned efforts to further the interaction be-
> tween physics and mathematics have been rare. They ought to be one
> of the major concerns of the international mathematical congresses.

Those who boast of creating mathematics untainted by the physical
world and maintain under pressure that others will some day find justification for their currently pointless endeavors may be left to work in
their own mental grooves. But they contradict the entire course of history. Their confidence that a mathematics freed from bondage to

science will produce richer, more varied, and more fruitful themes that will be applicable to far more than the older mathematics is not backed by anything but words.

The proponents of pure mathematics can and do make other claims for the value of their work, namely, intrinsic beauty and intellectual challenge. That there are such values cannot be denied. However, that they justify the enormous output of pure mathematics can indeed be challenged. Regardless of any judgments in these matters, these values do not contribute to what has been the major importance of mathematics, namely, the study of nature. Beauty and intellectual challenge are mathematics for mathematics' sake. Admittedly the worth of these intrinsic values warrants more recognition than the present discussion of the isolation of mathematics permits.

The defenders and critics of pure mathematics are clearly at odds with each other. All controversies prompt some humorous or sarcastic remarks. The applied mathematicians are not concerned with rigorous proof. For them agreement of their deductions with physical events is the chief concern. Typical of these men was Oliver Heaviside, who used what seemed to the pure mathematicians to be totally unjustified and outlandish techniques. Consequently, he was severely criticized. Heaviside was contemptuous of what he called the logic-choppers. He replied, "Logic can be patient for it is eternal." Somewhat later he confounded the pure mathematicians still more. At that time, what are called divergent series were outlawed. Heaviside said of a particular one, "Ha, the series diverges; now we can do something with it." It turns out that Heaviside's techniques have all been rigorized and even have supplied new mathematical themes. To nettle the purists, the applied mathematicians have remarked that the pure mathematicians can find the difficulty in any solution, but the applied men can find the solution to any difficulty.

The applied mathematicians make another gibe at the purists. Applied problems are set by physical phenomena and the mathematicians concerned are obliged to solve them while the pure mathematicians can create their own problems. The latter, the applied people say, are like a man who, searching for a key lost in some dark street, moves to search for it under a street lamp because there is more light there.

To further belittle their opponents, the applied mathematicians tell another tale. A man has a bundle of clothes to be laundered and looks around for a laundry. He finds a store with a sign in the window, "Laundry Done Here," enters, and puts his bundle on the counter. The storekeeper looks a little astonished and asks, "What is this?" The man answers, "I brought these clothes in to be laundered." "But we don't launder clothes here," the storekeeper replies. This time it is the would-be customer who is astonished. He points to the sign and asks, "What about that sign?" "Oh," says the storekeeper, "we just make the signs."

The controversy between pure and applied mathematicians continues, and since the pure mathematicians now rule the roost, they can look down at their misguided brethren and even reprove them. As Professor Clifford E. Truesdell has noted, " 'Applied mathematics' is an insult directed by those who consider themselves 'pure' mathematicians at those whom they take for impure. . . . but 'pure' mathematics as a parricide tantrum denying growth from human sensation, as a shibboleth to cast out the impure, is a disease invented in the last century. . . ." It has become an end in itself with no thought of what objective it might serve. It is not, in itself, a state of beatitude. Mathematics aims to discover something worth knowing. As matters stand now, research begets research, which begets research. In the halls of mathematics today, one dare not ask for meaning and purpose. Mathematics must not be tainted by reality. The ivy has grown so thick that the researchers within can no longer see the world outside. These sequestered minds are content in their isolation.

Mathematicians may disagree among themselves, but physicists and other scientists can only deplore that they have been left in the lurch. Let us listen to Professor John C. Slater, a distinguished professor until recently at the Massachusetts Institute of Technology:

> The physicist finds very little help from the mathematician. For every mathematician like von Neumann who realizes these problems [described earlier], and contributes practically to them, there are twenty who have no interest in them, who either work in fields of remote interest to physics, or who stress the older and more familiar parts of mathematical physics. Is it any wonder that in such a situation the physicist, looking at the mathematicians, feels that they have strayed from the path which has led to the past greatness of mathematics, and feels that they will not regain this path until they again resolutely enter the main current of progress of mathematical physics, the current which in the past has led to the most fruitful development of mathematics. . . . That, the physicist firmly feels, is the only path through which the mathematician of the present can achieve greatness.

The neglect of science was the subject of a major lecture by Professor Freeman J. Dyson to mathematicians in 1972. Dyson, a very distinguished physicist, pointed out opportunities, past and present, for mathematicians to take up significant and outstanding problems of science, but which they failed to take up. Some of these problems or fragments of them have somehow seeped into mathematics, but the mathematicians do not know their origin and physical significance. Hence they head in arbitrary directions or do not perceive what they have accomplished. As Dyson puts it, the marriage of mathematics and physics has ended in divorce.

The break from science has been accelerated during this century. Today it is common to hear and read mathematicians' declarations of independence from science. Mathematicians no longer hesitate to speak freely of their interest solely in mathematics proper and their indifference to science. Though no precise statistics are available, about ninety percent of the mathematicians active today are ignorant of science and are quite content to remain in that blissful state. Despite the history and some opposition, the trend to abstraction, to generalization for the sake of generalization, and to the pursuit of arbitrarily chosen problems has continued. The reasonable need for study of an entire class of problems in order to learn more about concrete cases and for abstraction in order to get at the essence of a problem has become an excuse for tackling generalities and abstractions in and for themselves.

Over the centuries man created such grand structures as Euclidean geometry, the Ptolemaic theory, the heliocentric theory, Newtonian mechanics, electromagnetic theory, and in recent times relativity and quantum theory. In all of these and in other significant and powerful bodies of science, mathematics as we now know is the method of construction, the framework, and indeed the essence. Mathematical theories have enabled us to know something of nature, to embrace in comprehensive intelligible accounts varieties of seemingly diverse phenomena. Mathematical theories have revealed whatever order and plan man has found in nature and have given us mastery or partial mastery over vast domains.

But most mathematicians have abandoned their traditions and heritage. The pregnant messages that nature sends to the senses now fall on closed eyes and inattentive ears. Mathematicians are living on the reputation earned by their predecessors and still expect the acclaim and support that the older work warranted. The pure mathematicians have gone further. They have expelled the applied mathematicians from their fraternity in the hope that by cornering the honorable title of mathematician they alone will gain the glory accorded to their predecessors. They have thrown away their rich source of ideas and are now spending their previously accumulated wealth. They have followed a gleam that has led them out of this world. It is true that some, aware of the noble tradition that motivated mathematical research in the past and warranted the honor accorded to the Newtons and Gausses, still claim potential value of their mathematical work for science.

They speak of creating models for science. But in truth they are not concerned with this goal. In fact, since most modern mathematicians know no science, they can't be creating models. They prefer to remain virgins rather than to bed with science. On the whole mathematics is now turned inward; it feeds on itself; and it is extremely unlikely, if one may judge by what happened in the past, that most of the modern mathematical research will ever contribute to the advancement of science; mathematics may be doomed to grope in darkness. Mathematics is now an almost entirely self-contained enterprise. Moving in directions determined by its own criteria of relevance and excellence, it is even proud of its independence from outside problems, motivations, and inspirations. It no longer has unity and purpose.

The isolation of most mathematicians today is deplorable for many reasons. The scientific and technological uses of mathematics are expanding at an enormous rate. Until recently, the vision of Descartes that mathematics represents the supreme achievement of the human intellect, the triumph of reason over empiricism, and the eventual pervasion of all fields of science by mathematically-based methodologies seemed closer to realization. But just when the mathematical approach to so many fields was spreading, the mathematicians withdrew into a corner. Whereas one hundred years ago and farther back mathematics and the physical sciences were intimately involved with each other (on a Platonic basis of course), since that time they have separated, and the separation has become marked in our time. The fact that mathematics is valuable because it contributes to the understanding and mastery of nature has been lost sight of. Most present-day mathematicians wish to isolate their subject and offer only an aloof study. The schism between those who would cleave to the ancient and honorable motivations for mathematical activity, the motivations that in the past supplied the most fruitful themes, and those who, sailing with the wind, investigate what strikes their fancy is pronounced. Blinded by a century of ever purer mathematics, most mathematicians have lost the skill and the will to read the book of nature. They have turned to fields such as abstract algebra and topology, to abstractions and generalizations such as functional analysis, to existence proofs for differential equations that are remote from applications, to axiomatizations of various bodies of thought, and to arid brain games. Only a few still attempt to solve the more concrete problems, notably in differential equations and allied fields.

Does the abandonment of science by most mathematicians mean that science will be deprived of mathematics? Not at all. A few discerning mathematicians have noted that the Newtons, Laplaces, and Hamiltons of the future will create the mathematics they will need just as they have in the past. These men, though honored as mathematicians, were physicists. In an obituary of Franz Rellich that Richard Courant wrote in 1957, he said that if present trends continue, "There exists the danger that the 'applied' mathematics of the future will be developed by physicists and engineers, and the professional mathematicians of rank will have no connection with the new development." Courant put quotes around the word applied because he really meant all significant mathematics. He did not distinguish pure and applied mathematics.

Courant's prophesy is being fulfilled. Because the establishment in the mathematical community favors pure mathematics, the best applied work is now being done by scientists in electrical engineering, computer science, biology, physics, chemistry, and astronomy. Like the mathematicians Gulliver met on his voyage to Laputa, the purists live on an island suspended in the air above the earth. They leave to others the problems of society on the earth itself. These mathematicians will live for a while in the atmosphere which mathematicians of the past supplied to the subject but they are doomed ultimately to expire in a vacuum.

Talleyrand once remarked that an idealist cannot last long unless he is a realist and a realist cannot last long unless he is an idealist. When applied to mathematics, this observation speaks for the need to idealize real problems and to study them abstractly but it also says that the work of the idealist who ignores reality will not survive. Mathematics should have its feet on the ground and its head in the clouds. It is the interplay between abstraction and concrete problems that produces live, vital, significant mathematics. Mathematicians may like to soar into the clouds of abstract thought, but like birds they must return to earth for food. Pure mathematics may be like cake for dessert. It pleases the appetite and even nourishes the system somewhat, but the body cannot survive on cake without the meat and potatoes of real problems for basic nourishment.

The danger lies in too much attention to artificial questions. If the present emphasis on pure mathematics continues, the mathematics of the future will not be the subject we have valued in the past, though it bears the same name. Mathematics is a marvelous invention, but the marvel lies in the human mind's capacity to construct understandable models of complex and seemingly inscrutable natural phenomena and thereby give man some enlightenment and power.

However, the individual must be free to choose his own course. Homer said in the *Odyssey*, "Different men take delight in different actions," and in the century after Homer the poet Archilochus put it, "Each man must have his heart cheered in his own way." Goethe expressed the same thought: "To the individual remains the freedom to occupy himself with what attracts him, what gives him pleasure, what seems to him likely to be useful." However, Goethe added, "the proper study of mankind is man." For our purposes we might paraphrase, the proper study of mathematicians is nature. As Francis Bacon put it in his *Novum organon* (New Instrument [of reasoning]). "But the real and legitimate goal of the sciences is the endowment of human life with new inventions and riches."

In the final analysis, sound judgment must decide what research is worth pursuing. The distinction with which the mathematical world should be concerned is not between pure and applied mathematics but between mathematics that is undertaken with sound objectives and mathematics that satisfies personal goals and whims, between pointed mathematics and pointless mathematics, between significant and insignificant mathematics, and between vital and bloodless mathematics.

XIV

Whither Mathematics?

Humble thyself, impotent reason!
PASCAL

Our lengthy account of the successively increasing perplexities which mathematicians confronted in attempting to determine what sound mathematics is and what foundation they should adopt in pursuing the creation of new mathematics has revealed the current plight. Even the one solace mathematicians derived from their work, namely, its remarkably effective applicability to science, can no longer be a comfort because most mathematicians have abandoned applications. How do mathematicians react to their dilemma and to what can they look forward? What is the essence of mathematics?

Let us first review how mathematics became involved in its present plight and just what the fundamental issues are. The Egyptian and Babylonian mathematicians, who first began building mathematics, were not at all able to foresee what kind of structure they would erect. Hence they did not lay a deep foundation. Rather they built directly on the surface of the earth. At that time the earth seemed to offer a secure base, and the material with which they commenced construction, facts about numbers and geometrical figures, was taken from simple earthly experiences. This historical origin of mathematics is signalized by our continuing use of the word geometry, the measurement of land.

But as the structure began to rise above the ground it became obvious that it was shaky and that further additions might imperil it. The Greeks of the classical period not only saw the danger but supplied the necessary rebuilding. They adopted two measures. The first was to select firm strips of ground along which one could run the walls. These strips were the self-evident truths about space and the whole numbers. The second was to put steel into the framework. The steel was deductive proof of each addition to the structure.

Insofar as mathematics was developed in Greek times, the structure, consisting mainly of Euclidean geometry, proved to be stable. One fault did show up, namely, that certain line segments—such as the diagonal of an isosceles right triangle whose arms are 1 unit long—would have to have a length of $\sqrt{2}$ units. Since the only numbers the Greeks recognized were the ordinary whole numbers, they would not accept such entities as $\sqrt{2}$. They resolved the dilemma by ostracizing these irrational "numbers," and abandoning the idea of assigning numerical lengths to line segments, areas, and volumes. Hence, they built no additions to arithmetic and algebra beyond the whole numbers and what could be incorporated into the structure of geometry. It is true that some Alexandrian Greeks, notably Archimedes, did operate with irrational numbers, but these were not incorporated into the logical structure of mathematics.

The Hindus and Arabs piled new floors on to the building with little concern for stability. First of all, about the year A.D. 600 the Hindus introduced negative numbers. Then the Hindus and Arabs, less fastidious than the Greeks, not only accepted irrational numbers but even developed rules for operating with them. The Renaissance Europeans, who adopted the mathematics of the Greeks, Hindus, and Arabs, at first balked at accepting these foreign elements. However, the needs of science prevailed and the Europeans overcame their concern for the logical soundness of mathematics.

By expanding the mathematics of number, the Hindus, Arabs, and Europeans added floor after floor: complex numbers, more algebra, the calculus, differential equations, differential geometry, and many more subjects. However, in place of steel they used wooden columns and beams composed of intuitive and physical arguments. But these supports proved unequal to their load and cracks began to show in the walls. By 1800, the structure was once again in peril and mathematicians hastened to replace the wood with steel.

While the superstructure was being strengthened, the ground—the axioms chosen by the Greeks—caved in under the walls. The creation of non-Euclidean geometry revealed that the axioms of Euclidean geometry were not truly solid strips of earth but only seemed so on superficial inspection. Nor were the axioms of the non-Euclidean geometries a firmer ground. What the mathematicians took to be the reality of nature, believing that their minds gave unfailing support for this cognition, proved to be unreliable sense data. To add to their troubles, the creation of new algebras forced the mathematicians to realize that the properties of number were no more firmly grounded in reality than those of geometry. Thus, the entire structure of mathematics, geometry, and arithmetic with its extentions to algebra and analysis, was in grave peril. The by now lofty building was in danger of collapsing and sinking into a quagmire.

To maintain the structure of mathematics called for strong measures, and the mathematicians rose to the challenge. It was clear that there was no solid earth on which to base mathematics, for the seemingly firm ground of nature had proved to be deceptive. But perhaps the structure could be made stable by erecting a solid foundation of another kind. This would consist of sharply worded definitions, complete sets of axioms, and explicit proof of all results no matter how obvious they might seem to the intuition. Moreover, in place of truth there was to be logical consistency. The theorems were to be carefully interwoven, so that the entire structure would be solid (Chapter VII). Through the axiomatic activity of the late 19th century, the mathematicians seemingly achieved the solidity of the structure. And so, though mathematics lost its grounding in reality, another crisis in the history of mathematics was resolved.

Unfortunately, the cement used in the foundation of the new structure did not harden properly. Its consistency had not been guaranteed by the builders, and when the contradictions of set theory appeared, mathematicians realized that an even graver crisis threatened their handiwork. Of course, they were not going to sit by and see centuries of effort crumble into ruins. Since consistency was dependent on the bases chosen for the reasoning, it seemed clear that only a reconstruction of the entire foundation of mathematics would suffice. The base on which the reconstructed mathematics rested, the logical and mathematical axioms, had to be strengthened, and so the workers decided to dig more deeply. Unfortunately, they disagreed about how and where to strengthen the foundations and so, while each maintained that he would ensure solidity, each set about rebuilding in his own way. The resulting structure was not tall and firmly grounded but sprawling and insecure, with each wing claiming to be the sole temple of mathematics and each housing what it regarded as the gems of mathematical thought.

When we were young, all of us undoubtedly read the story of the seven blind men and the elephant. Each man touched a different part of the elephant and drew his own conclusions about what the elephant was. So mathematics, perhaps a more graceful structure than an elephant, represents different bodies of knowledge to the foundationalists who contemplated it from different points of view.

Thus mathematics reached the stage where men held conflicting views of what may properly be designated mathematics—logicism, intuitionism, formalism, and set theory. Moreover within each construction there are somewhat diverging superstructures. Thus intuitionists differ in what they accept as fundamental, sound intuitions—the whole numbers only or also some irrationals, the law of excluded middle applied to finite sets only or to countable sets, and concepts of constructive methods. The logicists rely solely on logic and yet have misgivings about the axioms of reducibility, choice, and infinity. The set-theorists can proceed in any one of several distinct directions, depending on their acceptance or rejection of the axiom of choice and the continuum hypothesis. Even the formalists can pursue many paths. Some differ on the principles of metamathematics that should be used to establish consistency. The finitistic principles advocated by Hilbert do not suffice to prove the completeness of even the predicate (first order) logic, and certainly do not suffice to establish the consistency of Hilbert's formal mathematical systems. Hence non-finitistic methods have been used (Chapter XII). Further, under the limitations imposed by Hilbert, Gödel showed that any significant formal system contains undecidable propositions, propositions that are independent of the axioms. One could then take either such a proposition or its denial as an additional axiom. However, after this choice, the enlarged system must, according to Gödel's result, also contain undecidable propositions and so a choice again becomes possible. The process is in fact unending.

The logicists, formalists, and set-theorists rely on axiomatic foundations. In the first few decades of this century this type of foundation was hailed as the choice basis on which to build mathematics. But Gödel's theorem showed that no one system of axioms embraces all of the truths that belong to any one structure, and the Löwenheim-Skolem theorem showed that each embraces more than was intended. Only the intuitionists can be indifferent to the problems posed by the axiomatic approach.

To top all the disagreements and uncertainties about which foundation is the best, the lack of a proof of consistency still hangs over the heads of all mathematicians like the sword of Damocles. No matter which philosophy of mathematics one adopts, one proceeds at the risk of arriving at a contradiction.

The major fact that emerges from the several conflicting approaches to mathematics is that there is not one body of mathematics but many. The word mathematics should be understood in the plural sense and perhaps mathematic should be used for any one approach. The philosopher George Santayana once said, "There is no God and Mary is His mother." One might say today, there is no universally accepted body of mathematics and the Greeks were its founder. In fact, the multiplicity of choices mathematicians can now make can well be described in the words of Shelley:

> . . . Here
> In this interminable wilderness
> Of worlds, at whose immensity
> Even soaring fancy staggers.

Apparently we shall have to live, certainly for the foreseeable future, with no criterion for the desirable approach to mathematics proper.

Any hope of reconciling the divergent views on what is correct mathematics—or at least the proper course for mathematics—rests on recognizing the issues that oblige mathematicians to adopt different views. The basic issue is what is proof and, primarily as a consequence of differences of opinion on this issue, there are differences on what legitimate mathematics is.

Mathematical proof had supposedly always been a clear, indisputable process. True, it had been ignored for centuries (Chapters V–VIII), but mathematicians were on the whole conscious of this fact. The concept was there and served as the paradigm and standard to which mathematicians more or less conscientiously adhered.

What has caused concern and even conflict about proof? The older view of logic, that its principles as codified by Aristotle are absolute truths, was accepted for two thousand years. Confidence in them had been bolstered by long and seemingly reliable usage. But mathematicians came to the realization that the principles of logic were as much a product of experience as the axioms of Euclidean geometry. Hence some uneasiness about what the sound principles are did develop. Thus the intuitionists felt justified in restricting the application of the law of excluded middle. If the principles of logic have not proved to be immutable in the past, are we likely to find that principles acceptable now will be so in the future?

A second issue concerning proof, which arose when the logistic school was founded, is what do the principles of logic embrace. Although Russell and Whitehead had no hesitation in introducing the axioms of infinity and choice in the first edition of their *Principia Mathematica,* they certainly backtracked later, not only in acknowledging that the primary laws of logic were not absolute truths but also in recognizing that these two axioms are not axioms of logic. In the second edition of their *Principia,* these axioms were not listed at the outset and their use where needed to prove certain theorems was specifically mentioned.

Beyond any differences on what are acceptable principles of logic, there are differences on how far logic itself can serve. As we know, the logicists contend that it suffices for all of mathematics, though, as just noted, they did later equivocate on the axioms of infinity and choice. The formalists believe that logic alone does not suffice and axioms of mathematics must be added to axioms of logic in order to found mathematics. The set-theorists are rather casual about logical principles and some do not specify them. The intuitionists in principle dispense with logic.

Another issue is the concept of existence. For example, a proof that every polynomial equation must have at least one root establishes an existence theorem. Any proof, if consistent, is acceptable to the logicists, formalists, and set-theorists. However, even if a proof does not employ the law of excluded middle, the proof may give no method for calculating the existent object. Hence such existence proofs are unacceptable to the intuitionists. The unwillingness of intuitionists to accept transfinite cardinals and ordinals—because these are not clear to human intuition and cannot be constructively arrived at in the intuitionists' sense of construction or calculability—is another example of different standards of what constitutes existence. The issue, in what sense not only individual entities such as a root of an equation but all of mathematics exists, is a major one, and we shall say more about it later in this chapter.

Concern about what proper mathematics is arises from still another source. What are acceptable mathematical axioms? An outstanding example is whether one can use the axiom of choice. Here mathematicians are caught on the horns of a dilemma. Not to use it or to deny it means to forgo large portions of mathematics. To use it leads, as we have seen, not to contradictions but to intuitively unreasonable conclusions (Chapter XII).

Sullying the whole ideal of mathematics is the mathematicians' inability to prove its consistency. Contradictions appeared where they were unexpected. Though they have been resolved in more or less acceptable fashion, the danger that new ones may be discovered certainly has made some mathematicians sceptical of the extraordinary efforts required for rigor.

What then is mathematics if it is not a unique, rigorous, logical structure? It is a series of great intuitions carefully sifted, refined, and organized by the logic men are willing and able to apply at any time. The more they attempt to refine the concepts and systematize the deductive structure of mathematics, the more sophisticated are its intuitions. But mathematics rests upon certain intuitions that may be the product of what our sense organs, brains, and the external world are like. It is a human construction and any attempt to find an absolute basis for it is probably doomed to failure.

Mathematics grows through a series of great intuitive advances, which are later established not in one step but by a series of corrections of oversights and errors until the proof reaches the level of accepted

proof for that time. No proof is final. New counterexamples undermine old proofs. The proofs are then revised and mistakenly considered proven for all time. But history tells us that this merely means that
the time has not yet come for a critical examination of the proof. Such
an examination is often willfully delayed. Not only is there no glory in
finding errors but the mathematician who could have reason to question the proof of a theorem may want to cite it in behalf of his own
work. Mathematicians are far more concerned with establishing their
own theorems than with finding flaws in existing results.

Several of the schools have tried to enclose mathematics within the
confines of man's logic. But intuition defies encapsulation in logic. The
concept of a safe, indubitable, and infallible body of mathematics built
upon a sound foundation stems of course from the dream of the classical Greeks, embodied in the work of Euclid. This ideal guided the
thinking of mathematicians for more than twenty centuries. But apparently mathematicians were misled by the "evil genius" Euclid.

Actually the mathematician does not rely upon rigorous proof to the
extent that is normally supposed. His creations have a meaning for him
that precedes any formalization, and this meaning gives the creations
an existence or reality *ipso facto*. The attempt to determine precise
metes and bounds of a result by deriving it from an axiomatic structure
may help in some ways but does not actually enhance its status.

Intuition can be even more satisfying or reassuring than logic. When
a mathematician asks himself why some result should hold, the answer
he seeks is some intuitive understanding. In fact, a rigorous proof
means nothing to him if the result doesn't make sense intuitively. If it
doesn't, he will examine the proof very critically. If the proof seems
right, he will then try hard to find what is wrong with his intuition.
Mathematicians wish to know the inner reason for the success of a
series of syllogisms. Poincaré said: "When a somewhat long argument
leads us to a simple and striking result, we are not satisfied until we
have shown that we could have foreseen, if not the entire result, at least
its principal features."

Many mathematicians have placed far more reliance upon intuition.
The philosopher Arthur Schopenhauer expressed this attitude: "To
improve the method in mathematics it is necessary to demand above all
that one abandon the preconception that consists in believing that demonstrated truth is superior to intuitive knowledge." Pascal coined the
phrases *esprit de géométrie* and *esprit de finesse*. By the former, Pascal
meant force and rectitude of mind such as those exhibited by powerful
logical reasoning. By the latter, he meant amplitude of mind, the capacity to see more deeply and insightfully. To Pascal, even in science
esprit de finesse is a level of thought above and beyond logic and incom-

mensurable with it. Even what is incomprehensible to reason may nonetheless be true.

Long ago other mathematicians also asserted that intuitive conviction surpasses logic as the brilliance of the sun surpasses the pale light of the moon. Descartes relied upon fundamental intuitions and apropos of logic remarked: "I have found that, as for Logic, its syllogisms and the majority of its precepts are useful rather in the communication of what we already know or . . . in speaking without judgment about things of which one is ignorant." However, he was willing to supplement intuition with deductive reasoning (Chapter II).

Great mathematicians know before a logical proof is ever composed that a theorem must be true, and they sometimes are content with no more than an indication of a proof. In fact, Fermat in his vast and classic work on the theory of numbers and Newton in his work on third degree curves gave not even indications. Certainly, mathematical creation is furthered most by men who are distinguished by their power of intuition rather than by their capacity to make rigorous proofs.

The concept of proof then, large as it has loomed in the public mind and in the publications of mathematicians, has not played the role commonly assumed. The rise of conflicting philosophies of mathematics, each insisting on its own standards of proof, has fomented much scepticism about the value of proof, and attacks on the notion of proof began to appear even before the several philosophies had been clearly defined and their conflicting views were widespread. As far back as 1928, Godfrey H. Hardy spoke out with his usual bluntness:

> There is strictly speaking no such thing as mathematical proof; . . . we can, in the last analysis, do nothing but point; . . . proofs are what Littlewood and I call *gas,* rhetorical flourishes designed to affect psychology, pictures on the board in lectures, devices to stimulate the imagination of pupils.

To Hardy, proofs were a façade rather than the supporting columns of the mathematical structure.

In 1944 with all the more justification, the prominent American mathematician Raymond L. Wilder downgraded the status of proof. Proof, he said, is no more than the

> testing of the products of our intuition. . . . Obviously we don't possess, and probably will never possess, any standard of proof that is independent of time, the thing to be proved, or the person or school of thought using it. And under these conditions, the sensible thing to do seems to be to admit that there is no such thing, generally, as absolute truth [proof] in mathematics, whatever the public may think.

The value of proof was attacked by Whitehead in a lecture entitled "Immortality":

> The conclusion is that Logic, conceived as an adequate analysis of the advance of thought, is a fake. It is a superb instrument, but it requires a background of common sense. . . . My point is that the final outlook of philosophic thought cannot be based upon the exact statements which form the basis of special sciences. The exactness is a fake.

Proof, absolute rigor, and their ilk are will-o'-the wisps, ideal concepts, "with no natural habitat in the mathematical world." There is no rigorous definition of rigor. A proof is accepted if it obtains the endorsement of the leading specialists of the time or employs the principles that are fashionable at the moment. But no standard is universally acceptable today. This is not the finest hour of mathematical rigor. Certainly the previously accepted feature of mathematics, unquestionable proof from explicit axioms, now seems passé. Logic has all the fallibility and uncertainty that limit human minds. One must wonder how many fundamental assumptions we are habitually making in mathematics without ever recognizing them.

The philosopher Nietzsche once said that "jokes are the epitaphs of emotions." To relieve their dejection, mathematicians have resorted to joshing about the logic of their subject. "The virtue of a logical proof is not that it compels belief but that it suggests doubts. The proof tells us where to concentrate our doubts." "Respect but suspect mathematical proof." "We can no longer hope to be logical; the best we can hope for is not to be illogical." "More vigor and less rigor." The mathematician Henri Lebesgue, an intuitionist, said in 1928, "Logic can make us reject certain proofs but it cannot make us believe any proof." In an article of 1941, he added that logic does not serve to convince, to create confidence. We have confidence in what accords with our intuition. Lebesgue did grant that this intuition becomes more sophisticated as we learn more about mathematics.

Even Bertrand Russell, despite his thoroughly logistic program, could not refrain from a caustic remark about logic. In his *Principles of Mathematics* (1903), he wrote, "It is one of the chief merits of proofs that they instill a certain skepticism about the result proved." He also said in his *Principles* of 1903 that it follows from the very nature of an attempt to base mathematics on a system of undefined concepts and primitive propositions that results can be disproved by the discovery of a contradiction but can never be proved. All depends in the end upon immediate perception. Somewhat later in his work (1906), disturbed by the current paradoxes, he spoke more truly than he did in after years.

When the antinomies had shown that the logical proof of the time was not infallible, he said, "An element of uncertainty must always remain, just as it remains in astronomy. It may with time be immensely diminished; but infallibility is not granted to mortals. . . ."

To these gibes at proof we may add the words of a leading student of the logic of mathematics, Karl Popper: "There are three levels of understanding of a proof. The lowest is the pleasant feeling of having grasped the argument; the second is the ability to repeat it; and the third or top level is that of being able to refute it."

Rather ironic is the counterargument made by Oliver Heaviside, who was contemptuous about the mathematicians' concern for rigor: "Logic is invincible because one must use logic to defeat logic."

Though Felix Klein, the head of mathematics during the first quarter of this century at what was then the world center for mathematics, the University of Göttingen, was not primarily concerned with the foundational problems, he perceived something about the growth of mathematics that has, thus far at least, been borne out by history. In his *Elementary Mathematics from an Advanced Standpoint* (1908), Klein described the growth of mathematics thus:

> In fact, mathematics [has] grown like a tree, which does not start at its tiniest rootlets and grow merely upward, but rather sends its roots deeper and deeper at the same time and rate that its branches and leaves are spreading upward. . . . We see, then, that as regards the fundamental investigation in mathematics, there is no final ending, and therefore on the other hand, no first beginning.

Though in a somewhat different connection, Poincaré expressed a similar view: there are no solved problems; there are only problems more or less solved.

Mathematicians have been worshipping a golden calf—rigorous, universally acceptable proof, true in all possible worlds—in the belief that it was God. They now realize it was a false god. But the true god refuses to reveal himself, and now they must question whether God exists. The Moses who might relay the word has yet to appear. There is reason to question reason.

There are sound critics of the foundations who are even more impatient with the fine distinctions these mathematical foundationists make. If, they say, mathematics rests ultimately on intuitions, then to cite Imre Lakatos (1922–1974), why do we probe deeper and deeper?

> Why then don't we stop earlier, why not say that "the ultimate test whether a method is admissible in arithmetic must of course be whether it is intuitively convincing." . . . Why not honestly admit mathematical fallibility, and try to defend the dignity of fallible knowl-

edge from cynical scepticism, rather than delude ourselves that we can
invisibly mend the latest tear in the fabric of our "ultimate" intuitions?

The value of intuition vis-à-vis proof is aptly described by a story. A
physicist had a horseshoe hanging on the door of his laboratory. A visi-
tor taken aback asked him whether this brought luck to him and his
work. "No," the physicist answered, "I don't believe in superstition. But
it seems to work nevertheless."

Arthur Stanley Eddington once said, "Proof is an idol before whom
the mathematician tortures himself." Why should they continue to do
so? We might well ask what mathematicians accomplish with their stress
on reasoning if they no longer know that their subject is consistent and
if, especially, they no longer agree on what correct proof is. Should
they rather become indifferent to rigor, throw up their hands and say
that mathematics as a soundly established body of knowledge is an
illusion? Should they abandon deductive proof and resort merely to
convincing, intuitively sound arguments? After all, the physical sciences
use such arguments, and even where they use mathematics they are not
too concerned with the mathematician's passion for rigor. Abandonment
is not the advisable path. Anyone who has looked into the contributions
of mathematics to human thought would not sacrifice the concept of
proof.

One must grant that logic plays a part. If intuition is master and logic
the servant, the servant has some power over the master. It restricts
unbridled intuition. Though intuition does play the major role, it can
lead to assertions that are far too general. The proper limiting conditions are imposed by logic. Intuition throws caution to the winds but
logic teaches restraint. It is true that the adherence to logic involves
long assertions qualified by many hypotheses and usually requires
many theorems and proofs, in mincing steps, to arrive at what a powerful intuition often conquers in one swoop. But the bold bridgeheads
seized by intuition must be secured by thorough scouring for hostile
bands that might surround and destroy the bridgeheads.

Intuition can be deceptive. Throughout most of the 19th century,
mathematicians, including Cauchy, the founder of rigor, believed that
a continuous function must have a derivative. But Weierstrass astonished the mathematical world by exhibiting a continuous function
that does not have a derivative at any point. Such a function was not
and is not accessible to intuition. Mathematical reasoning not only sup-
plements intuition to confirm or correct it, but even occasionally surpasses it.

What mathematics accomplishes with its reasoning may be more evident from an analogy. Suppose a farmer takes over a tract of land in a

wilderness with a view to farming it. He clears a piece of ground but notices wild beasts lurking in a wooded area surrounding the clearing who may attack him at any time. He decides therefore to clear that area. He does so but the beasts move to another area. He therefore clears this one. And the beasts move to still another spot just outside the new clearing. The process goes on indefinitely. The farmer clears away more and more land but the beasts remain on the fringe. What has the farmer gained? As the cleared area gets larger the beasts are compelled to move farther back and the farmer becomes more and more secure at least as long as he works in the interior of his cleared area. The beasts are always there and one day they may surprise and destroy him but the farmer's relative security increases as he clears more land. So, too, the security with which we use the central body of mathematics increases as logic is applied to clear up one or another of the foundational problems. Proof, in other words, gives us relative assurance. We become quite convinced that a theorem is correct if we prove it on the basis of reasonably sound statements about numbers or geometrical figures which are intuitively more acceptable than the one we prove. In the words of Raymond L. Wilder, proof is a testing process that we apply to the thoughts suggested by our intuition.

Unfortunately, the proofs of one generation are the fallacies of the next. As E. H. Moore, a foremost American mathematician, put it as far back as 1903, "All science, logic and mathematics included, is a function of the epoch—all science in the ideals as well as its achievements." "Sufficient unto the day is the rigor thereof." Today the concept of proof depends also on the school of thought to which one adheres. Wilder himself would be satisfied with a proof that, as far as one can see, does not involve a contradiction and is mathematically useful. He would, for example, use the continuum hypothesis as an axiom. Downgrading the importance of proof, he criticizes the divisiveness of the various schools of thought. Is not their insistence on one school rather than another like the fanaticism of religionists who claim to represent the true God and reject all other sects?

We are now compelled to accept the fact that there is no such thing as an absolute proof or a universally acceptable proof. We know that, if we question the statements we accept on an intuitive basis, we shall be able to prove them only if we accept others on an intuitive basis. Nor can we probe these ultimate intuitions too far without running into paradoxes or other unresolved difficulties, some lying in the realm of logic itself. Sometime about 1900, the famous French mathematician Jacques Hadamard said, "The object of mathematical rigor has been only to sanction and legitimatize the conquests of intuition." We can no longer accept this judgment. It is more appropriate to say with Hermann Weyl, "Logic is the hygiene which the mathematician practices to keep his ideas healthy and strong." Proof does play a role; it minimizes the risk of contradictions.

We must recognize that absolute proof is not an actuality but a goal, a goal to be sought but very likely never to be attained. It may be no more than a phantom, constantly pursued but forever elusive. We should make constant efforts to strengthen what we have without ever expecting to perfect it. The moral of the history of proof is that even though we pursue an unattainable goal, we may still produce the wonderful values that mathematics has furnished in the past. If then we reorient our attitude toward mathematics, we shall be more content to pursue the subject despite our disillusionment.

The recognition that intuition plays a fundamental role in securing mathematical truths and that proof plays only a supporting role suggests that, in a larger sense, mathematics has turned full circle. The subject started on an intuitive and empirical basis. Proof became a goal with the Greeks and, though honored in the breach until the 19th century, it seemed late in that century to be achieved. But the efforts to pursue rigor to the utmost have led to an impasse in which, like a dog chasing its tail, logic has defeated logic. As Pascal put it in his *Pensées,* "Reason's last step is the recognition that there are an infinite number of things that are beyond it."

Kant, too, recognized the limitations of reason. In his *Critique of Pure Reason* he acknowledged:

> Our reason has this peculiar fate that, with reference to one class of its knowledge, it is always troubled with questions that cannot be ignored, because they spring from the very nature of reason, and that cannot be answered, because they transcend the power of human reason.

Or as Miguel de Unamuno stated in *The Tragic Sense of Life,* "The supreme triumph of reason is to cast doubt on its own validity."

More pessimistic about the role of logic was Weyl. In 1940 he said, "In spite, or because, of our disposed critical insight, we are today less sure than at any previous time of the ultimate foundations on which mathematics rests." In 1944 he elaborated:

> The question of the ultimate foundations and the ultimate meaning of mathematics remains open; we do not know in what direction it will find its final solution or even whether a final objective answer can be expected at all. "Mathematizing" may well be a creative activity of man, like language or music, of primary originality, whose historical decisions defy complete objective rationalization.

As Weyl stated, mathematics is an activity of thought, not a body of exact knowledge. It is best viewed historically. The rational constructions and reconstructions of the foundations appear now only as a travesty of the history.

The most extreme view was expressed by Karl Popper, a notable philosopher of science, in *The Logic of Scientific Discovery.* Mathematical reasoning is never verifiable but only falsifiable. Mathematical theorems are not guaranteed in any way. One may continue to use the existent theory in the absence of a better one, just as Newton's theory of mechanics was used for two hundred years before relativity, or as Euclidean geometry was before Riemannian geometry. But assurance of correctness is not attainable.

History supports the view that there is no fixed, objective, unique body of mathematics. Moreover, if history is any guide, there will be new additions to mathematics that will call for new foundations. In this respect, mathematics is like any one of the physical sciences. Theories must be modified as new observations or new experimental results conflict with previously established theories and compel formulation of new ones. No timeless account of mathematical truth is possible. The attempts to erect mathematics on an unshakable foundation have ended in failure. The successive attempts to provide a solid foundation, from Euclid through Weierstrass to the modern foundational schools, do not give any indication of an evolutionary advance which promises eventual success.

This account of the roles of intuition and proof represents the outlook for mathematics today. But it does not reflect all opinions about the future. The case for logic has been reaffirmed by the group of mathematicians who write under the pseudonym of Nicolas Bourbaki. In the introduction to the first volume of their *Elements of Mathematics,* they observe:

> Historically speaking, it is of course quite untrue that mathematics is free from contradiction; non-contradiction appears as a goal to be achieved, not as a God-given quality that has been granted to us once for all. Since the earliest times, all critical revisions of the principles of mathematics as a whole, or of any branch of it, have almost invariably followed periods of uncertainty, where contradictions did appear and had to be resolved. . . . There are now twenty-five centuries during which the mathematicians have had the practice of correcting their errors and thereby seeing their science enriched, not impoverished; this gives them the right to view the future with serenity.

There may be some comfort in this appeal to history, but history also tells us that new crises will appear. However, this prospect does not diminish Bourbaki's optimism.

Jean Dieudonné, one of the leading French mathematicians and a Bourbakist, is confident that problems of logic that arise will always be resolved:

> One may add that if some day it will be shown that mathematics is contradictory, it is probable that we shall know which rule to attribute the result to, and that the contradiction will be avoided by leaving this rule out or modifying it suitably. In short, mathematics will take a change in direction but will not disappear as a science. This is not entirely a speculation; it is almost what happened after the discovery of irrationals. Far from deploring it as having revealed a contradiction in Pythagorean mathematics, we consider it today as one of the great victories of the human spirit.

Dieudonné might well have added the case of Leibniz's approach to the calculus (Chapter VII). After all the criticism it received in the 18th century, a new formulation—non-standard analysis (Chapter XII)—has rigorized it on a basis consistent with the logicist, formalist, and set-theoretic foundations.

Beyond the confidence which men such as the Bourbakists profess about adjustments to the logic of mathematics, there are mathematicians who believe that there is a unique, correct, eternally existing body of mathematics, though it may or may not apply to the physical world. Not all of this eternally existing body of ideas may be known to man, but it nevertheless exists. Their contention is that the disagreements and uncertainties of proof are due only to the limitations of man's reason. Moreover, man's current differences are merely a temporary obstacle that will gradually be overcome.

Some of these men, Kantians in this respect, see mathematics as so deeply embedded in human reason that there can be no question of what must be correct. For example, William Rowan Hamilton, though he invented the very objects—quaternions—that led to the questioning of the physical truth of arithmetic, maintained in 1836 a position very much like Descartes's:

> Those purely mathematical sciences of algebra and geometry are sciences of the pure reason, deriving no weight and assistance from experiment, and isolated or at least isolable from all outward and accidental phenomena. . . . They are, however, ideas which seem so far born within us that the possession of them in any conceivable degree is only the development of our original power, the unfolding of our proper humanity.

Arthur Cayley, one of the leading 19th-century algebraists, said in an address to the British Association for the Advancement of Science (1883) that "We are . . . in possession of cognitions a priori, independent, not of this or that experience, but absolutely so of all experience. . . . These cognitions are a contribution of the mind to the interpretation of experience."

Whereas men such as Hamilton and Cayley saw mathematics as embedded in man's mind, others see it existing in a world outside of man. That the belief in a unique objective world of mathematical truths independent of man should have been held prior to 1900 is certainly understandable. The belief goes back to Plato (Chapter I) and was reaffirmed many times, notably by Leibniz, who distinguished between truths of reason and truths of fact, the former holding in all possible worlds. Even Gauss, who first appreciated the significance of non-Euclidean geometry, maintained the truth of number and analysis (Chapter IV).

The fine 19th-century analyst Charles Hermite (1822–1901) also expressed a belief that there is an objective real world of mathematics. He said in a letter to the mathematician Thomas Jan Stieltjes:

> I believe that the numbers and functions of analysis are not the arbitrary product of our spirits; I believe that they exist outside of us with the same character of necessity as the objects of objective reality; and we find or discover them and study them as do the physicists, chemists, and zoologists.

On another occasion he said, "We are servants rather than masters in mathematics."

Despite the controversies on the foundations, many 20th-century men maintained the same position. Georg Cantor, the creator of the theory of sets and transfinite numbers, believed that the mathematician does not invent but discovers concepts and theorems. They exist independently of human thoughts. Cantor regarded himself as a reporter and secretary. Though Godfrey H. Hardy was sceptical of man's proof, he said in an article of 1929:

> It seems to me that no philosophy can possibly be sympathetic to a mathematician which does not admit, in one manner or other, the immutability and unconditional validity of mathematical truth. Mathematical theorems are true or false; their truth or falsity is absolutely independent of our knowledge of them. In *some* sense, mathematical truth is part of objective reality.

He expressed the same view in his book *A Mathematician's Apology:*

> I will state my own position dogmatically in order to avoid minor misapprehensions. I believe that mathematical reality lies outside us, that our function is to discover or *observe* it, and that the theorems which we prove, and which we describe grandiloquently as our "creations," are simply our notes of our observations.

The leading French mathematician of this century, Jacques Hadamard (1865–1963), affirmed in his *Psychology of Invention in the Mathe-*
matical Field: "Although the truth is not yet known to us, it *pre-exists,*
and inescapably imposes on us the path we must follow."

Gödel, too, maintained that there is a transcendent world of mathematics. Apropos of set theory, he affirmed that it is legitimate to consider all sets as real objects:

> It seems to me that the assumption of such objects is quite as legiti-
> mate as the assumption of physical objects and there is quite as much
> reason to believe in their existence. They are in the same sense neces-
> sary to obtain a satisfactory theory of mathematics as physical bodies are
> necessary for a satisfactory theory of our sense perceptions and in
> both cases it is impossible to interpret the propositions one wants to as-
> sert about these entities as propositions about the "data," i.e., in the
> latter case the actually occurring sense perceptions.

Some of these affirmations come from 20th-century men who were
not much if at all concerned with foundations. What is surprising is
that even some of the leaders in the work on foundations—Hilbert,
Alonzo Church, and the members of the Bourbaki school—affirm that
the mathematical concepts and properties exist in some objective sense
and that they can be apprehended by human minds. Thus mathematical truth is discovered not invented. What evolves is not mathematics
but man's knowledge of mathematics.

The men who hold these views are often called Platonists. Though
Plato did believe that mathematics exists in some ideal world independent of human beings, his doctrines included much that does not apply
to the current views, and the use of the appellation Platonist is more
unsuitable than helpful.

These assertions about the existence of an objective, unique body of
mathematics do not explain where mathematics resides. They say
merely that mathematics exists in some extra-human world, a castle in
the air, and is merely detected by man. The axioms and theorems are
not purely human creations, rather they are like riches in a mine
which have to be brought to the surface by patient digging. But their
existence is as independent of man as the planets appear to be.

Is then mathematics a collection of diamonds hidden in the depths of
the universe and gradually unearthed, or is it a collection of synthetic
stones manufactured by man, yet so brilliant nevertheless that they
bedazzle those mathematicians who are already partially blinded by
pride in their own creations?

Moreover, if there is a world of entities that are supra-sensible and
transcendentally absolute, and if our propositions in logic and mathe-

matics are mere registers of observations of these entities, then do not the contradictions and false propositions exist in exactly the same sense as true propositions? The noxious weeds of falsehood and inconsistency may flourish side by side with the good, the true, and the beautiful. Perhaps the Devil sows his seeds and raises his harvest along with the God of truth. The Platonists can of course retort that false propositions and contradictions arise only because man's efforts to grasp the truth are inadequate.

The second view—that mathematics is entirely a product of human thought—is of course held by the intuitionists and can be traced back to Aristotle. However, whereas some assert that truth is guaranteed by the mind, others maintain that mathematics is the creation of fallible human minds rather than a fixed body of knowledge. A classic statement to this effect was made by Pascal in his *Pensées* long before the modern controversies arose. "Truth is so subtle a point that our instruments are too blunt to touch it exactly. When they do reach it, they crush the point and bear down around it, more on the false than on the true." Arend Heyting, a leading intuitionist, affirmed that no one today can speak of the true mathematics, that is, true in the sense of a correct, unique body of knowledge.

Hermann Hankel, Richard Dedekind, and Karl Weierstrass all believed that mathematics is a human creation. Dedekind in a letter to Heinrich Weber affirmed: "I advise moreover that we understand by number not the class itself, but something new . . . which the mind creates. We are of a divine race and we possess . . . the power to create." Weierstrass endorsed this thought with the words, "The true mathematician is a poet." And Ludwig Wittgenstein (1889–1951), a student of Russell and an authority in his own right, believed that the mathematician is an inventor not a discoverer. All these men and others conceive of mathematics as something far beyond bondage to empirical findings or rational deductions. In favor of their position is the fact that such elementary concepts as irrational numbers and negative numbers are neither deductions from empirical findings nor entities obviously existing in some external world.

Hermann Weyl, too, was rather ironic about eternal truths. In his *Philosophy of Mathematics and Natural Science,* he said:

> Gödel, with his basic trust in transcendental logic, likes to think that our logical optics is only slightly out of focus and hopes that after some minor correction of it we shall see *sharp,* and then everybody will agree that we see *right.* But he who does not share this trust will be disturbed by the high degree of arbitrariness in a system like Z [Zermelo's], or even in Hilbert's system. . . . No Hilbert will be able to assure us of consistency forever; we must be content if a simple axiomatic sys-

tem of mathematics has met the test of our elaborate mathematical experiments so far. It will be early enough to change the foundations when, at a later stage, discrepancies appear.

The Nobel prize-winning physicist Percy W. Bridgman, in *The Logic of Modern Physics* (1946), rejected flatly any objective world of mathematics. "It is the merest truism, evident at once to unsophisticated observation, that mathematics is a human invention." Theoretical science is a game of mathematical make-believe. All these men contend that mathematics is not only man-made but very much influenced by the cultures in which it was developed. Its "truths" are as dependent on human beings as is the perception of color or the English language. Only the relatively universal acceptance of mathematics as opposed to political, economic, and religious doctrines may have lured us into believing that it is a body of truths existing objectively outside of man. It may exist independently of any one human but not of the culture in which he resides. To paraphrase Hermann Weyl, mathematics is not an isolated technical accomplishment but only a part of human existence in its totality, and thereby it finds its justification.

The men who favor the view that mathematics is man-made are in essence Kantians, for he placed the source of mathematics in the organizing power of the mind. However, the modernists say that it is not in the morphology or physiology of the mind that mathematics originates; rather it is in the activity of the mind. It organizes by methods that are evolutionary. The creative activity of the mind constantly evolves newer and higher forms of thought. In mathematics the human mind is able to see clearly that it is free to create a body of knowledge that it finds interesting or useful. Moreover the field of creation is not a closed one. Notions that apply to existing and newly arising fields of thought will be created. The mind has the power to devise structures that will embrace the data of experience and provide a mode of arranging them. The source of mathematics is the progressive development of the mind itself.

The present conflicts about the nature of mathematics itself and the fact that mathematics is not today a universally accepted, indisputable body of knowledge certainly favor the view that mathematics is man-made. As Einstein put it, "Whoever undertakes to set himself up as a judge in the field of Truth and Knowledge is shipwrecked by the laughter of the gods."

It is rather ironic that the intellectuals of the Age of Reason, pointing to mathematics as evidence of man's rational powers and his ability to obtain truths, confidently asserted that reason would solve all of man's problems. Intellectuals of the 20th century, however confident some may be of the power of reason, certainly cannot point to mathematics as the standard and paradigm. This turn of events is not far short of an intellectual disaster. It is still true that mathematics is man's most extensive and most profound effort to achieve precise and effective thinking and what it accomplishes measures the capacity of the human mind. It represents the upper limit of what we can hope to attain in all rational domains. But today one cannot derive much comfort from the current confusion about what valid mathematics is. This is why Hilbert sought so desperately to restore truth in the sense of objective, unassailable reasoning. As he put it in his paper of 1925 "On the Infinite": "And where else would reliability and truth be found if even mathematical thinking fails?"

He repeated this concern in a talk he gave at the International Congress in Bologna (1928):

> For how would it be above all with the truth of our knowledge and with the existence and progress of science if there were no truth in mathematics? Indeed there often appears today in professional writings and public lectures skepticism and despondency about knowledge; this is a certain kind of occultism which I regard as damaging.

A continual, never-ending search for absolutes may seem second best to actual attainment of absolutes, but a long time ago Goethe pointed out that this is man's saving grace:

> Wer immer strebend sich bemüht
> Den können wir erlösen.*

Though not as confident of the existence of absolute truths, André Weil, one of the leading mathematicians of our time, maintains that the pursuit of mathematics must be furthered even though it is not the majestic tower of human reason. As he put it:

> For us, whose shoulders sag under the weight of the heritage of Greek thought and who walk in the paths traced out by the heroes of the Renaissance, a civilization without mathematics is unthinkable. Like the parallel postulate, the postulate that mathematics will survive has been stripped of its "evidence;" but, while the former is no longer necessary, we would not be able to get on without the latter.

The future of mathematics has never been of greater promise; the nature of it has never been less clear. The subtle analysis of the obvious has produced a spiral of never ending complications. But mathematicians will continue to struggle with foundational problems. As Des-

---

* Whoever strives constantly can be saved.

cartes put it, "I shall persevere until I find something that is certain—or, at least, until I find for certain that nothing is certain."

According to Homer, the gods condemned Sisyphus, king of Corinth, after his death perpetually to roll a big rock uphill, only to see it fall back to the bottom each time he neared the summit. He had no illusion that some day his labors would end. Mathematicians have the will and the courage that comes almost instinctively to complete and solidify the foundations of their subject. Their struggle too may go on forever; they too may never succeed. But the modern Sisyphuses will persist.

XV

The Authority of Nature

> And this prayer I make,
> Knowing that Nature never did betray
> The heart that loved her . . .
>
> WILLIAM WORDSWORTH

Mathematicians can pursue many conflicting directions to derive new results. In the absence of internal criteria that favor or justify one direction rather than another, a choice must be based on external considerations. Of these, certainly the most important is the traditional and still most justifiable reason for the creation and development of mathematics, its value to the sciences. The uncertainties now evident about the proper foundation for mathematics and the questions about the solidity of its logic can be parried, though not resolved, by stressing application to nature. In the words of Emerson, let us "build in matter home for mind." One cannot determine on a priori grounds whether mathematical theorems produced will necessarily apply directly, or whether, if used in conjunction with sound physical principles, the deductions from these principles will lead to physically correct results. However, application does provide a pragmatic test. The theorems that lead to correct results time after time may be used with increasing confidence. If continued use of the axiom of choice, for example, leads to sound physical results, then certainly doubts concerning its acceptability are at least diminished.

From the historical standpoint, the appeal to applications is not as radical as it may seem to present-day mathematical purists. The concepts and axioms came from observation of the physical world. Even the laws of logic are now generally granted to be the product of experience. The problems which led to theorems and even suggestions about methods of proof came from this same source. And the value or significance of the results deduced from the axioms was judged, at least until seventy-five or so years ago, by what they affirmed about the physical world. Why not test correctness by how well mathematics continues to describe and predict physical happenings? Insofar as one judges the correctness of mathematics by its applicability, there can, of course, be no absolute test. A theorem may work in n cases and fail in the $(n + 1)$th case. One disagreement disqualifies the theorem. But modification may lead, and historically has led, to corrections that will continue to ensure usefulness.

An empirical foundation and test for mathematics had been advocated by John Stuart Mill (1806–1873). He admitted that mathematics is more general than the several physical sciences. But what "justifies" mathematics is that its propositions have been tested and confirmed to a greater extent than those of the physical sciences. Hence men came to think incorrectly of mathematical theorems as qualitatively different from confirmed hypotheses and theories of other branches of science. The theorems were taken as certain whereas physical theories were thought of as very probable or merely corroborated by experience.

Mill based his assertions on philosophical grounds, long before the modern foundational controversies arose. With all the more reason many recent and current workers in the foundations have become pragmatic. As Hilbert remarked, "By their fruits shall ye know them." Hilbert also said in 1925, "In mathematics as elsewhere success is the supreme court to whose decisions everyone submits."

Andrzej Mostowski, one of the prominent and active workers in the foundations, agrees. At a congress held in Poland in 1953, he stated:

> The only consistent point of view, which is in accord not only with healthy human understanding but also with mathematical tradition, is rather the assumption that the source and last raison d'etre of the number concept—not only the natural but also the real numbers—lie in experience and practical applicability. The same is true of the concepts of set theory insofar as they are needed in the classical domains of mathematics.

Mostowski goes further. He says that mathematics is a natural science. Its concepts and methods have their origin in experience, and any attempts to found mathematics without regard to its origin in the natural sciences, its applications, and even its history are doomed to fail.

Perhaps more surprising is that Weyl, an intuitionist, also agreed that soundness may be judged by application to the physical world. Weyl had contributed much to mathematical physics and, firm as he was in support of the intuitionistic principles, he was not willing to sacrifice useful results by unyielding adherence to these principles. In his *Philosophy of Mathematics and Natural Science* (1949), he conceded:

How much more convincing and closer to facts are the heuristic arguments and the subsequent systematic constructions in Einstein's general relativity theory, or the Heisenberg-Schrödinger quantum mechanics. A truly realistic mathematics should be conceived, in line with physics, as a branch of the theoretical construction of the one real world, and should adopt the same sober and cautious attitude toward hypothetic extensions of its foundations as is exhibited by physics.

Weyl is certainly advocating treating mathematics as one of the sciences. Its theorems, like those of physics, may be tentative and precarious. They may have to be recast but correspondence with reality is one sure test of soundness.

Haskell B. Curry, a prominent and active formalist, is willing to go further. In his *Foundations of Mathematical Logic* (1963), he argued:

But does mathematics need absolute certainty for its justification? In particular, why do we need to be sure a theory is consistent or that it can be derived by an absolutely certain intuition of pure time, before we use it? In no other science do we make such demands. In physics all theorems are hypothetical; we adopt a theory so long as it makes useful predictions and modify or discard it as soon as it does not. This is what happened to mathematical theories in the past, where the discovery of contradictions has led to modification in the mathematical doctrines accepted up to the time of that discovery. Why should we not do the same in the future?

Willard Van Orman Quine, an active logicist who made many unsuccessful efforts to simplify the Russell-Whitehead *Principia,* has also been willing, at least for the present, to settle for physical soundness. In an article of 1958, part of the collection entitled *The Philosophical Bearing of Modern Logic,* he said:

We may more reasonably view set theory, and mathematics generally, in much the way in which we view theoretical portions of the natural sciences themselves; as comprising truths or hypotheses which are to be vindicated less by the light of pure reason than by the indirect systematic contribution which they make to the organizing of empirical data in the natural sciences.

Von Neumann, who made fundamental contributions to formalism and to set theory, was also willing to surmount the present impasse in the same way. In a famous article "The Mathematician" (to be found in Robert B. Heywood's *The Works of the Mind,* 1947), he argued that though the several foundational schools have not succeeded in justifying classical mathematics, most mathematicians use it anyway:

After all classical mathematics was producing results which were both elegant and useful, and, even though one could never again be absolutely certain of its reliability, it stood on at least as sound a foundation as, for example, the existence of the electron. Hence if one was willing to accept the sciences, one might as well accept the classical system of mathematics.

The status of mathematics, then, is no better than that of physics.

Even Russell, who in 1901 claimed that the edifice of mathematical truth, logical and physical, remained unshakable, admitted in an essay of 1914 that "our knowledge of physical geometry is synthetic but not a priori." It is not deducible from logic alone. In the second edition of his *Principia* (1926), he conceded still more. Logic and mathematics like Maxwell's equation of electromagnetic theory "are believed because of the observed truth of certain of their logical consequences."

Perhaps more surprising is Gödel's statement of 1950 that

> the role of the alleged "foundations" is rather comparable to the function discharged, in physical theory, by explanatory hypotheses. . . . The so-called logical or set-theoretical foundation for number theory or of any other well established mathematical theory is explanatory, rather than foundational, exactly as in physics where the actual function of axioms is to *explain* the phenomena described by the theorems of this system rather than to provide a genuine foundation for such theorems.

What these leaders are acknowledging is that the attempt to establish a universally acceptable, logically sound body of mathematics has failed. Mathematics is a human activity and is subject to all the foibles and frailties of humans. Any formal, logical account is a pseudo-mathematics, a fiction, even a legend, despite the element of reason.

Many other prominent workers in the foundations have accepted as a practical solution the same test of what sound mathematics is. Mathematics can be firmly, if not absolutely, secured by its applicability even if occasional corrections are required. As Wordsworth put it, "To the solid ground of nature trusts the Mind that builds for aye."

In urging the pragmatic test, the applicability of mathematics to science, the foundationalists may seem to be abandoning their own principles and convictions. But, whether they realized it or not, they were affirming only what had always been the test of mathematical soundness. In view of the illogical development (Chapters V-VIII), why did the mathematicians of those centuries believe in mathematics? Failing to recognize that their proofs were faulty, they thought that they had proved some results. But surely they knew that no logic supported negative, irrational, and complex numbers, or algebra, or the calculus. They relied upon applicability.

The resort to scientific applicability, or, one can say, empirical evidence, has an implication that is worth noting. The Euclidean ideal presupposed that one starts with axioms that are truths and that from these axioms one deduces further truths by valid reasoning. Reliance upon physical applicability reverses the entire concept of mathematics. If the deductions apply, then the axioms are at least reasonable, though not necessarily the only ones which can give rise to the conclusions. Truth, in the sense of useful or applicable mathematics, does not flow upward.

Actually the leaders of foundational schools did at least for long periods of time put aside their own convictions. Thus Kronecker, one of the founders of the intuitionist school, did some fine work in algebra that did not conform to his own standards because, as Poincaré remarked, Kronecker forgot his own philosophy. Brouwer, too, after proclaiming his philosophy of intuitionism in his thesis of 1907, spent the next decade on research and proofs in topology wherein he ignored the intuitionist doctrines.

The upshot of these views is that sound mathematics must be determined not by any one foundation which may some day prove to be right. The "correctness" of mathematics must be judged by its applicability to the physical world. Mathematics is an empirical science much as Newtonian mechanics. It is correct only to the extent that it works and when it does not, it must be modified. It is not a priori knowledge even though it was so regarded for two thousand years. It is not absolute or unchangeable.

If mathematics is to be treated as one of the sciences, it is important to be fully aware of how science operates. It makes observations and experiments and constructs a theory, a theory of motion, or of light, sound, heat, electricity, chemical combinations, and so forth. These theories are man-made and are tested by checking their predictions with further observation and experiment. If the predictions are verified at least within experimental error, the theory is maintained. But it may be overthrown later and must always be regarded as a theory and not as truth imbedded in the design of the physical world. We are accustomed to this view of scientific theories because we have had many examples of scientific theories being overthrown and rejected for new ones. The only reason men did not accept this view of mathematics is, as Mill pointed out, that basic arithmetic and Euclidean geometry were effective for so many centuries that people mistook it for truth. But we must now see that any branch of mathematics offers only a theory that works. As long as it works we shall hold to it, but a better one may be needed later. Mathematics does mediate between man and nature, between his inner and outer worlds. It is a bold and formidable bridge between ourselves and the external world. It is tragic to have to recognize that the bridge is not firmly anchored in reality or in human minds.

Reason possesses insights only into that which reason itself constructs according to its own plan, and though reason may take the lead with its own proposals, it must then by experiment elicit from nature the wisdom of these proposals. There is a time for theory and a time to decide the disposition of that theory by nature's behavior.

There is one quality that distinguishes most of mathematics from physical theories. Whereas in science there have been radical changes in theories, in mathematics most of the logic, number theory, and classical analysis has functioned for centuries. They have been and still are applicable. To this extent mathematics differs from the other sciences. Whether or not these portions of mathematics are absolutely reliable, they have served well. They can be called quasi-empirical.

Support for this view can be drawn especially from the history of the calculus. Despite the open debates about the logic of the calculus, it has been successful as a methodology. Ironically non-standard analysis (Chapter XII) has justified Leibniz's theory of infinitesimals but not all of the techniques of the calculus.

One can apply the test of applicability even to the axiom of choice. As Zermelo himself said in his paper of 1908, "How does Peano arrive at his own fundamental principles . . . since after all he cannot prove them either? Evidently by analyzing the modes of inference that in the course of history came to be recognized as valid and by pointing out that the principles are intuitively evident and necessary for science. . . ." Zermelo in defending his use of the axiom of choice pointed to the successes achieved by using the axiom. He cited in his 1908 paper how useful the axiom had been (even by then) in the theory of transfinite numbers, in Dedekind's theory of finite numbers, and in more technical problems of analysis.

The recommendation of various leaders that applications to science be utilized as a guide and a test of what is reliable is motivated by more than a desire to choose among the several foundations. These men recognize that the power of mathematics to master physical phenomena has increased enormously and this service to mankind cannot be abandoned while foundational issues are thrashed out. Though many mathematicians, for reasons more meretricious than meritorious, did abandon science about one hundred years ago, the greatest of the recent leaders, Poincaré, Hilbert, von Neumann, and Weyl, have constantly pursued physical applications.

Unfortunately, most mathematicians today do not work on applications (Chapter XIII). Instead they continue to produce new results in pure mathematics at an ever increasing pace. Some indication of the current volume of research (pure and applied) may be gained from the journal *Mathematical Reviews,* which reviews briefly the new and presumably significant results. There are about 2500 in each monthly issue or about 30,000 per year.

One would think that the present quandary concerning correct mathematics, which school of thought is soundest, and the multiple directions mathematics can pursue even within any one school of thought, would give pure mathematicians pause and move them to concentrate on foundational problems before creating new mathematics that may indeed prove to be logically incorrect. How then can they proceed so blithely to produce new results in areas of mathematics that are not applied?

There are several answers to this question. Many mathematicians are ignorant of the work in foundations. The manner in which mathematicians have been working since 1900 is typical of the way in which human beings face many of their problems. Almost all of them maintain offices in the tall structure of mathematics. While the foundation workers dig farther and farther down to secure the structure, the tenants of the building continue to occupy it and do their work. The people working on the foundations delve so far below ground that they are completely out of sight and the tenants are not aware that there is any concern about the support of the structure or that there is a danger of collapse. Hence they continue to use the conventional mathematics. They do not know of any challenge to the prevailing orthodoxy and so are happy to work within it.

Other contemporary mathematicians are aware of the uncertainties in the foundations but prefer to take an aloof attitude toward what they characterize as philosophical (as opposed to purely mathematical) questions. They find it hard to believe that there can be any serious concern about the foundations, or at least about their own mathematical activity. They prefer to be suckled in a creed outworn. For these men the unwritten code states: Let us proceed as though nothing has happened in the last seventy-five years. They speak of proof in some universally accepted sense even though there is no such animal, and write and publish as if the uncertainties were non-existent. What matters for them is new publications, the more the better. If they respect sound foundations at all, it is only on Sundays and on that day they either pray for forgiveness or they desist from writing new papers in order to read what their competitors are doing. Personal progress is a must—right or wrong.

Are there then no authorities who might urge restraint on the ground that foundational issues remain to be resolved? The editors of journals could refuse papers. But the editors and referees are peers who take the same position as mathematicians at large. Hence papers that maintain some semblance of rigor, the rigor of 1900, are accepted and published. If the emperor has no clothes and the court also has none, nudity is no longer astonishing, nor does it cause any embarassment. As Laplace once wrote, human reason has less difficulty in making progress than in investigating itself.

In any case, foundational questions are relegated by many mathematicians to the hinterland. The mathematical logicians do devote their energies to the foundational problem, but this group is often regarded as outside the pale of mathematics proper.

One cannot condemn all of the mathematicians who ignore foundational problems and proceed as though they were non-existent. Some are seriously concerned with utilizing mathematics, and they call upon historical support for their *modus vivendi*. As we have seen (Chapters V and VI), despite the lack of logical foundations for the number system and operations with it and for the calculus, about which heated arguments took place for over a century, mathematicians proceeded to utilize the material and to produce new results which are certainly effective. The proofs were crude or even non-existent. When contradictions were discovered, mathematicians reexamined their reasoning and modified it. Often the reasoning was improved but was still not rigorous even by late-19th-century standards. Had the mathematicians waited until they attained that standard, they would have made no progress. As Emile Picard remarked, if Newton and Leibniz had been aware that continuous functions were not necessarily differentiable, the calculus would never have been invented. In the past, audacity and prudence produced the important progress.

The philosopher George Santayana in his book *Scepticism and Animal Faith* pointed out that while scepticism and doubt are important for thinking, animal faith is important for behavior. The values of much mathematical research are superb and if these values are to be nourished, research must go on. Animal faith supplies the confidence to act.

A few mathematicians have expressed their concern about foundational issues that impugn their work. Emile Borel, René Baire, and Henri Lebesgue stated explicitly their doubts about the validity of set-theoretic methods but continued to use them with some reservation about the reliability of what they yielded. Borel said in 1905 that he was willing to indulge in reasoning about the Cantorian transfinite numbers because they are helpful in vital mathematical work. However, the course adopted by Borel and others has not been light-hearted. Let us listen to the words of Hermann Weyl, one of the deepest and certainly the most erudite of modern mathematicians:

> We are less certain than ever about the ultimate foundations of mathematics and logic. Like everybody and everything in the world today,

we have our "crisis." We have had it for nearly fifty years [as of 1946]. Outwardly it does not seem to hamper our daily work, and yet I for one confess that it has had a considerable practical influence on my mathematical life; it directed my interests to fields I considered relatively "safe," and has been a constant drain on the enthusiasm and determination with which I pursued my research work. This experience is probably shared by other mathematicians who are not indifferent to what their scientific endeavors mean in the context of man's whole caring and knowing, suffering and creative existence in the world.

To test the soundness of mathematics by its applicability immediately raises a question. How well does mathematics work? As for the mathematics created and applied before 1800, we have already had occasion (Chapter III) to demonstrate through several examples how remarkably well mathematics describes and predicts what does happen in the physical world. However, in the 19th century, mathematicians introduced concepts and theories that, however commendable the motivation, were not drawn directly from nature and even seemed at variance with nature, for example, infinite series, non-Euclidean geometries, complex numbers, quaternions, strange algebras, infinite sets of various sizes, and other creations we have not treated. On a priori grounds there is no reason to expect that these concepts and theories should apply. Let us first assure ourselves that this modern mathematics does work and indeed marvelously well.

The greatest scientific creations of the past hundred years are electromagnetic theory, the theory of relativity, and quantum theory, all of which employ modern mathematics extensively. We shall consider only the first of these because its applicability is familiar to all of us. During the first half of the 19th century, numerous investigations of electricity and magnetism were undertaken by a number of physicists and mathematicians, and a few mathematical laws about the behavior of the two phenomena were obtained. During the 1860s James Clerk Maxwell undertook to assemble these laws and to examine their compatibility. He found that mathematical compatibility required the addition to the equations of a term, which he called displacement current. The only physical meaning he could find for this term was that from a source of electricity (roughly a wire carrying current) an electromagnetic field or electromagnetic wave must spread out into space. These electromagnetic waves can be of various frequencies and include what we now receive over our radios and television sets, as well as X rays, light, infrared rays, and ultra-violet rays. Thus by purely mathematical considerations Maxwell predicted the existence of a great variety of phenomena unknown until then and rightly concluded that light is an electromagnetic phenomenon.

What is especially remarkable about electromagnetic waves—and reminiscent of gravitation (Chapter III)—is that we have not the slightest physical knowledge of what electromagnetic waves are. Only mathematics vouches for their existence, and only mathematics enabled engineers to invent the marvels of radio and television.

The same observation applies to all sorts of atomic and nuclear phenomena. Mathematicians and theoretical physicists speak of fields—the gravitational field, the electromagnetic field, the field of electrons, and others—as though they were material waves which spread out into space and exert their effects somewhat as water waves pound against ships and shores. But these fields are fictions. We know nothing of their physical nature. They are only distantly related to observables such as sensations of light, sound, motions of objects, and the now perhaps too familiar radio and television. Berkeley once described the derivative as the ghost of departed quantities. Modern physical theory is the ghost of matter. But by formulating mathematically the laws of these fictional fields, which have no apparent counterparts in reality, and by deducing consequences of these laws, we obtain conclusions, which when suitably interpreted in physical terms can be checked against sense perceptions.

The modern fictional character of science was emphasized by Einstein in 1931:

> According to Newton's system, physical reality is characterized by the concepts of space, time, material point, and force (reciprocal action of material points). . . .
> After Maxwell they conceived physical reality as represented by continuous fields, not mechanically explicable, which are subject to partial differential equations. This change in the conception of reality is the most profound and fruitful one that has come to physics since Newton. . . .
> The view I have just outlined of the purely fictitious character of the fundamentals of scientific theory was by no means the prevailing one in the eighteenth and nineteenth centuries. But it is steadily gaining ground from the fact that the distance in thought between the fundamental concepts and laws on one side and, on the other, the conclusions which have to be brought into relation with our experience grows larger and larger, the simpler the logical structure becomes— that is to say, the smaller the number of logically independent conceptual elements which are found necessary to support the structure.

Modern science has been praised for eliminating humors, devils, angels, demons, mystic forces, and animism by providing rational explanations of natural phenomena. We must now add that modern science is gradually removing the intuitive and physical content, both of which appeal to the senses; it is eliminating matter; it is utilizing purely synthetic and ideal concepts such as fields and electrons about which all we know are mathematical laws. Science retains only a small but nevertheless vital contact with sense perceptions after long chains of mathematical deduction. Science is rationalized fiction, rationalized by mathematics.

Heinrich Hertz, a great physicist and the man who first confirmed experimentally Maxwell's prediction that electromagnetic waves can travel through space, was so impressed with the power of mathematics that he could not restrain his enthusiasm. "One cannot escape the feeling that these mathematical formulas have an independent existence and intelligence of their own, that they are wiser than we are, wiser even than their discoverers, that we get more out of them than was originally put into them."

Sir James Jeans (1877–1946) underscored the role of mathematics in the investigation of nature. In *The Mysterious Universe* he affirmed: "The essential fact is simply that all the pictures which science now draws of nature, and which alone seem capable of according with observational fact, are *mathematical* pictures. . . . We go beyond the mathematical formulas at our own risk." Physical concepts and mechanisms are conjectured to construct the mathematical account and then, paradoxically, the physical aids are seen to be hardly more than fantasies, but the mathematical equations remain as the only sure hold on the phenomena.

In *Between Physics and Philosophy* Jeans reaffirmed the thought. Nature does not function in a way that can be comprehensible to the human mind through models or pictures that the senses can grasp. We can never understand what events are, but must limit ourselves to describing the patterns of events in mathematical terms. The final harvest in physics will always be a collection of mathematical formulas. The real essence of material substance is forever unknowable.

Thus the role of mathematics in modern science is now seen to be far more than that of a major tool. The role has often been described as one of summarizing and systematizing in symbols and formulas what is physically observed or physically established by experimentation and then deducing from the formulas additional information that is not accessible to observation or experimentation nor to information more readily obtained. However, this account of the role of mathematics falls far short of what it achieves. Mathematics is the essence of scientific theories, and the applications made in the 19th and 20th centuries on the basis of purely mathematical constructs are even more powerful and marvelous than those made earlier when mathematicians operated with concepts suggested directly by physical happenings. Though credit for the achievements of modern science—radio, television, airplanes, the telephone, the telegraph, high fidelity phonographs and recording instruments, X rays, transistors, atomic power (and bombs), to mention a few that are familiar—cannot be accorded only to mathematics, the role of mathematics is more fundamental and less dispensable than any contribution of experimental science.

Francis Bacon in the 17th century was sceptical of theories such as the Copernican and Keplerian astronomical theories. He feared that they were fashioned by philosophic or religious beliefs—such as God's predilection for simplicity or His design of a mathematically ordered nature—rather than demanded by observation or experimentation. Bacon's attitude was certainly reasonable, but modern mathematical theories have come to dominate physical science solely because they are so effective. Of course, conformity to observations is a requisite for the acceptance of any mathematical theory of science.

Hence any question about whether mathematics works can be answered with a resounding yes. But the question of why it works is not so readily answered. In Greek times and for many centuries thereafter, mathematicians believed that they had clear indications of where to search for gold—mathematics was truth about the physical world and the logical principles were also truths—and so they dug for it arduously, vigorously, and eagerly. They succeeded gloriously. But now we know that what was taken for gold was not gold but precious metal nevertheless. This precious metal has continued to describe the working of nature with remarkable accuracy. Why it has served so well warrants analysis. Why should one expect the construction of an independent, abstract, a priori body of "exact" thought to bear on man's physical world?

One might answer that mathematical concepts and axioms are suggested by experience. Even the laws of logic have been acknowledged to be suggested by and therefore in accord with experience. But such an explanation is far too simplistic. It may suffice to explain why fifty cows and fifty cows make one hundred cows. In the areas of number and geometry, experience may indeed suggest the proper axioms, and the logic employed may be no more than what experience has taught. But human beings have created mathematical concepts and techniques in algebra, the calculus, differential equations, and other fields that are not suggested by experience.

Beyond these examples of non-empirical mathematics, we should take into account that the mathematical straight line consists of a non-denumerable set of points. The calculus uses a concept of time composed of instants that are "crowded together" like the numbers of the real number system. The notion of a derivative (Chapter VI) may be suggested by the physical concept of velocity at some infinitesimal period of time, but the derivative, when it represents velocity, represents it at an instant. The variety of infinite sets is certainly not suggested by experience but is used in mathematical reasoning and is as necessary for a satisfactory mathematical theory as physical bodies are for sense perceptions. Mathematics has also contributed concepts such as the electromagnetic field whose physical nature is certainly not known to us.

Moreover, even if the laws of logic and some physical principles are derived from experience, in the course of an extensive mathematical proof of a physically significant conclusion the laws are used dozens of times and the proof rests solely on logic. Purely mathematical reasoning led to predictions such as the existence of Neptune. Does nature then conform to logical principles? Is there in other words a body of logic, however obtained, which tells us how nature must behave? The fact that major theories which involve hundreds of theorems and thousands of deductions about abstractions still cleave as closely to reality as the axioms, speaks for a power of mathematics to represent and predict real phenomena with an accuracy which is incredible. Why should long chains of pure reasoning produce such remarkably applicable conclusions? This is the greatest paradox in mathematics.

Hence man is left with a twofold mystery. Why does mathematics work even where, although the physical phenomena are understood in physical terms, hundreds of deductions from axioms prove to be as applicable as the axioms themselves? And why does it work in domains where we have only mere conjectures about the physical phenomena but depend almost entirely upon mathematics to describe these phenomena? One cannot dismiss these questions lightheartedly. Far too much of our science and technology depends upon mathematics. Is there perhaps some magical power in the subject which, though it used to fight under the invincible banner of truth, has actually achieved its victories through some inner mysterious strength?

The problem has been posed repeatedly, notably by Albert Einstein in his *Sidelights on Relativity* (1921):

> Here arises a puzzle that has disturbed scientists of all periods. How is it possible that mathematics, a product of human thought that is independent of experience, fits so excellently the objects of physical reality? Can human reason without experience discover by pure thinking properties of real things? . . .
>
> As far as the propositions of mathematics refer to reality they are not certain; and as far as they are certain they do not refer to reality.

He went on to explain that the axiomatization of mathematics made this distinction clear. Though Einstein understood that the axioms of mathematics and principles of logic derive from experience, he was questioning why long and intricate pure reasoning, which is independent of experience and involves concepts created by the human mind, can produce such remarkably applicable conclusions.

One modern explanation stems from Kant. Kant maintained (Chapter IV) that we do not and cannot know nature. Rather we have sense perceptions. Our minds, endowed with established structures (intuitions in Kant's terminology) of space and time, organize these perceptions in accordance with what these built-in mental structures dictate. Thus we organize spatial perceptions in accordance with the laws of Euclidean geometry because our minds require this. Being so organized, the spatial perceptions continue to obey the laws of Euclidean geometry. Of course, Kant was wrong in insisting upon Euclidean geometry, but his point that man's mind determines how nature behaves is a partial explanation. The mind shapes our concepts of space and time. We see in nature what our minds predetermine for us to see.

A view somewhat similar to Kant's but extending beyond it was advocated by Arthur Stanley Eddington (1882–1944), one of the great physicists of our time. According to Eddington the human mind decides how nature must behave:

> We have found that where science has progressed the farthest, the mind has but regained from nature that which the mind has put into nature. We have found a strange footprint on the shores of the unknown. We have devised profound theories, one after another, to account for its origin. At last we have succeeded in reconstructing the creature that made the footprint. And Lo! It is our own.

The Kantian explanation of why mathematics works has in recent years been elaborated at length by Alfred North Whitehead and espoused even by Brouwer in a paper of 1923. The key idea is that mathematics is not something independent of and applied to phenomena taking place in an external world but rather an element in our way of conceiving the phenomena. The natural world is not objectively given to us. It is man's interpretation or construction based on his sensations, and mathematics is a major instrument for organizing the sensations. Almost automatically then mathematics describes the external world insofar as it is known to man. That many human beings accept the same mathematical organization is explained by the proposition that human minds may indeed function alike or by the fact that they are born into a culture and language that condition them to accept a particular mathematical scheme. The dominance of Euclidean geometry, though it is not the final word about space, supports the latter view. The same may be said of the heliocentric theory because it was not at the outset motivated by any discrepancies of observation with Ptolemaic theory.

Moreover, if Ptolemaic theory had been retained and refined to agree with more recent observations, it would undoubtedly serve equally well at the expense only of mathematical complexity.

The substance of the above thought may be expressed thus. We attempt to abstract from the complexity of phenomena some simple systems whose properties are susceptible of being described mathematically. This power of abstraction is responsible for the amazing mathematical description of nature. Moreover, we see only what our mathematical "optics" permits us to see. This thought was also expressed, for example, by the philosopher William James in his *Lectures on Pragmatism*: "All the magnificent achievements of mathematical and physical science . . . proceed from our indomitable desire to cast the world into a more rational shape in our minds than the shape into which it is thrown there by the crude order of our experience."

In the somewhat more poetic language of a recent writer: "Reality is the most alluring of all courtesans, for she makes herself what you would have her at the moment; but she is no rock on which to anchor your soul, for her substance is of the stuff of shadows; she has no existence outside your dreams and is often no more than the reflection of your own thoughts shining upon the face of nature."

However, this Kantian explanation that we see in nature what our minds predetermine for us to see does not fully answer the question of why mathematics works. Developments since Kant's time such as electromagnetic theory can hardly be endowments of the human mind or the mind's organization of sensations. Radio and television do not exist because the mind organized some sensations in accordance with some internal structure which then enabled us to experience radio and television as consequences of the mind's conception of how nature must behave.

There are mathematicians who believe that mathematics is autonomous (Chapter XIV). That is, whether its axioms are the products of pure reason or suggested by experience, thereafter the entire body of mathematics is built up independently of experience. How then under this view can mathematics apply to the physical world and especially to physical phenomena? There are several answers. One is that mathematical axioms use undefined terms and these can be differently interpreted to suit the physical situation. Thus, as a minor example, elliptic non-Euclidean geometry applies to straight lines in the usual sense and it applies to the geometry of a sphere where straight lines are great circles.

This type of explanation was offered by Poincaré. He was willing to grant that mathematics was a purely deductive science which merely deduced the implications of its axioms. Thus man, using plausible axioms, perhaps suggested by sensations, constructs Euclidean geometry

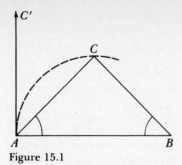

Figure 15.1

or a non-Euclidean geometry. The axioms and theorems of these geometries are neither empirical nor a priori truths. They are neither true
nor false any more than the use of polar coordinates rather than rectangular is true or false. Poincaré called them conventions for ordering
and measuring bodies, or disguised definitions of the concepts. We use
the geometry that is most convenient. However, he insisted that we
would always use Euclidean geometry with the usual interpretation of
straight line, that is, the stretched string or the ruler's edge, because it
is simplest. Why then should the deductions still apply? Poincaré's answer is that we modify the physical laws to make the mathematics fit.

To illustrate Poincaré's thesis, let us consider how surveyors determine distances. They begin by adopting a convenient base line AB (Fig.
15.1) whose length is measured by actually applying a tape measure.
To determine the distance AC, a surveyor measures angle A by sighting
point C in his telescope stationed at A and then swings the telescope
around until he sights point B. On his theodolite he has a scale which
tells him how much he has rotated his telescope, and hence he knows
angle A. In a similar manner, he measures angle B. The surveyor proceeds on the assumption that the light rays which travel from C to A
and from B to A follow the straight-line (stretched string) paths between those pairs of points, and, since the axioms of Euclidean geometry fit stretched strings, he applies Euclidean geometry or trigonometry
to calculate AC and BC. However, the surveyor's results may be wrong.
Why? The light ray from C to A may have followed the broken-line
path shown in Figure 15.1, and the surveyor at A would have to point
his telescope tangentially to the light ray in order to receive the light.
Hence the telescope would really be pointed to C' although the surveyor sees the point C in his telescope. Consequently, the angle he actually measures is $C'AB$ and not CAB. Then the subsequent use of Euclidean geometry may have led to erroneous results for AC and BC.

There is some question of what path the light rays followed. Some-

MATHEMATICS: THE LOSS OF CERTAINTY

times the paths are truly straight-line; sometimes they are bent by the refractive effect of the atmosphere. Suppose the surveyor does get incorrect results for AC and BC. Even though he may have no reason to believe that the paths of the light rays are curved, he should treat them as such. He can then correct the measurements of the angles at A and B and apply Euclidean geometry to obtain the correct values of AC and BC.

As another example of Poincaré's thesis that mathematics can be made to fit physical reality, let us see what he had to say about the question of whether the earth rotates. He argued that we should accept rotation as a physical fact because it enables us to devise a simpler mathematical theory of astronomy. And in fact simplicity of the mathematical theory was the only argument Copernicus and Kepler could advance in favor of their heliocentric theory as opposed to the older Ptolemaic theory.

Poincaré's philosophy of science has merit. We do try to use the simplest mathematics and alter physical laws if necessary to make our reasoning conform to physical facts. However, the criterion used by mathematicians and scientists today is the simplicity of the *whole* of mathematical and physical theory. And if we must use a non-Euclidean geometry—as Einstein did in his theory of relativity—to produce the simplest combined theory, we do so.

Though Poincaré was more explicit in explaining how mathematics is made to work, he did agree somewhat with the Kantian explanation in that he believed the accord between mathematics and nature is fashioned by human minds. In *The Value of Science* he affirmed:

> Does the harmony which human intelligence thinks it discovers in Nature exist apart from such intelligence? Assuredly no. A reality completely independent of the spirit that conceives it, sees it or feels it, is an impossibility. A world so external as that, even if it existed, would be forever inaccessible to us. What we call "objective reality" is, strictly speaking, that which is common to several thinking beings and might be common to all; this common part, we shall see, can only be the harmony expressed by mathematical laws.

There is a somewhat vague, perhaps a too simplistic, explanation of why mathematics works. According to this view there is an objective physical world and man constantly strives to fit his mathematics to it. We modify the mathematics when applications reveal misrepresentation or downright errors in the mathematics. This view was expressed by Hilbert in his address at the Second International Congress of Mathematics (1900):

> But even while this creative activity of pure thought is going on, the external world once again reasserts its validity, and by thrusting new

questions upon us through the phenomena that occur, it opens up
new domains of mathematical knowledge; and as we strive to bring
these new domains under the dominion of pure thought we often find
answers to outstanding unsolved problems, and thus at the same time
we advance in the most effective way the earlier theories. On this ever-
repeated interplay of thought and experience depend, it seems to me,
the numerous and astonishing analogies and the apparently pre-
established harmony that the mathematician so often perceives in the
problems, methods, and concepts of diverse realms of knowledge.

Simpler explanations of why mathematics works, less credible in our
day, repeat what mathematicians believed from Greek times until about
1850. Some men still believe in the mathematical design of nature.
They may grant that many of the earlier mathematical theories of
physical phenomena were imperfect, but they point to continuing improvements that not only embrace more phenomena but offer far
more accurate agreement with observations. Thus Newtonian mechanics replaced Aristotelian mechanics, and the theory of relativity
improved on Newtonian mechanics. Does not this history imply that
there is design and that man is approaching closer and closer to the
truth? Hermite offered this explanation of the accord between mathematics and science:

> There exists, if I am not deceived, a world which is the collection of
> mathematical truths, to which we have access only through our in-
> tellects, just as there is the world of physical reality; the one and the
> other independent of us, both of divine creation, which appear dis-
> tinct because of the weakness of our minds, but for a more powerful
> mode of thinking are one and the same thing. The synthesis of the
> two is revealed partially in the marvelous correspondence between ab-
> stract mathematics on the one hand and all the branches of physics on
> the other.

In a letter to Leo Königsberger, Hermite added "that these notions of
analysis have their existence apart from us, that they constitute a whole
of which only a part is revealed to us, incontestably although mysteriously associated with that other totality of things which we perceive
through the senses."

Sir James Jeans in *The Mysterious Universe* also accepted the older view
that "from the intrinsic evidence of his creation, the Great Architect of
the Universe now begins to appear as a pure mathematician." However, at first he granted that man's mathematics is "not yet in contact
with ultimate reality." But later in the book he became more dogmatic:

> Nature seems very conversant with the rules of pure mathematics, as
> our mathematicians have formulated them in their studies, out of
> their own inner consciousness and without drawing to any appreciable

extent on their experience of the outer world. . . . In any event, it can hardly be disputed that nature and our conscious mathematical minds work according to the same laws.

Later in life Eddington too became convinced that nature is mathematically designed and he affirmed categorically in *Fundamental Theory* (1946) that our minds can build up a pure science of nature from a priori knowledge. This science is the only one possible; any other one would contain logical inconsistencies. Not all details of science could be so obtained but general laws could be. Thus that light must travel at a finite velocity is obtainable by the mind. Even constants of nature—such as the ratio of the mass of a proton to the mass of an electron—can be determined a priori. This knowledge is independent of actual observation of the universe and more certain than experimental knowledge.

George David Birkhoff, the first of the great mathematicians in the United States, did not hesitate to repeat and support in 1941 the words of Eddington:

> . . . There is nothing in the whole system of laws of physics that cannot be deduced unambiguously from epistemological considerations. An intelligence, unacquainted with our universe but acquainted with our system of thought by which the human mind interprets to itself the content of its sensory experience, should be able to attain all the knowledge of physics that we have attained by experiment. . . . For example, he would infer the existence and properties of radium, but not the dimensions of the Earth.

Earlier in life (1918), Einstein expressed a somewhat inadequate but reasonable explanation of why mathematics fits reality:

> The development of physics has shown that at any given moment, out of all conceivable constructions, a single one has proved itself decidedly superior to all the rest. Nobody who has really gone deeply into the matter will deny that in practice the world of phenomena uniquely determines the theoretical system, in spite of the fact that there is no logical bridge between phenomena and their theoretical principles; this is what Leibniz described so happily as a "pre-established harmony."

In his mature position, presented in *The World As I See It* (1934), he asserted:

> Our experience hitherto justifies us in believing that nature is the realization of the simplest conceivable mathematical ideas. I am convinced that we can discover by purely mathematical constructions the concepts and the laws connecting them with each other, which furnish the key to the understanding of natural phenomena. Experience may

suggest the appropriate mathematical concepts, but they most certainly cannot be deduced from it. Experience remains, of course, the sole criterion of the utility of a mathematical construction. But the creative principle resides in mathematics. In a certain sense, therefore, I hold it true that pure thought can grasp reality as the ancients dreamed.

In another passage Einstein reaffirmed his belief through the now famous phrase about God: "I, at any rate, am convinced that He does not throw dice." And if He does, then, as Ralph Waldo Emerson once suggested, "The dice of God are always loaded." Einstein is not affirming here that the mathematical laws we now have are the correct ones but that there are such, and we can hope to come closer and closer to them. As he put it, "God is subtle; He is not malicious."

Like Einstein, Pierre Duhem, one of the great historians and philosophers of science, in *The Aim and Structure of Physical Theory* passed through doubts to positive affirmation. He first described a physical theory as "an abstract system whose aim is to summarize and classify logically a group of experimental laws without claiming to explain these laws." Theories are approximate, provisional, and "stripped of all objective references." Science is acquainted only with sensible appearances, and we should shed the illusion that in theorizing we are "tearing the veil from these sensible appearances." And when a scientist of genius brings mathematical order and clarity into the confusion of appearances, he achieves his aim only at the expense of replacing relatively intelligible concepts by symbolic abstractions which reveal nothing of the true nature of the universe. But Duhem ended by declaring that "It is impossible for us to believe that this order and this organization [produced by mathematical theory] are not the reflected image of a real order and organization." The world is mathematically designed by a great architect. God does eternally geometrize and man's mathematics describes that design.

Hermann Weyl was sure that mathematics reflects the order of nature. In a talk he said:

> There is inherent in nature a hidden harmony that reflects itself in our minds under the image of simple mathematical laws. That then is the reason why events in nature are predictable by a combination of observation and mathematical analysis. Again and again in the history of physics this conviction, or should I say this dream, of harmony in nature had found fulfillments beyond our expectation.

Perhaps, however, the wish was father to the thought, for, in his book *Philosophy of Mathematics and Natural Science,* he added:

> And yet science would perish without a supporting transcendental
> faith in truth and reality, and without the continuous interplay be-
> tween its facts and constructions on the one hand and the imagery of
> ideas on the other.

Though one cannot dismiss lightly the views of Jeans, Weyl, Eddington, and Einstein, their views of the relation of mathematics and nature
are not the dominant ones. It is true the successes of the mathematical
description of nature have been so astounding that the explanations
they offer seem to be reasonable, just as Euclidean geometry seemed to
be indubitable truth to mathematicians of many centuries. But today
the belief in the mathematical design of nature seems far-fetched.

Still another account of the relationship of mathematics to the physical world allows for some correspondence but not the kind usually understood. In the past hundred years there has arisen the statistical view
of nature. Somewhat ironically, it was initiated by Laplace who firmly
believed that the behavior of nature is strictly determined in accordance with mathematical laws but that causes of the behavior of nature
are not always known and observations are only approximately correct.
Hence one should apply the theory of probability to determine the
most likely causes and the data most likely to be correct. His *Analytical
Theory of Probability* (3rd edition, 1820) is the classic on this subject. The
history of probability and statistics is extensive and perhaps it is unnecessary to include it here. But in less than a century, it led to the view
that the behavior of nature is not at all determined, but rather chaotic.
However, there is a most probable behavior, an average behavior, and
this is what we observe and say is determined by mathematical laws.
Just as human beings live to all ages—some die as infants and some at a
hundred—there is not only a life expectancy for all men and women
but even a life expectancy at any given age. And by using these data,
insurance companies build a very successful business. The statistical
view of nature has received enormous support in recent years because
of the development of quantum mechanics, according to which there
are no rigid, discrete, localized particles of matter. They exist only with
a certain probability in every part of space, but most probably in some
one place.

At any rate, according to the statistical view, the mathematical laws of
nature describe at best how nature will most probably behave, but they
do not preclude that the earth may suddenly wander off into space.
Nature can make up her own mind and decide not to do what is most
probable. The conclusion which some philosophers of science adopt
today is that the inexplicable power of mathematics remains inexplicable. This was first expressed by the philosopher Charles Sanders

Peirce: "It is probable that there is some secret here which remains to be discovered."

More recently (1945) Erwin Schrödinger in *What Is Life* said that the miracle of man's discovering laws of nature may well be beyond human understanding. Another physicist, the highly distinguished Freeman Dyson agrees: "We are probably not close yet to understanding the relation between the physical and the mathematical worlds." To which one can add Einstein's remark, "The most incomprehensible thing about the world is that it is comprehensible."

In 1960 Eugene P. Wigner, another Nobel prize-winner in physics, discussed the problem of the unreasonable effectiveness of mathematics in the natural sciences in an article with that very title and offered no more explanation than to restate the issue:

> The miracle of the appropriateness of the language of mathematics for the formulation of the laws of physics is a wonderful gift which we neither understand nor deserve. We should be grateful for it, and hope that it will remain valid in future research and that it will extend, for better or for worse, to our pleasure even though perhaps also to our bafflement, to wide branches of learning.

These last few "explanations" are apologetic. They say rather little, though often in impressive language that tempers the admission that they have no answer to why mathematics is effective.

However satisfactory or unsatisfactory any explanation of why mathematics works may be, it is important to recognize that nature and the mathematical representation of nature are not the same. The difference is not merely that mathematics is an idealization. The mathematical triangle is assuredly not a physical triangle. But mathematics departs even further. In the 5th century B.C., Zeno of Elea posed a number of paradoxes. Whatever his purpose, even the first of his paradoxes on motion illustrates the difference between mathematical conceptualization and experience. Zeno's first paradox states that a runner can never get to the end of a race-course because first he must cover ½ the distance, then ½ of the remaining distance, then ½ of the remaining distance, and so on. Hence the runner must cover

$$\frac{1}{2}+\frac{1}{4}+\frac{1}{8}+\frac{1}{16}+\cdots.$$

Then, Zeno argued, the time required to cover an infinite number of distances must be infinite.

One physical resolution of Zeno's paradox and the most obvious is that a runner will cover the distance in a finite number of steps. However, even if one accepts Zeno's mathematical analysis, the time required may be ½ minute, plus ¼ minute, plus ⅛ minute, and so forth, and the sum of all these infinite numbers of time intervals is just one minute. This analysis departs from the physical process, but the result agrees nevertheless.

It may be that man has introduced limited and even artificial concepts and only in this way has managed to institute some order in nature. Man's mathematics may be no more than a workable scheme. Nature itself may be far more complex or have no inherent design. Nevertheless, mathematics remains the method *par excellence* for the investigation, representation, and mastery of nature. In those domains where it is effective, it is all we have; if it is not reality itself, it is the closest to reality we can get.

Though it is a purely human creation, the access it has given us to some domains of nature enables us to progress far beyond all expectations. Indeed it is paradoxical that abstractions so remote from reality should achieve so much. Artificial the mathematical account may be; a fairy tale perhaps, but one with a moral. Human reason has a power even if it is not readily explained.

The mathematical successes are bought at a price, the price of viewing the world in terms of measure, mass, weight, duration, and other similar concepts. Such an account cannot be a complete representation of the rich and variegated experiences any more than a person's height is the person. At best mathematics describes some processes of nature, but its symbols do not contain all of it.

Moreover, mathematics deals with the simplest concepts and phenomena of the physical world. It does not deal with man but with inanimate matter. The behavior of matter is repetitive and mathematics can describe it. But in economics, political theory, psychology, as well as biology, mathematics is far less useful. Even in the physical realm, mathematics deals with simplifications which merely touch reality as a tangent touches a curve at one point. Is the path of the earth around the sun an ellipse? No. Only if the earth and sun are regarded as points and only if all other bodies in the universe are ignored. Do the four seasons on earth repeat themselves year after year? Hardly. Only in their grossest aspects, which are about all that man can perceive anyway, do they repeat.

Should we reject mathematics because we do not understand its unreasonable effectiveness? Heaviside once remarked: Should I refuse my dinner because I do not understand the process of digestion? Experience refutes the doubters. Rational explanations are disregarded by the confident. With all due respect to religion, social science, and philosophy and with explicit acknowledgment of the fact that mathematics does not treat those aspects of our lives, the fact is that mathematics is infinitely more successful in giving us knowledge. This knowledge is not based on mere assertions of its correctness. It is tested daily in the working of every radio set and atomic power plant, in the prediction of eclipses, and in thousands of events in the laboratory and in daily life.

Mathematics may treat the simpler problems—the physical world—but in its sphere it is the most successful development. Some of the hope that man has any significance at all springs from the power he has acquired through mathematics. It has harnessed nature and lightened man's burden. One can take heart from its victories.

The question of why mathematics works is not solely academic. In the use of mathematics in engineering, to what extent can one rely upon mathematics to predict and design? Should one design a bridge using theory involving infinite sets or the axiom of choice? Might not the bridge collapse? Fortunately some engineering projects use theorems so well supported by past experience that the theorems may be used with confidence. Many engineering projects are deliberately over-designed. Thus a bridge uses materials such as steel, but our knowledge of the strength of materials is not accurate. Hence engineers use stronger cables and beams than theory calls for. However, in the case of a type of project that has never been built before, one must be concerned about the reliability of the mathematics used. In such cases, prudence would dictate the use of small-scale models or other tests before the construction proper is undertaken.

The concern of this chapter has been to seek some resolution of the quandary in which mathematics and mathematicians find themselves. There is no universally accepted body of mathematics and the multitude of paths that are advocated by various groups cannot all be pursued, for to do so would hinder the major goal of mathematics, namely, the advancement of science. Hence we have advocated using that goal as a standard. We have also discussed the problems and issues that this course entails.

However, while emphasis on application to science does seem the wisest course, this program does not exclude other worthy and even wise pursuits in the realm of mathematics. We did point out (Chapter XIII) that even the pursuit of applied mathematics requires a variety of supporting activities, abstraction, generalization, rigorization, and improvements in methodology. Beyond these, one can justify the pursuit of foundational approaches that do not bear directly on the mathematics that has proved useful in scientific inquiries. The constructivist program of the intuitionists, though intended by them to replace meaningless existence theorems, does produce methods of calculating quantities which the pure existence theorems merely tell us exist. To use for the sake of simplicity an old example, Euclid proved that the ratio of the area of any circle to the square of its radius is the same for all circles. This ratio is of course π. Hence Euclid proved a pure existence theorem. But to know the value of π is obviously important if one is to calculate the area of any given circle. Fortunately Archimedes' approximate calculation and several later series expansions enabled us to calculate π long before the challenge to pure existence theorems was posed by the intuitionists. Certainly the ability to calculate π was important. Similarly other quantities whose existence alone is now established should be calculated. And so the constructivist program should be pursued.

Another potential value in pursuing foundational approaches is the possibility of arriving at a contradiction. Consistency is not established, and so finding a contradiction or a patently absurd theorem would at least eliminate an alternative that now absorbs the time and energy of some mathematicians.

Our account of the status of mathematics is certainly not a comforting one. Mathematics has been shorn of its truth; it is not an independent, secure, solidly grounded body of knowledge. Most mathematicians have renounced their devotion to science, an act that would be deplorable at any period of history but especially when applications might supply some guide to the sound direction mathematics should pursue. And the remarkable confirmation and power of what has been applied remains to be explained.

Despite these defects and limitations mathematics has much to offer. It is man's supreme intellectual achievement and the most original creation of the human spirit. Music may rouse or pacify the soul, painting may delight the eye, poetry may stir the emotions, philosophy may satisfy the mind, and engineering may improve the material life of man. But mathematics offers all these values. Moreover, in the direction of what reasoning can accomplish, mathematicians have exercised the greatest care that the human mind is capable of to secure the soundness of their results. It is not accidental that mathematical precision is a byword. Mathematics is still the paradigm of the best knowledge available.

The accomplishments of mathematics are the accomplishments of the human mind, and this evidence of what human beings can achieve has given man the courage and confidence to tackle the once seemingly impenetrable mysteries of the cosmos, to overcome fatal diseases to which man is subject, and to question and to improve the political systems under which man lives. In these endeavors mathematics may or may not be effective but our unquenchable hope for success derives from mathematics.

The values are there, values at least as great as any human creation can offer. If all are not readily or widely perceptible or appreciated, fortunately they are utilized. If the climb to reach them is more arduous than in music, say, the rewards are richer, for they include almost all the intellectual, aesthetic, and emotional values that any human creation can offer. To ascend a high mountain may be more strenuous than to climb a low hill but the view from the top extends to far more distant horizons. Values there are in abundance and the only question one may raise is the order of importance. But this question each must answer for himself; individual judgments, opinions, and tastes enter.

Insofar as certainty of knowledge is concerned, mathematics serves as an ideal, an ideal toward which we shall strive, even though it may be one that we shall never attain. Certainty may be no more than a phantom constantly pursued and interminably elusive. However, ideals do have a force and a value. Justice, democracy, and God are ideals. It is true that people have murdered in the name of God and miscarriages of justice are notorious. Nevertheless, these ideals are the major product of thousands of years of civilization. So is mathematics, even if it is only an ideal. Perhaps contemplation of the ideal will make us more aware of the direction we must pursue to obtain truths in any field.

The plight of man is pitiable. We are wanderers in a vast universe, helpless before the devastations of nature, dependent upon nature for food and other necessities, and uninformed about why we were born and what we should strive for. Man is alone in a cold and alien universe. He gazes upon this mysterious, rapidly changing, and endless universe and is confused, baffled, and even frightened by his own insignificance. As Pascal put it:

> For after all what is man in nature? A nothing in relation to infinity, all in relation to nothing, a central point between nothing and all and infinitely far from understanding either. The ends of things and their beginnings are impregnably concealed from him in an impenetrable secret. He is equally incapable of seeing the nothingness out of which he was drawn and the infinite in which he is engulfed.

Montaigne and Hobbes said the same thing in other words. The life of man is solitary, poor, nasty, brutish, and short. He is the prey of contingent happenings.

Endowed with a few limited senses and a brain, man began to pierce the mystery about him. By utilizing what the senses reveal immediately or what can be inferred from experiments, man adopted axioms and applied his reasoning powers. His quest was the quest for order; his goal, to build systems of knowledge as opposed to transient sensations, and to form patterns of explanation that might help him attain some mastery over his environment. His chief accomplishment, the product of man's own reason, is mathematics. It is not a perfect gem and continued polishing will probably not remove all flaws. Nevertheless, mathematics has been our most effective link with the world of sense perceptions and though it is discomfiting to have to grant that its foundations are not secure, it is still the most precious jewel of the human mind and must be treasured and husbanded. It has been in the van of reason and no doubt will continue to be even if new flaws are discovered by more searching scrutiny. Alfred North Whitehead once wrote, "Let us grant that the pursuit of mathematics is a divine madness of the human spirit." Madness, perhaps, but surely divine.

Selected Bibliography

Note: *A. M. M.* stands for *The American Mathematical Monthly*

Barker, S. F.: *Philosophy of Mathematics*, Prentice-Hall Inc., Engelwood Cliffs, N.J., 1964.

Baum, Robert J.: *Philosophy and Mathematics from Plato to the Present*, Freeman, Cooper & Co., San Francisco, 1973.

Bell, E. T.: "The Place of Rigor in Mathematics," *A. M. M.*, 41 (1934), 599–607.

Benacerraf, Paul and Hilary Putnam: *Philosophy of Mathematics, Selected Readings*, Prentice-Hall Inc., Englewood Cliffs, N.J., 1964.

Beth, Evert W.: *The Foundations of Mathematics*, North-Holland Publishing Co., N.Y., 1959; paperback, Harper and Row, N.Y. 1966.

——: *Mathematical Thought: An Introduction to the Philosophy of Mathematics*, D. Reidel, Dordrecht, Holland, 1965; Gordon and Breach, N.Y., 1965.

Bishop, Errett, et al.: "The Crisis in Contemporary Mathematics," *Historia Mathematica*, 2 (1975), 505–33.

Black, Max: *The Nature of Mathematics*, Harcourt, Brace, Jovanovich, N.Y., 1935; Routledge & Kegan Paul, London, 1933.

Blumenthal, L. M.: "A Paradox, A Paradox, A Most Ingenious Paradox," *A. M. M.*, 47 (1940), 346–53.

Bochenski, I. M.: *A History of Formal Logic*, Chelsea Publishing Co., N.Y., reprint, 1970.

Bourbaki, Nicholas: "The Architecture of Mathematics," *A. M. M.*, 57 (1950), 221–32; also in F. Le Lionnais, *Great Currents of Mathematical Thought*, Dover Publications, Inc. N.Y., 23–36.

Brouwer, L. E. J.: "Intuitionism and Formalism," *American Mathematical Society Bulletin*, 20 (1913–14), 81–96.

Burington, A. S.: "On the Nature of Applied Mathematics," *A. M. M.*, 56 (1949), 221–41.

Calder, Allan: "Constructive Mathematics," *Scientific American* (Oct. 1979), 146–71.

Cantor, Georg: *Contributions to the Founding of the Theory of Transfinite Numbers*, Dover Publications, Inc. N.Y., 1955.

Cohen, Morris R.: *A Preface to Logic*, Holt, Rinehart and Winston, N.Y., 1944; Dover reprint, N.Y., 1977.

Cohen, Paul J. and Reuben Hersh: "Non-Cantorian Set Theory," *Scientific American* (Dec. 1967), 104–16.

Courant, Richard: "Mathematics in the Modern World," *Scientific American* (Sept. 1964), 40–49.

Dauben, J. W.: *Georg Cantor: His Mathematics and Philosophy of the Infinite,* Harvard University Press, Cambridge, 1978.

Davis, M. and R. Hersh: "Nonstandard Analysis," *Scientific American* (June 1972), 78–86.

Davis, P. J.: "Fidelity in Mathematical Discourse: Is One and One Really Two?" *A. M. M.,* 79 (1972), 252–63.

De Long, Howard: *A Profile of Mathematical Logic,* Addison-Wesley, Reading, Mass., 1970.

————: "Unsolved Problems in Arithmetic," *Scientific American* (March 1971), 50–60.

Desua, Frank: "Consistency and Completeness—A Résumé," *A. M. M.,* 63 (1956), 293–305.

Dieudonné, Jean: "Modern Axiomatic Methods and the Foundations of Mathematics," in Le Lionnais: *Great Currents of Mathematical Thought,* vol. I, Dover Publications, N.Y., 1971, 251–66.

————: "The Work of Nicholas Bourbaki," *A. M. M.,* 77 (1970), 134–45.

Dresden, Arnold: "Brouwer's Contributions to the Foundations of Mathematics," *American Mathematical Society Bulletin,* 30 (1924), 31–40.

————: "Some Philosophical Aspects of Mathematics," *American Mathematical Society Bulletin,* 34 (1928), 438–52.

Eves, Howard and Carroll V. Newsom: *An Introduction to the Foundations and Fundamental Concepts of Mathematics,* rev. ed., Holt, Rinehart and Winston, N.Y., 1965.

Fraenkel, Abraham A.: "On the Crisis of the Principle of the Excluded Middle," *Scripta Mathematica,* 17 (1951), 5–16.

————: "The Recent Controversies About the Foundations of Mathematics," *Scripta Mathematica,* 13 (1947), 17–36.

————, Y. Bar-Hillel, and A. Levy: *Foundations of Set Theory,* 2nd rev. ed., North-Holland Publishing Co., N.Y., 1973.

Gödel, K.: "What is Cantor's Continuum Problem?" *A. M. M.,* 54 (1947), 515–25; also in P. Benacerraf and H. Putnam, with additions, 258–73.

Goodman, Nicolas D.: "Mathematics as an Objective Science," *A. M. M.,* 86 (1979), 540–51.

Goodstein, R. L.: *Essays in the Philosophy of Mathematics,* Leicester University Press, 1965.

Hahn, Hans: "The Crisis in Intuition," in J. R. Newman, *The World of Mathematics,* vol. III, 1956–76.

Halmos, Paul R.: "The Basic Concepts of Algebraic Logic," *A. M. M.,* 63 (1956), 363–87.

Hardy, G. H.: "Mathematical Proof," *Mind,* 38 (1928), 1–25; also in Collected Papers, vol. VII, 581–606.

————: *A Mathematician's Apology,* Cambridge University Press, 1941.

Heijenoort, Jean van, ed.: *From Frege to Gödel, A Source Book in Mathematical Logic, 1879–1931*, Harvard University Press, Cambridge, 1967.

Hempel, Carl G.: "Geometry and Empirical Science," *A. M. M.*, 52 (1945), 7–17.

——: "On the Nature of Mathematical Truth," *A. M. M.*, 52 (1945), 543–56; also in P. Benacerraf and H. Putnam.

Hersh, Reuben: "Some Proposals For Reviving the Philosophy of Mathematics," *Advances in Mathematics*, 31 (1979), 31–50.

Hilbert, David: "On the Infinite," *Mathematische Annalen*, 95 (1925), 161–90; also in P. Benacerraf and H. Putnam, 134–51 and in J. van Heijenoort, 367–92.

Kline, Morris: *Mathematical Thought from Ancient to Modern Times*, Oxford University Press, N.Y., 1972.

——: *Mathematics in Western Culture*, Oxford University Press, N.Y., 1953.

Kneale, William and Martha Kneale: *The Development of Logic*, Oxford University Press, N.Y., 1962.

Kneebone, G. T.: *Mathematical Logic and the Foundations of Mathematics*, D. Van Nostrand, N.Y., 1963.

Körner, S.: *The Philosophy of Mathematics*, Hutchinson University Library, London, 1960.

Lakatos, Imre: *Mathematics, Science and Epistemology*, 2 vols., Cambridge University Press, N.Y., 1978.

——, ed.: *Problems in the Philosophy of Mathematics*, vol. I, North-Holland Publishing Co., N.Y., 1972.

——: *Proofs and Refutations*, Cambridge University Press, N.Y., 1976.

Langer, Susanne K.: *An Introduction to Symbolic Logic*, 2nd ed., Dover Publications, N.Y., 1953.

Le Lionnais, F., ed.: *Great Currents of Mathematical Thought*, 2 vols., Dover Publications, N.Y., 1971.

Lewis, C. I.: *A Survey of Symbolic Logic*, Dover Publications, N.Y., 1960.

Luchins, E. and A.: "Logicism," *Scripta Mathematica*, 27 (1965), 223–43.

Luxemburg, W. A. J.: "What is Non-Standard Analysis?" *A. M. M.*, 80 (1973), part II, 38–67.

Mackie, G. L.: *Truth, Probability and Paradox*, Oxford University Press, N.Y., 1973.

Mendelson, Elliott: *Introduction to Mathematical Logic*, 2nd ed., D. Van Nostrand, N.Y., 1979.

Monk, J. D.: "On the Foundations of Set Theory," *A. M. M.*, 77 (1970), 703–11.

Myhill, John: "What is a Real Number?" *A. M. M.*, 79 (1972), 748–54.

Nagel, Ernest and J. R. Newman: "Gödel's Proof," *Scientific American* (June 1956), 71–86.

—— and ——: *Gödel's Proof*, New York University Press, N.Y., 1958.

Neumann, John von: "The Mathematician," in Robert B. Heywood, *The Works of the Mind*, University of Chicago Press, Chicago (1947), 180–96; also in J. R. Newman: *The World of Mathematics*, vol. IV (1956), 2053–68; also in *Collected Works*, vol. I (1961), 1–9.

Newman, James R.: *The World of Mathematics*, 4 vols., Simon and Schuster, N.Y., 1956.

Pierpont, James: "Mathematical Rigor Past and Present," *American Mathematical Society Bulletin*, 34 (1928), 23–52.

Poincaré, Henri: *The Foundations of Science*, The Science Press, Lancaster, Pa., 1946.

———: *Last Thoughts*, Dover Publications, N.Y., 1963.

Putnam, Hilary: "Is Logic Empirical?" *Boston Studies in Philosophy of Science*, 5 (1969), 216–41.

———: *Mathematics, Matter and Method, Philosophical Papers*, vol. I, Cambridge University Press, N.Y., 1975.

Quine, W. V.: "The Foundations of Mathematics," *Scientific American* (Sept. 1964), 112–27.

———: *From a Logical Point of View*, 2nd ed., Harvard University Press, Cambridge, 1961.

———: "Paradox," *Scientific American* (April 1962), 84–96.

———: *The Ways of Paradox and Other Essays*, Random House, N.Y., 1966.

Richmond, D. E.: "The Theory of the Cheshire Cat," *A. M. M.*, 41 (1934), 361–68.

Robinson, Abraham: *Non-Standard Analysis*, 2nd ed., North-Holland Publishing Co., N.Y., 1974.

Rotman, B. and G. T. Kneebone: *The Theory of Sets and Transfinite Numbers*, Oldbourne Book Co., London, 1966.

Russell, Bertrand: *The Autobiography of Bertrand Russell: 1872 to World War I*, Bantam Books, N.Y., 1965.

———: *Introduction to Mathematical Philosophy*, George Allen & Unwin, London, 1919.

———: *Mysticism and Logic*, Longmans, Green, London, 1925.

———: *The Principles of Mathematics*, 2nd ed., George Allen & Unwin, London, 1937.

Schrödinger, Erwin: *Nature and the Greeks*, Cambridge University Press, N.Y., 1954.

Sentilles, D.: *A Bridge to Advanced Mathematics*, Williams & Wilkins, Baltimore, 1975.

Snapper, Ernst: "What is Mathematics?" *A. M. M.*, 86 (1979), 551–57.

Stone, Marshall: "The Revolution in Mathematics," *A. M. M.*, 68 (1961), 715–34.

Tarski, Alfred: *Introduction to Logic and to the Methodology of Deductive Sciences*, 2nd ed., Oxford University Press, N.Y., 1946.

———: "Truth and Proof," *Scientific American* (June 1969), 63–77.

Waismann, F.: *Introduction to Mathematical Thinking*, Harper & Row, N.Y., 1959.

Wavre, Robin: "Is There a Crisis in Mathematics?" *A. M. M.*, 41 (1934), 488–99.

Weil, André: "The Future of Mathematics," *A. M. M.*, 57 (1950), 295–306.

Weyl, Hermann: "A Half-Century of Mathematics," *A. M. M.*, 58 (1951), 523–53.

———: "Mathematics and Logic," *A. M. M.*, 53 (1946), 2–13.

————: *Philosophy of Mathematics and Natural Science,* Princeton University Press, Princeton, 1949.

White, Leslie A.: "The Locus of Mathematical Reality: An Anthropological Footnote," *Philosophy of Science,* 14 (1947), 289–303; also in James R. Newman (above): vol. IV, 2348–64.

Whitehead, Alfred North and Bertrand Russell: *Principia Mathematica,* 3 vols., Cambridge University Press, N.Y., 1st ed., 1910–13; 2nd ed., 1925–27.

Wigner, Eugene P.: "The Unreasonable Effectiveness of Mathematics," *Communications on Pure and Applied Mathematics,* 13 (1960), 1–14.

Wilder, Raymond L.: *Introduction to the Foundations of Mathematics,* 2nd ed., John Wiley, N.Y., 1965.

————: "The Nature of Mathematical Proof," *A. M. M.,* 51 (1944), 309–23.

————: "The Role of the Axiomatic Method," *A. M. M.* 74 (1967), 115–27.

————: "The Role of Intuition," *Science,* 156 (1967), 605–10.

Index